T0143347

THE
TRANSFORM
AND DATA
COMPRESSION
HANDBOOK

THE ELECTRICAL ENGINEERING
AND SIGNAL PROCESSING SERIES
Edited by Alexander Poularikas and Richard C. Dorf

Handbook of Antennas in Wireless Communications
Lal Chand Godara

Propagation Data Handbook for Wireless Communications
Robert Crane

The Digital Color Imaging Handbook
Guarav Sharma

Handbook of Neural Network Signal Processing
Yu Hen Hu and Jeng-Neng Hwang

Handbook of Multisensor Data Fusion
David Hall

The Advanced Signal Processing Handbook:
Theory and Implementation for Radar, Sonar,
and Medical Imaging Real Time Systems
Stergios Stergiopoulos

The Transform and Data Compression Handbook
K.R. Rao and P.C. Yip

The Encyclopedia of Signal Processing
Alexander Poularikas

Applications in Time Frequency Signal Processing
Antonia Papandreou-Suppappola

THE
TRANSFORM
AND DATA
COMPRESSION
HANDBOOK

Edited by

K.R. RAO
University of Texas at Arlington

AND

P.C. YIP
McMaster University

CRC Press
Taylor & Francis Group
Boca Raton London New York

CRC Press is an imprint of the
Taylor & Francis Group, an **informa** business

Library of Congress Cataloging-in-Publication Data

The transform and data compression handbook / editors, P.C. Yip, K.R. Rao.
    p. cm.--(Electrical engineering and signal processing series)
    Includes bibliographical references and index.
    ISBN 0-8493-3692-9 (alk. paper)
    1. Data transmission systems--Handbooks, manuals, etc.. 2. Data compression
(Telecommunication)--Handbooks, manuals, etc. I. Yip, P.C. (Pat C.) II. Rao, K.
Ramamohan (Kamisetty Ramamohan) III. Series

TK5105 .T72 2000
621.382--dc21   00-057149

CRC Press
Taylor & Francis Group
6000 Broken Sound Parkway NW, Suite 300
Boca Raton, FL 33487-2742

© 2001 by Taylor & Francis Group, LLC
CRC Press is an imprint of Taylor & Francis Group, an Informa business

No claim to original U.S. Government works

ISBN-13: 9780849336928 (hbk)

Visit the Taylor & Francis Web site at
http://www.taylorandfrancis.com

and the CRC Press Web site at
http://www.crcpress.com

Preface

While this handbook is an exposition of different discrete transforms and their ever-expanding applications in the general area of signal processing, the overriding task is to maintain the continuity and connectivity among the chapters. This task is accomplished by the common theme of data compression. The handbook seeks to provide the reader with a wealth of information regarding the transforms (some have been widely used while others have great potential) as well as a demonstration of their power and practicality in data compression. Such compression is a necessary and desirable ingredient in today's world of massive data storage and data transmission. By providing a plethora of Web sites, ftp locations, and references to general review papers, the chapter authors have expanded the usefulness of this handbook for the common reader. The clear and concise presentations of the ideas and concepts, as well as the detailed descriptions of the algorithms, provide important insights into the applications and their limitations. With the understanding of these concepts, readers can apply the techniques presented in this handbook to their own areas of interest and improve on the performance by marrying this with their own expertise. We are confident that this handbook will be a valuable addition to the bookshelf of anyone actively engaged in or studying the art and science of signal processing.

The Transform and Data Compression Handbook is aimed at providing a description of various discrete transforms and their applications in different disciplines. In view of the proliferation of digital data (images, video, text, documents, audio, music, graphics, etc.), it is imperative that the data be mapped from the data domain (in which there are usually redundancies) to a different one (the transform domain) for efficient and economical storage and/or transmission. Transforms by themselves do not provide any compression. However, by reallocation of the energy in the data, transforms provide the possibilities for compression. Techniques such as adaptive quantization and entropy coding applied to the transform coefficients can result in significant reduction in bit rates. Depending on the quality levels required by the end user, other parameters such as human visual/acoustic sensitivity, adaptive scanning, statistical modeling, and variable length coding would further contribute to the bit rate reduction. Generally transforms, wavelet transforms in particular, are well suited for scalable coding (in spatial or temporal domains, or in SNR). This concept facilitates data transmission in embedded bit-stream format, providing for multi-resolution (spatial/temporal) and multiquality (SNR) end products, subject to bandwidth limitation, processing power, and cost constraints.

Many international standards relating to audio, video, and data, such as JPEG, H.261, H.262, MPEG-1, MPEG-2, MPEG-4, HDTV, and JPEG-2000, utilize transforms in their overall compression schemes. A number of consumer and commercial products, such as video-CD, DVD, videophone, set-top boxes, digital TV, and digital camera/VCR, have been made possible because of signal compression. Other electronic innovations, such as MP3, video-streaming, and wireless PCS, are completely dependent on the reduction of bit rates made possible by compression. It is not exaggerating to say that data compression is one of the main contributing factors in the explosive growth in information technology.

While different coding schemes can accomplish an amazing amount of compression, the cornerstone is still undoubtedly the underlying transform. It is for this reason that the definitions and properties for each of the transforms dealt with in this handbook are presented with such care and detail. The bibliography sections and Web sites provide further sources of information.

Outline of Chapters

Chapter 1   The Karhunen-Loève Transform

The first transform described in this handbook is the Karhunen-Loève transform (KLT). It takes its rightful place as the leadoff transform to be discussed. Dony does an excellent job of interpreting this statistically optimal transform. The simple and yet elegant explanation of rotation of axes in the data domain to achieve the "principal components" representation underscores the significant energy compaction provided by this transform. Other properties of the transform follow, and the chapter is rounded off with descriptions of applications in chest radiographs and other monochrome and color images. Web sites and software download locations are listed as well.

Chapter 2   The Discrete Fourier Transform

Discrete Fourier transform (DFT), the best known and arguably the most universally applied transform, is presented by Selesnick and Schuller. Following an exposition of the definitions and properties of the DFT, it is shown that by a symmetric extension of the sequence, the DFT can lead to the discrete cosine transform (DCT), another favorite transform described in Chapter 4. The authors then go on to develop the fast Fourier transform (FFT) algorithms, a catalyst for all DFT applications. A novel feature of this chapter is the linkage provided by the authors between DFT and filter-banks, which are used extensively in audio coders. Cosine-modulated filter-banks and complex DFT-based filter-banks are the byproducts of the DFT that are used in Moving Picture Expert Group (MPEG) audio coders. There is an extensive list of Web sites providing information for available software, algorithms, and applications, as well as other related links.

Chapter 3   Comparametric Transforms for Transmitting Eye Tap Video with Picture Transfer Protocol (PTP)

This is a unique, challenging, and provocative chapter written by Mann, the inventor of the wearable computer (WearComp), the Eye Tap camera, and reality mediator. This chapter takes us to the forefront of the multimedia revolution with a new computational/communications device that subsumes the functionality of the videophone, digital camera, and other wireless personal electronics innovations. Mann's invention functions as a true extension of the mind and body and causes the eye to function as if it were a camera. His invention has given rise to a whole new philosophical and mathematical approach to image compression and image storage, and it gives a refreshingly new definition of functionality in image transmission and processing. The new Eye Tap genre of video is best processed and compressed by comparametric equations, essentially equations representing projections and tone scale adjustments of images. Traditionally image compression has been directed to ensure a certain minimum quality or reliability (e.g., worst case scenario). The author instead makes a compelling argument in favour of "best case" scenario; Mann argues that being able to broadcast even intermittent still images to the Internet can provide a measure of security unmatched by conventional "robust" security systems. These arguments are based on a definition of "fear of functionality," a completely novel approach to the idea of security. The author has set up a Web site from which computer programs can be freely downloaded. Such a generous spirit is to be commended. It is also interesting to note that this chapter was typeset using LaTex running on a small wearable computer designed and built by the author.

Chapter 4   Discrete Cosine and Sine Transforms

Next to the DFT, discrete cosine transform (DCT) is probably the most used transform in digital signal processing work. DCT is one of a family of trigonometric transforms including the discrete sine transform (DST). In this chapter, Britanak presents a unified treatment of the family of DCTs and DSTs starting with the definitions, properties, and fast algorithms. This chapter is particularly relevant as the DCT has been adopted in several international standards for image/video coding. In modified form, both DCT and DST have been used in MDCT/MDST audio coding. Computer programs in C (listed in Sections 4.3 and 4.4) that can be implemented to perform the transforms are very useful in all signal processing applications. The chapter concludes with a specific application in a Joint Photographic Experts Group (JPEG) base line system (Fig. 4.3) using the standard test image of Lena.

Chapter 5   Lapped Transforms for Image Compression

Lapped transforms (LTs), developed originally to eliminate or reduce the blocking artifacts of block transforms such as DCT in low bit rate image/video coding, are presented by de Queiroz and Tran. Several versions of the LTs, such as orthogonal and nonorthogonal LTs, tree-structured hierarchical, symmetric, bi-orthogonal, and variable length LTs, are defined, and their properties and factorization schemes are described. Generalized versions of the lapped orthogonal transform (LOT), called GenLOT, are developed in Sections 5.6.3–4 while cosine-modulated LTs, otherwise known as MLT or ELT, are discussed in Section 5.8. To demonstrate the promise and potential for LTs in image coding, well known image compression algorithms are applied to standard test images, with DCT or the wavelet transform replaced by LTs. Comparative analysis shows the elimination of ringing and blocking artifacts that are characteristic of the DCT based coders and also performance rivaling that of the wavelet transforms.

Chapter 6   Wavelet-Based Image Compression

This is another highly valuable chapter as it addresses wavelet-based image compression. Wavelet-based transforms give a time-frequency decomposition of the signal, which has multi-resolution characteristics. The transforms have superior energy compaction and compatibility with Human Visual System (HVS). They make possible the embedded bit-stream coding corresponding to various subbands (the basis for fast browsing of images or databases over the Internet). Discrete wavelet transforms (DWT) and its variants have been adopted both by the FBI in the use of fingerprint image compression and the international standards groups (JPEG-2000 and MPEG-4 still frame image coding). It is highly possible that wavelets may eventually replace DCT in all the coders. Walker and Nguyen provide a clear explanation of the multiresolution aspects of DWT and its implementation using a 2-channel filter bank. Some of the recent enhancements of the basic DWT, such as EZW, SPIHT, WDR, and ASWDR, are enumerated, followed by their implementation in image coding and subsequent evaluation. Various Web sites that provide software, literature, simulation results, and innumerable other details further strengthen the chapter's utility.

Chapter 7   Fractal-Based Image and Video Compression

The concepts and techniques of fractal-based image/video compression are introduced in this chapter by Lu. The seminal work by Mandelbrot forms the basis of many treatises of fractal applications, made popular by movie scenes generated graphically by the use of fractals. Fractal-based signal analysis is currently at the forefront of research. Although compression techniques based on affine transforms or iterated function systems (IFS) may not have caught the attention of every researcher, their attractive properties making possible high compression ratios and asymmetric coding certainly deserve further study. With the advent of super HDTV, wireless cellular multimedia phones, and interactive services on the Internet, fractal transform and its variants such as IFS, QPIFS, and PIFS will find their rightful place in the compression arena. Starting with the basic properties of fractals, Lu demonstrates the compression property of fractals using the encoding/decoding procedures. The capabilities of fractals are illustrated using images and video. As with the other chapters, Web and ftp sites, mostly maintained by universities, provide access to software, literature, products, R&D, and applications to the interested readers.

Chapter 8   Compression of Wavelet Transform Coefficients

The concluding chapter presents a philosophical and thoughtful argument for the effectiveness of transforms in general and wavelets in particular for bandwidth reduction. The superiority of wavelet transform over others, including the widely used DCT, is clearly demonstrated by the characteristics of the DWT. From the chapter's title, the reader may get a wrong impression of duplication with Chapter 6. On the contrary, this chapter complements the topics in Chapter 6 by a clear exposition of the superior performance of the DWT over other transforms. The subband decomposition inherent in dyadic wavelet transform, preservation of spatial signal features in subbands of different scales, and self similarities among subbands of the spatial orientation are some of the reasons for this superiority. These self-similarities are conducive to statistical context modeling and adaptive entropy coding of wavelet coefficients. By a lucid presentation of these concepts aided by implementation on test images, Wu convincingly demonstrates the validity of the DWT adopted in JPEG-2000 and MPEG-4 and the bright future it has in other applications.

Acknowledgements

The editors have been entrusted with the organizational and administrative process in compiling this handbook. Needless to say, without the expertise and efforts of the individual chapter authors, this handbook would never have seen the light of day. The editors sincerely acknowledge the energetic contributions from the chapter authors, whose uniform excellence has made this an outstanding volume. The editors thank the authors for their prompt and timely responses in spite of their heavy commitments in their daily academic or professional lives. It is hoped that the completion of this handbook will elicit a sense of pride and accomplishment, a well-earned and well-deserved reward for their efforts. The editors would also like to thank their families for the patience and perseverance they showed during the months of preparation of this handbook.

List of Acronyms

AFB	Analysis filter bank
ASPEC	Audio spectral perceptual entropy coding
ASWDR	Adaptively scanned wavelet difference reduction
bpp	Bits per pixel
CREW	Compression by reversible embedded wavelets
DCT	Discrete cosine transform
DFT	Discrete Fourier transform
DPCM	Differential pulse code modulation
DSP	Digital signal processing
DST	Discrete sine transform
DTFT	Discrete time Fourier transform
DWP	Discrete wavelet packet
DWT	Discrete wavelet transform
ECECOW	Embedded conditional entropy coding of wavelet
ECG	Electrocardiogram
ELT	Extended lapped transform
EZC	Embedded zerotree coding
EZW	Embedded zerotree wavelet
FAQ	Frequently asked questions
FFT	Fast Fourier transform
FIR	Finite impulse response
FLT	Fast lapped transform
FoF	Fear of functionality
FPGA	Field programmable gate array
GenLOT	Generalized LOT
GNU	GNU's Not Unix
GNUX	GNU-Linux
H.261	Standard for compression of videotelephony and teleconferencing
H.263	Standard for visual communication via telephone lines
HDTV	High definition TV
HLT	Hierarchical lapped transform
HSI	Hue, saturation, intensity
HV	Horizontal vertical
HVS	Human visual system
IDFT	Inverse discrete Fourier transform
IFS	Iterated function systems

ISO	International Standards Organization
ITU	International Telecommunication Union
JBIG	Joint Binary Image Group
JPEG	Joint Photographic Experts Group
JPEG-LS	JPEG-Lossless
KLT	Karhunen-Loève transform
LBT	Lapped bi-orthogonal transform
LOT	Lapped orthogonal transform
LT	Lapped transform
LZC	Layered zero coding
MC	Motion compensated
MDCT	Modified discrete cosine transform
MDST	Modified discrete sine transform
MIMO	Multi-input multi-output
MLT	Modulated lapped transform
MOS	Mean opinion score
MP3	MPEG-Layer 3
MPEG	Moving Pictures Expert Group
MPEG-AAC	MPEG advanced audio coder
MSE	Mean squares error
PAC	Perceptual audio coder
PCA	Principal component analysis
PIFS	Partitioned iterated function systems
PR	Perfect reconstruction
PSD	Personal safety device
PSNR	Peak signal to noise ratio
PTM	Polyphase transfer matrix
PTP	Picture transfer protocol
QCLS	Quadratic-constrained least squares
QM	Cute sound
QPIFS	Quadtree partitioned iterated function systems
RGB	Red, green, and blue
RLC	Run-length coding
RLD	Run-length decoder
ROI	Region of interest
RTT	Round trip time
SDF	Symmetric delay factorization
SFB	Synthesis filter bank
SPIHT	Set partitioning of hierarchical tree
STW	Spatial orientation tree wavelet
SVD	Singular value decomposition
TDAC	Time domain aliasing cancellation
TF	Time-frequency
VLC	Variable-length coding
VLD	Variable-length decoder
VQ	Vector quantization
WDR	Wavelet difference reduction
YIQ	Luminance, in-phase, and quadrature-phase chrominance

Contributors

Vladimir Britanak   Institute of Control Theory and Robotics, Slovak Academy of Sciences, Bratislava, Slovak Republic

Ricardo L. de Queiroz   Digital Imaging Technology Center, Xerox Corporation, Webster, New York

R.D. Dony   School of Engineering, University of Guelph, Guelph, Ontario, Canada

Guojun Lu   Gippsland School of Computing and Information Technology, Monash University, Churchill, Victoria, Australia

Steve Mann   Department of Electrical and Computer Engineering, University of Toronto, Toronto, Ontario, Canada

Truong Q. Nguyen   Department of Electrical and Computer Engineering, Boston University, Boston, Massachusetts

Gerald Schuller   Bell Labs, Lucent Technologies, Murray Hill, New Jersey

Ivan W. Selesnick   Department of Electrical Engineering, Polytechnic University, Brooklyn, New York

Trac D. Tran   Department of Electrical and Computer Engineering, The Johns Hopkins University, Baltimore, Maryland

James S. Walker   Department of Mathematics, University of Wisconsin-Eau Claire, Eau Claire, Wisconsin

Xiaolin Wu   Department of Computer Science, University of Western Ontario, London, Ontario, Canada

Contents

Chapter 1

Karhunen-Loève Transform

R.D. Dony

University of Guelph

1.1 Introduction

The goal of image compression is to store an image in a more compact form, i.e., a representation that requires fewer bits for encoding than the original image. This is possible for images because, in their "raw" form, they contain a high degree of redundant data. Most images are not haphazard collections of arbitrary intensity transitions. Every image we see contains some form of structure. As a result, there is some correlation between neighboring pixels. If one can find a reversible transformation that removes the redundancy by decorrelating the data, then an image can be stored more efficiently. The Karhunen-Loève Transform (KLT) is the linear transformation that accomplishes this.

In Section 1.2 we show how pixels are correlated in typical images. With the pixel values forming the axes of a vector space, a rotation of this space can remove this correlation. The basis vectors of the new space define the linear transformation of the data. The basis vectors of the KLT are the eigenvectors of the image covariance matrix. Its effect is to diagonalize the covariance matrix, removing the correlation of neighboring pixels.

As presented in Section 1.3, the KLT minimizes the theoretical bound on bit rate as given by the signal entropy. The entropy for both discrete random variables and continuous random processes is defined. The KLT also maximizes the coding gain defined as the ratio of the arithmetic mean of the coefficient variances to their geometric mean. Further, the effects of truncation, block size, and interblock correlation are also presented. Section 1.4 presents the results of using the KLT for a number of examples.

1.2   Data Decorrelation

Data from neighboring pixels are highly correlated for most images. Fig. 1.1 shows a typical gray scale image. The image is 512×512 pixels in size with each gray level brightness value of pixel being represented by an 8-bit value for a range of [0–255]. This particular image is commonly used in evaluations and is often referred to as the Lena image. Even with a large degree of detail in many regions, the gray level value of any given pixel tends to be similar to its neighboring pixels. To illustrate this relationship, one can plot the gray level values of pairs of adjacent pixels as shown in Fig. 1.2. Each dot represents a pixel in the image with the x coordinate being its gray level value and the y coordinate being the gray level value of its neighbor to the right. The strong diagonal relationship about the $x = y$ line clearly shows the strong correlation between neighboring pixels.

If we were to block the image into nonoverlapping 1×2 pixel blocks as shown in Fig. 1.3, we can represent an image by a collection of two-dimensional vectors $\mathbf{x}_i$. The scatter plot of this collection is equivalent to Fig. 1.2. Looking at the distributions of the values for each of the two components as shown in Fig. 1.4, we see that they are relatively wide and cover most of the 0–255 range. In fact, the distributions of each component would be quite similar to the overall distribution of individual pixels in the image.

Now, what would happen if we rotated the distribution shown in Fig. 1.2 by $45°$ about the center? The result is shown in Fig. 1.5. The two components are now decorrelated, i.e., knowing the value of the first component does not help in estimating the value of the second. The distributions of the new components are shown in Fig. 1.6. The first component, save for the shift and a scaling factor of $\sqrt{2}$, is still quite similar to the previous distributions — quite broad and covering most of the dynamic range of the original individual pixels. The second component, however, is quite different. It is much narrower, with a strong peak at 0. Because it has a smaller dynamic range, we could encode its value with fewer bits. So even with a decorrelation by a simple rotation of the axis, we can reduce the number of bits required for encoding an image.

In general, a process is decorrelated when, for zero mean random variables x_i and x_j, the expectation of their product, the covariance, is zero if $i \neq j$, i.e.,

$$E\left(x_i x_j\right) = \begin{cases} 0 & i \neq j, \\ \sigma_i^2 & i = j, \end{cases} \tag{1.1}$$

where $E(\cdot)$ is the expectation operator. Using vector notations, we may define the vector of the values of an image block of N pixels as

$$\mathbf{x} = [x_1 \; x_2 \; \ldots \; x_N]^T . \tag{1.2}$$

We can then define the covariance matrix as

$$[\mathbf{C}]_x = E\left[(\mathbf{x} - \mathbf{m})(\mathbf{x} - \mathbf{m})^T\right], \tag{1.3}$$

FIGURE 1.1
Example "Lena" image. Reproduced by Special Permission of *Playboy* magazine. Copyright ©1972, 2000 by Playboy.

where $\mathbf{m} = E(\mathbf{x})$ is the mean. For notational convenience, we will assume zero mean input for the rest of this chapter. In practice, the mean can simply be removed from the data before processing.

We wish to find a linear transformation matrix, $[\mathbf{W}]$, whose transpose, $[\mathbf{W}]^T$, will rotate $\mathbf{x}$ to produce a diagonal covariance matrix for the transformed variable $\mathbf{y}$,

$$\mathbf{y} = [\mathbf{W}]^T \mathbf{x} . \tag{1.4}$$

Each column vector, $\mathbf{w}_i$, of $[\mathbf{W}]$ is a basis vector of the new space. So, alternatively, each element, y_i, of $\mathbf{y}$ is calculated as

$$y_i = \mathbf{w}_i^T \mathbf{x} . \tag{1.5}$$

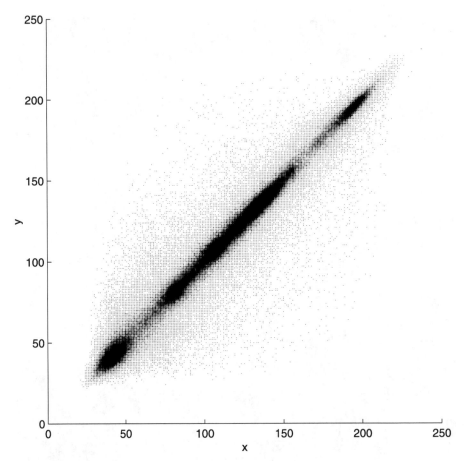

FIGURE 1.2
Scatter plot of adjacent pixel value pairs.

For simple rotations with no scaling, the matrix $[\mathbf{W}]$ must be orthonormal, that is

$$[\mathbf{W}]^T [\mathbf{W}] = [\mathbf{I}] = [\mathbf{W}][\mathbf{W}]^T \tag{1.6}$$

where $[\mathbf{I}]$ is the identity matrix. This means that the column vectors of the matrix are mutually orthogonal and are of unit norm. From Eq. (1.6), it follows that the inverse of an orthonormal matrix is simply its transpose, $[\mathbf{W}]^T = [\mathbf{W}]^{-1}$. The inverse transformation is then calculated as

$$\mathbf{x} = [\mathbf{W}]\mathbf{y} . \tag{1.7}$$

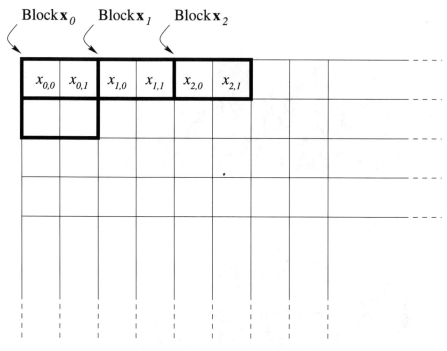

FIGURE 1.3
Image blocking with 1×2 pixel nonoverlapping blocks.

Further, the total energy under the transformation is preserved

$$
\begin{aligned}
\|\mathbf{y}\|^2 &= \mathbf{y}^T \mathbf{y} \\
&= \left([\mathbf{W}]^T \mathbf{x}\right)^T \left([\mathbf{W}]^T \mathbf{x}\right) \\
&= \mathbf{x}^T [\mathbf{W}][\mathbf{W}]^T \mathbf{x} \\
&= \mathbf{x}^T \mathbf{x} \\
&= \|\mathbf{x}\|^2 ,
\end{aligned} \tag{1.8}
$$

where $\|\mathbf{x}\|$ is the norm of the vector $\mathbf{x}$ defined as

$$
\begin{aligned}
\|\mathbf{x}\| &= \sqrt{\mathbf{x}^T \mathbf{x}} \\
&= \sqrt{\sum_{i=1}^{N} x_i^2} .
\end{aligned} \tag{1.9}
$$

For the above example where $N = 2$, by inspection, the matrix $[\mathbf{W}]$ is simply a

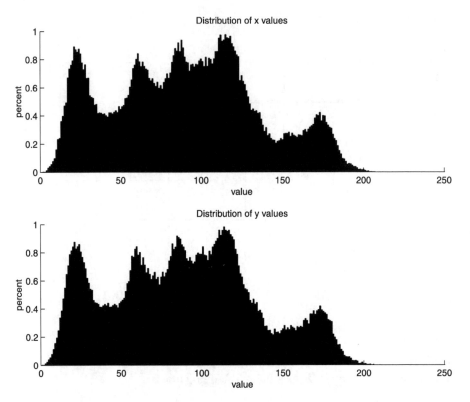

FIGURE 1.4
Distributions for each component.

rotation by 45°

$$[\mathbf{W}] = \begin{bmatrix} \cos 45° & -\sin 45° \\ \sin 45° & \cos 45° \end{bmatrix}. \qquad (1.10)$$

For an arbitrary covariance matrix, the problem of finding the appropriate transformation is the orthonormal eigenvector problem. Since the covariance matrix is real and symmetric, we can find its real eigenvalues and corresponding eigenvectors. Let $[\mathbf{C}]_y$ be the desired diagonal covariance matrix of the transformed variable $\mathbf{y}$ which will be of the form

$$[\mathbf{C}]_y = \begin{bmatrix} \lambda_1 & & 0 \\ & \ddots & \\ 0 & & \lambda_N \end{bmatrix}, \qquad (1.11)$$

where the diagonal elements are the variances of the transformed data. The diagonal

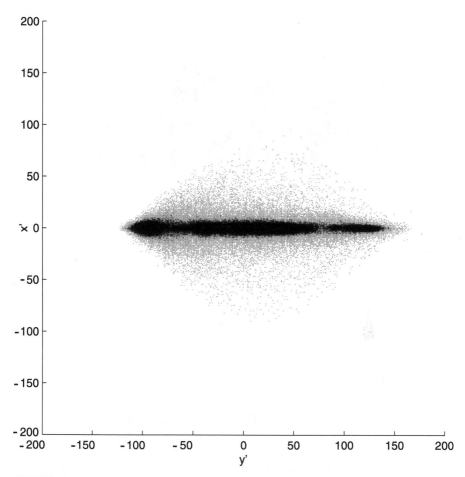

FIGURE 1.5
Scatter plot of pixel value pairs rotated by 45°.

matrix can be calculated from the original covariance matrix, $[\mathbf{C}]_x$, as

$$
\begin{aligned}
[\mathbf{C}]_y &= E\left[\mathbf{y}\mathbf{y}^T\right] \\
&= E\left[\left([\mathbf{W}]^T\mathbf{x}\right)\left([\mathbf{W}]^T\mathbf{x}\right)^T\right] \\
&= E\left[[\mathbf{W}]^T\left(\mathbf{x}\mathbf{x}^T\right)[\mathbf{W}]\right] \\
&= [\mathbf{W}]^T[\mathbf{C}]_x[\mathbf{W}] ,
\end{aligned}
\tag{1.12}
$$

or equivalently,

$$
[\mathbf{C}]_x[\mathbf{W}] = [\mathbf{W}][\mathbf{C}]_y .
\tag{1.13}
$$

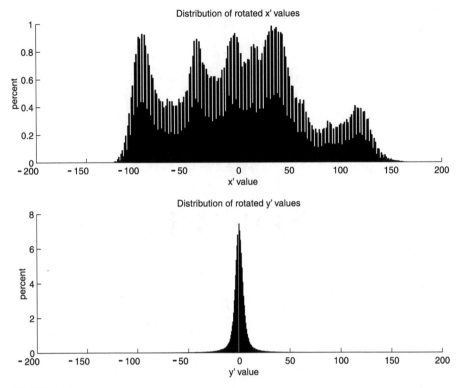

FIGURE 1.6
Distributions for each component of the rotated pixel value pairs.

Since the desired $[\mathbf{C}]_y$ is diagonal, Eq. (1.13) can be rewritten for each column vector, $\mathbf{w}_i$, of $[\mathbf{W}]$ as

$$[\mathbf{C}]_x \mathbf{w}_i = \lambda_i \mathbf{w}_i . \tag{1.14}$$

The solutions for λ_i and $\mathbf{w}_i$ with $i = 1, \ldots , N$ in Eq. (1.14) are the N eigenvalue, eigenvector pairs of the matrix $[\mathbf{C}]_x$ of dimension $N \times N$. That is, each column vector of $[\mathbf{W}]$ is an eigenvector of the covariance matrix, $[\mathbf{C}]_x$, of the original data. To ensure that $[\mathbf{W}]$ is orthonormal, Gram-Schmidt orthogonalization may be applied to the eigenvectors as they are obtained.

This transformation defined by the eigenvalues of the covariance matrix is the Karhunen-Loève transform (KLT), named after Karhunen [17] and Loève [19] who developed the continuous version of the transformation for decorrelating signals. Earlier, Hotelling [15] had developed a "method of principal components" for removing the correlation from the discrete elements of a random variable. As a result, the method is also referred to as the Hotelling transform or principal components analysis (PCA).

1.2.1   Calculation of the KLT

Estimation of Covariance

The calculation of the KLT is typically performed by finding the eigenvectors of the covariance matrix, which, of course, requires an estimate of the covariance matrix. If the entire signal is available, as is the case for coding a single image, the covariance matrix can be estimated from n data samples as

$$[\widehat{\mathbf{C}}]_x = \frac{1}{n} \sum_{i=1}^{n} \mathbf{x}_i \mathbf{x}_i^T , \qquad (1.15)$$

where $\mathbf{x}_i$ is a sample data vector. If only portions of the signal are available, care must be taken to ensure that the estimate is representative of the entire signal. In the extreme, if only one data vector is used then only one nonzero eigenvalue exists, and its eigenvector is simply the scaled version of the data vector. For typical images, it is rarely the case that their covariance matrix has any zero eigenvalues. For a data vector of dimension N, a good rule of thumb is that at least $10 \times N$ representative samples from the various regions within an image be used to ensure a good estimate if it is not feasible to use the entire image.

Calculation of Eigenvectors

While it is beyond the scope of this chapter to provide a detailed discussion of the algorithms for extracting the eigenvalues and eigenvectors, we will present a brief overview of the general methods commonly used. The reader is referred to [16, 28] for more detailed explanations. For actual implementations of the methods, many numerical packages such as LAPACK [22] (which is based on EISPACK [21] and LINPACK [23]), MATLAB [20], IDL [31], and Octave [11], and the routines in "cookbooks," such as that by Press et al. [28], provide routines for the solution of eigensystems.

A simple approach is the Jacobi method. It develops a sequence of rotation matrices, $[\mathbf{P}]_i$, that diagonalizes $[\mathbf{C}]$ as

$$[\mathbf{D}] = [\mathbf{V}]^T [\mathbf{C}][\mathbf{V}] , \qquad (1.16)$$

where $[\mathbf{D}]$ is the desired diagonal matrix and $[\mathbf{V}] = [\mathbf{P}]_1[\mathbf{P}]_2[\mathbf{P}]_3 \cdots$. Each $[\mathbf{P}]_i$ rotates in one plane to remove one of the off-diagonal elements. It is an iterative technique which is terminated when the off-diagonal values are close to zero within some tolerance. Upon termination, the matrix $[\mathbf{D}]$ contains the eigenvalues on the diagonals and the columns of $[\mathbf{V}]$ are the basis vectors of the KLT.

While this technique is quite simple, for larger matrices it can take a large number of calculations for convergence. A more efficient approach for larger, symmetric matrices divides the problem into two stages. The Householder algorithm can be applied to reduce a symmetric matrix into a tridiagonal form in a finite number of steps. Once the matrix is in this simpler form, an iterative method such as QL factorization can be used to generate the eigenvalues and eigenvectors. The advantage of this approach is that the factorization on the simplified tridiagonal matrix typically requires fewer iterations than the Jacobi method.

Recently, there has been some interest in iterative methods of principal components extraction that do not require the calculation of a covariance matrix [7, 14, 26]. These techniques update the estimate of the eigenvectors for each input training vector. One such method developed by Oja [25] is of the form

$$\hat{\mathbf{w}}(t+1) = \hat{\mathbf{w}}(t) + \alpha \left[y(t)\mathbf{x}(t) - y^2(t)\hat{\mathbf{w}}(t) \right] , \qquad (1.17)$$

where $\mathbf{x}$ is an input vector, $\hat{\mathbf{w}}(t)$ is the current estimate of the basis vector, $y = \mathbf{w}^T\mathbf{x}$ is the coefficient value, and α is a learning-rate parameter. Eq. (1.17) has been shown to converge to the largest principal component [14, 27]. This algorithm can be generalized through deflation to extract any or all of the principal components [7, 33]. Also, adaptive schemes have been based on this method [8]. While these algorithms have some advantages over covariance-based methods, there are still some concerns over stability and convergence [3, 4, 35].

Markov-1 Solution

The calculation of the eigenvectors for an arbitrary covariance matrix can still require a large number of computations. However, there is a special class of matrix that has an analytical solution for its eigenvectors and eigenvalues [29, 30]. If a process were to have a covariance function of the form

$$[\mathbf{C}]_{ij} = \sigma^2 \rho^{|i-j|} , \qquad (1.18)$$

where ρ is the correlation coefficient such that $0 < \rho < 1$, such a process is referred to as a first order stationary Markov process or simply Markov-1. The solution for the ith element of the jth basis vector for N-dimensional data is given by

$$w_{ij} = \left[\frac{2}{(N+\mu_j)} \right]^{1/2} \sin\left\{ r_j \left[(i+1) - \frac{(N+1)}{2} \right] + (j+1)\frac{\pi}{2} \right\} , \qquad (1.19)$$

where μ_j is the jth eigenvalue calculated as

$$\mu_j = \left(1 - \rho^2 \right) \left[1 - 2\cos\left(r_j \right) + \rho^2 \right] , \qquad (1.20)$$

and r_j is the jth real positive root of the transcendental equation

$$\tan\left(Nr \right) = -\frac{\left(1 - \rho^2 \right) \sin\left(r \right)}{\cos\left(r \right) - 2\rho + \rho^2 \cos\left(r \right)} . \qquad (1.21)$$

To extend this to two-dimensional data, one can assume a separable transform. The horizontal and vertical correlation coefficients, ρ_H and ρ_V, are estimated from the image to calculate a horizontal basis set, $w_{ij}^{(H)}$, and vertical basis set, $w_{ij}^{(V)}$, respectively.

Then, the i, j element of the kth two-dimensional basis vector, w_{ijk}, is calculated as the product of the two:

$$w_{ijk} = w_{ik}^{(H)} w_{jk}^{(V)} . \tag{1.22}$$

As many images exhibit a Markov-1 structure, this solution to the KLT can be quite useful due to its ease of generation.

1.3 Performance of Transforms

On its own, an orthonormal transformation does not effect data compression. The blocks of pixels are simply transformed from one set of values to another and, for reversible transformations, back again on reconstruction. To reduce the number of bits for representing an image, the coefficients are quantized, incurring some irreversible loss, and then encoded for more efficient representation. By decorrelating the data before these steps using the KLT, more data compaction can be achieved.

To examine the effects of this extra efficiency, we can make use of Shannon's information measures [34].

1.3.1 Information Theory

The information conveyed by an observation of some random process is related to its probability of occurrence. If an observation were all but certain to occur, i.e., its probability were close to 1, it would not be very informative. However, if it were quite unexpected, the observation would convey much more information. Shannon formalized this relationship between the probability of an event, $P(x)$, and its information content, $I(x)$, as

$$I(x) = -\log P(x) . \tag{1.23}$$

If the logarithm is taken with respect to base 2, the information, $I(x)$, is measured in units of *bits*.

A random variable, $\mathbf{x}$, is a collection of all possible events and their associated probabilities. The average information for a random variable can be calculated as

$$\begin{aligned} H(\mathbf{x}) &= \sum_i P(x_i) I(x_i) \\ &= -\sum_i P(x_i) \log P(x_i) , \end{aligned} \tag{1.24}$$

where the sum is taken through all possible events. The average information is called the entropy of the process.

Entropy is useful in determining theoretical performance measures of compression methods. Shannon showed that entropy gives a lower bound on the average number of bits required to encode the events of a random process without introducing error. In other words, one needs at least as many bits per event, on average, as the entropy to represent a set of observations.

However, these measures are not directly applicable to the coefficients of an arbitrary transformation. They are defined for discrete events whereas the coefficients, since they are floating-point values, must be considered real-valued samples of continuous distributions. Since the probability of any such real-valued sample is zero, the (discrete) entropy is undefined. Instead, we define the *differential entropy* [13] as

$$h(x) = -\int_{-\infty}^{\infty} p(s) \log p(s)ds . \tag{1.25}$$

For simple distributions such as the Gaussian, uniform, or Laplacian distributions the differential entropy is of the form

$$h(x) = \frac{1}{2} \log \sigma_x^2 + k , \tag{1.26}$$

where σ_x^2 is the variance of the random variable and k is a distribution-dependent constant (e.g., for a Gaussian, $k = \frac{1}{2} \log_2 2\pi e$) [1].

A good transformation, then, should minimize the sum of the differential entropies for the resulting coefficients. Due to the logarithmic term, this is equivalent to minimizing the product of the variances of the coefficients. However, recall that for any orthonormal transformation, the total energy is preserved, so the sum of the coefficient variances is fixed. One measure of the efficiency of the transform is the coding gain [10] defined as the ratio between the algebraic mean of the variances, which is independent of the transform, and the geometric mean of the variances, which is transform dependent:

$$G_W = \frac{\frac{1}{N}\sum_{i=1}^{N}\sigma_{y_i}^2}{\left(\prod_{i=1}^{N}\sigma_{y_i}^2\right)^{1/N}} . \tag{1.27}$$

For the raw signal, before any transformation, all the variances are approximately equal giving a unity coding gain. Any increase in one of the coefficient variances must be matched by an equal decrease in one or more of the other variances for an orthonormal transform. The arithmetic mean is therefore the same, but the geometric mean decreases resulting in a coding gain of greater than one.

For a given energy of the signal, minimizing the product of the variances maximizes the coding gain. Conversely, maximizing the coding gain minimizes the lower bound on the number of bits required to encode the image. So, to minimize the product of the variances given a fixed sum, one should maximize the variance of the first coefficient. Next, subject to the orthonormality constraint, maximize the variance of the second coefficient, and so on. This procedure is nothing more than extracting the principal components or, equivalently, generating the KLT. Therefore, the KLT, by decorrelating the data, produces a set of coefficients that minimizes the differential entropy of the data.

1.3.2 Quantization

In transform coding, the transform coefficients are quantized to effect the data reduction. While the transformation is reversible, quantization is not, and therefore introduces error. Let $\hat{\mathbf{y}}$ be the set of quantized coefficient values for a block. On reconstruction, the block is calculated as

$$\hat{\mathbf{x}} = [\mathbf{W}]\hat{\mathbf{y}} . \tag{1.28}$$

The squared error for the block is calculated as

$$
\begin{aligned}
\varepsilon^2 &= \left\| \hat{\mathbf{x}} - \mathbf{x} \right\|^2 \\
&= (\hat{\mathbf{x}} - \mathbf{x})^T (\hat{\mathbf{x}} - \mathbf{x}) \\
&= ([\mathbf{W}]\hat{\mathbf{y}} - [\mathbf{W}]\mathbf{y})^T ([\mathbf{W}]\hat{\mathbf{y}} - [\mathbf{W}]\mathbf{y}) \\
&= (\hat{\mathbf{y}} - \mathbf{y})^T [\mathbf{W}]^T [\mathbf{W}] (\hat{\mathbf{y}} - \mathbf{y}) \\
&= (\hat{\mathbf{y}} - \mathbf{y})^T (\hat{\mathbf{y}} - \mathbf{y}) \\
&= \left\| \hat{\mathbf{y}} - \mathbf{y} \right\|^2 .
\end{aligned}
\tag{1.29}
$$

So, the squared error on reconstruction is the same as the squared error of the coefficients for orthonormal transformations.

The quantized coefficients are typically encoded using a lossless method, such as arithmetic coding or Huffman coding. These methods can, at best, reduce the average number of bits to the entropy of the quantized coefficients.

To illustrate the advantage of performing the KLT before quantization, we calculate the total entropy for a number of quantization intervals on both the original data and the transformed data. For this example, a midstep, uniform quantizer is used where the quantized value is calculated as

$$\hat{y} = q \text{ round } (y/q) , \tag{1.30}$$

based on the width of the quantization interval, q, where the function round(x) returns the nearest integer to the real value x. The results are shown in Fig. 1.7. For a given squared error due to quantization, the entropy in bits per pixel is less for the transformed data than for the original data.

1.3.3 Truncation Error

Another approach to reducing the data and hence introducing error is the complete removal of a number of the coefficients before quantization. Say only M of the N coefficients were to be retained. The resulting expected squared error is calculated as

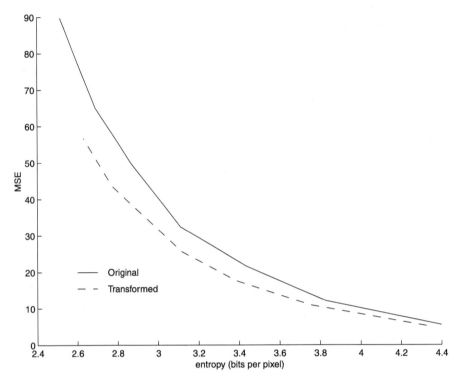

FIGURE 1.7
Plot of mean squared error (MSE) versus entropy in bits per pixel for a number
of quantization widths.

$$
\begin{aligned}
E\left[\varepsilon^2\right] &= E\left[\frac{1}{N}\sum_{i=1}^{N}\left(y_i - \hat{y}_i\right)^2\right] \\
&= \frac{1}{N}E\left[\sum_{i=1}^{M}(y_i - y_i)^2 + \sum_{i=M+1}^{N}(y_i - 0)^2\right] \quad (1.31) \\
&= \frac{1}{N}E\left[\sum_{i=M+1}^{N}y_i^2\right] \\
&= \frac{1}{N}\sum_{i=M+1}^{N}\sigma_i^2 \, .
\end{aligned}
$$

Recall that for the KLT the variances of the coefficients, σ_i^2, are the eigenvalues, λ_i, of the covariance matrix. To minimize the expected squared error, the M coefficients corresponding to the M largest eigenvalues should be kept.

Notice that the above minimization is valid for any transformation whose M basis vectors span the M-dimensional subspace defined by the M largest principal components (eigenvectors for the M largest eigenvalues). However, only the KLT ensures that the remaining coefficients can be coded with the minimum number of bits since it minimizes the differential entropy of the coefficients. To illustrate this point, let us generate the 64 KLT basis vectors for an 8×8 blocking of the test image and keep only the first four. The variances of the resulting coefficients are shown in the first column of Table 1.1. The MSE due to the removal of the 60 lowest variance coefficients is 96.1. Now, let us generate another set of 4 basis vectors by taking random linear combinations of the first 4 KLT basis vectors. The new set still spans the space defined by the original 4 KLT basis vectors. As a result, the MSE due to truncation and the sum of the remaining variances are identical to those of the KLT bases. However, the product of the variances is much higher, and, as a result, the coding gain is much smaller than for the KLT bases. This means that the representation is less efficient and will require more bits to encode the coefficients for the same degree of distortion.

Table 1.1 Performance Differences Between First Four Basis Vectors of KLT and a Random Combination of Them

	KLT bases	Random span
σ_1^2	113995	20876
σ_2^2	6880	18236
σ_3^2	2727	79310
σ_4^2	1691	6873
$\sum_{i=1}^{4} \sigma_i^2$	125294	125294
$\sum_{i=5}^{64} \sigma_i^2$	6147	6147
Truncation MSE	96.1	96.1
$\prod_{i=1}^{4} \sigma_i^2$	3.6×10^{15}	207.5×10^{15}
Coding gain	4.04	1.47

1.3.4 Block Size

The question remains of what size to use for the image blocks. The larger the block, the greater the decorrelation, hence the greater the coding gain. However, the number of arithmetic operations for the forward and inverse transformations increases linearly with the number of pixels in the block. Furthermore, the size of the covariance matrix is the square of the number of pixels. Not only does the calculation of the eigenvectors require more resources, but the number of samples to get a reasonable estimate of the covariance matrix increases significantly. As well, if the set of KLT basis vectors is to be kept with the image for reconstruction, the size of the basis set is also of concern. Therefore, there is a trade-off between computational requirements and the degree of decorrelation in determining the block size.

Fig. 1.8 shows the coding gain as a function of block size for the test image. It clearly shows that the use of larger block sizes results in larger coding gains. For example, increasing the block size from 4 × 4 to 8 × 8 increases the gain from 27 to 39. However, the number of floating point operations per pixel increases by a factor of four from 32 to 128.

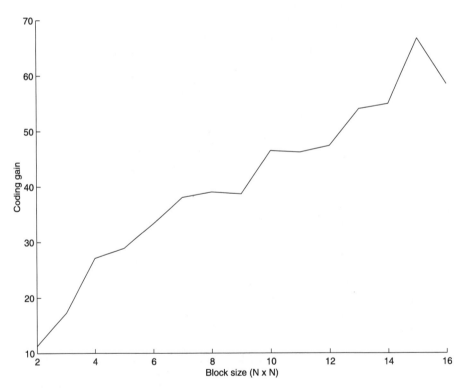

FIGURE 1.8
Coding gain as a function of block size for test image.

Of course, using a block the same size as the image results in a perfect coding gain since the entire image can be represented by a single component. Unfortunately, this representation is so image specific that the transform basis itself must also be included with the compressed image to enable reconstruction. Since the basis vector *is* the image, one is no further ahead. However, such full-frame transform coding may be appropriate for sequences or collections of similar images.

Interblock Correlation

The KLT produces decorrelated coefficients within the image blocks. There is no assurance, however, that the coefficients from block-to-block are also decorrelated. In fact, for most images there is a significant correlation between the first coefficients for adjacent blocks. For example, Fig. 1.9 shows the scatter plot of adjacent pairs of the first coefficient for the 8 × 8 KLT of the test image. Note the strong correlation between the adjacent values. In contrast, Fig. 1.10 shows little if any correlation between adjacent second coefficients.

A simple method of reducing such correlation is to encode only the difference between adjacent coefficients after initially encoding the first. This method is known as differential pulse code modulation (DPCM). The use of DPCM on the first coefficients significantly increases the overall coding efficiency by reducing the variance of the coefficient. For example, performing DPCM on the first coefficient of the above 8 × 8 KLT coefficients reduces the variance from 113995 to 51676. The resulting scatter plot of the adjacent pairs of differences is shown in Fig. 1.11. The use of DPCM has removed the correlation between adjacent values of the first coefficient.

1.4 Examples

1.4.1 Calculation of KLT

To calculate the KLT of an image, the covariance matrix is first estimated. The estimate is calculated from the set of sequential nonoverlapping blocks for the image. For the following examples, blocks of 8 × 8 pixels are used. For the "Lena" image, this results in 4096 blocks. The eigenvalues and the corresponding eigenvectors are extracted from the covariance matrix. Because the matrix is symmetric, the eigenvalues and eigenvectors can be calculated using the tridiagonalization and QL factorization approach.

The resulting 64 basis vectors are shown in Fig. 1.12 as two-dimensional basis images or blocks. The bases are in order from the largest variance at the top left to the lowest at the bottom right. Dark pixels represent negative values and light pixels represent positive values. The first basis is almost flat due to the similarity of pixel values within most blocks. As was the case for the two-dimensional scatter plot of Fig. 1.2, the 64-dimensional scatter plot would show a strong concentration of points along the diagonal line $x_1 = x_2 = \cdots = x_{64}$. As this is true for most images, the

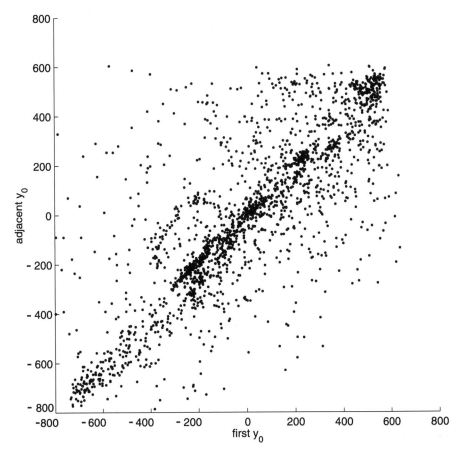

FIGURE 1.9
Scatter plot of adjacent pairs of the first coefficient.

first component of the KLT tends to be constant or d.c. As the variance increases, the degree of variation, or frequency, increases. This relationship generally agrees with the form of the KLT solution for a Markov-1 process as shown in Eq. (1.19) where the frequency increases as the basis index increases. Again, as most images have an approximate Markov-1 structure, the form of the KLT bases are similar.

1.4.2   Quantization and Encoding

Once the coefficients are calculated, they are quantized and then losslessly encoded. There are numerous such methods, but a discussion and comparison of them would be beyond the scope of this chapter. For illustrative purposes, we will use an encoding scheme similar to that adopted by the JPEG standard [36]. The coefficients are quantized by a midstep uniform quantizer as defined in Eq. (1.30). For simplicity, the

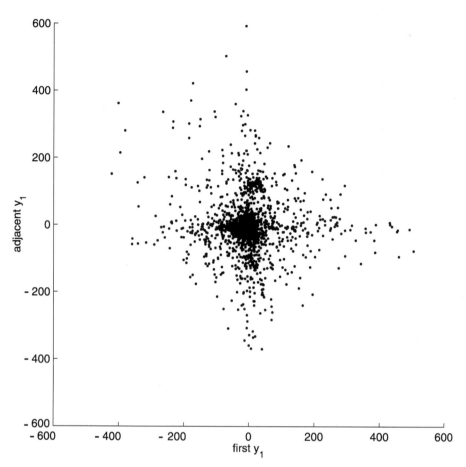

FIGURE 1.10
Scatter plot of adjacent pairs of the second coefficient.

same quantization step size, q, is used for all coefficients, unlike the JPEG standard that varies the degree of quantization for each coefficient according to the visibility of error as judged by human observers. Each quantized coefficient is encoded first by a Huffman encoded value for the number of bits required by the coefficient followed by the minimum number of bits for the coefficient value itself. Zero-valued coefficients from adjacent blocks are run-length encoded for further compaction.

The results for various degrees of quantization are shown in Table 1.2. As the coarseness of quantization increases, the size of the file decreases resulting in greater compression. The equivalent average number of bits per pixel is also shown. For comparison to show the efficiency of the coefficient encoding, the entropy of the quantized coefficient values is also shown. The actual bit rate and the entropy are very similar. At high compression the actual bit rate is slightly lower than the entropy because of the run-length encoding of zero values.

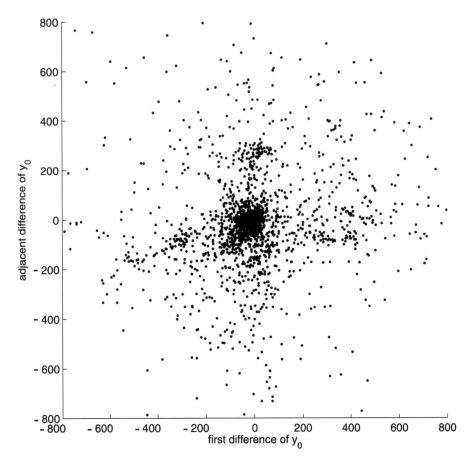

FIGURE 1.11
Scatter plot of adjacent pairs of differences of the first coefficient.

As the bit-rate decreases, distortion increases. Table 1.2 shows the distortion in two equivalent common measures [6]. The mean squared error (MSE) is defined as

$$\text{MSE} = E\left[(x - \hat{x})^2\right], \tag{1.32}$$

where x is the original pixel value and $\hat{x}$ is the reconstructed value. The peak signal-to-noise ratio (PSNR) is a logarithmic measure of distortion given in decibels (dB) and is defined as

$$\text{PSNR} = 10\log_{10}\frac{(255)^2}{E\left[(x - \hat{x})^2\right]}, \tag{1.33}$$

where 255 is the peak value of an 8-bit image. The larger the PSNR value, the better the accuracy of reconstruction. The plot of the distortion as PSNR versus the bit

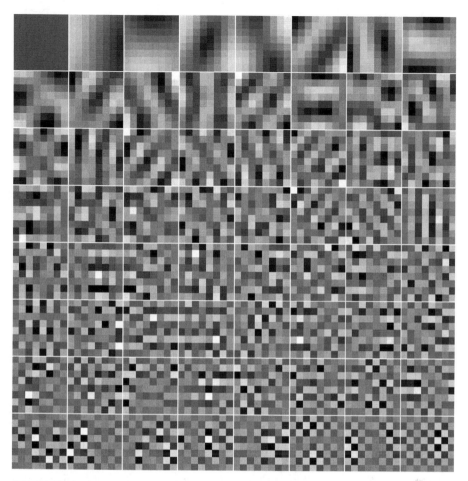

FIGURE 1.12
KLT basis images for "Lena" image.

rate is shown in Fig. 1.13. From rate-distortion theory, for a stationary memoryless Gaussian source, the bit rate, R, as a function of the squared error distortion, ε^2, is given by [1]

$$R(\varepsilon) = \begin{cases} \frac{1}{2}\log_2\left(\sigma^2/\varepsilon^2\right) & 0 \leq \varepsilon^2 < \sigma^2, \\ 0 & \sigma^2 \leq \varepsilon^2. \end{cases} \quad (1.34)$$

For high bit rates, the rate-distortion curve follows the logarithmic relationship between the squared error and the bit rate. As the quantization interval increases, the distortion overtakes the variance for more coefficients. As a result, the curve begins to drop sharply as the distortion increases without a corresponding further reduction in bit rate. In the limit as the quantization interval increases, the bit rate becomes zero

Table 1.2 Compression of "Lena" Image Using KLT

Quantizer Width	File Size (bytes)	Bits/pixel	Entropy (bits)	MSE	PSNR (dB)
2	139948	4.27	4.08	0.42	51.95
4	109141	3.33	3.11	1.42	46.62
8	78820	2.41	2.18	5.19	40.98
16	42245	1.29	1.28	15.01	36.37
24	27196	0.83	0.90	23.78	34.37
36	18375	0.56	0.64	36.27	32.54
48	13893	0.42	0.50	48.45	31.28
64	10548	0.32	0.39	64.70	30.02
92	7547	0.23	0.28	93.68	28.41
128	5492	0.17	0.21	130.19	26.98
192	3797	0.12	0.15	199.21	25.14
256	2831	0.09	0.11	273.42	23.76
512	1457	0.04	0.06	638.18	20.08

and the squared error is then simply the variance.

Fig. 1.14 shows the reconstructed image after a compression of 10:1 (0.8 bits per pixel). Overall, very little distortion is visible. Areas of constant brightness, edges, lines, and textured regions are all reproduced quite faithfully. Even on closer examination, little distortion is evident, as shown by comparing Figs. 1.15(a) and (b). At 10:1 compression, some minor distortion is seen as spurious texture in the background. As well, the lone feather piece in the center-left region is somewhat distorted. As the compression ratio increases, though, the distortion becomes more apparent, as shown by Figs. 1.15(c) and (d) for ratios of 20:1 and 40:1, respectively. The texture of the hat is lost in areas at 20:1, while artifacts in the background region are more pronounced. The edges of the hat, however, are still rather crisp and the textured region of the feathers on the brim does not seem as distorted as the hat texture. Because the set of bases is image specific, certain features, such as these, may be well represented and be somewhat resistant to distortion at moderate compression ratios. By 40:1, though, the image is quite distorted. This type of distortion is sometimes referred to as "block effect distortion" because the block boundaries used in block transform coding are visible.

1.4.3 Generalization

In theory, the transform basis set for the KLT is specific to a particular image. However, in practice the statistics of images at the block-size level of detail tend to be similar. As a result, the KLT computed from one set of image data performs quite well on another set. For example, the above results were based on the KLT computed from the covariance matrix of the set of sequential, nonoverlapping blocks from the image. These blocks are the exact data that are used to encode the image. If the covariance

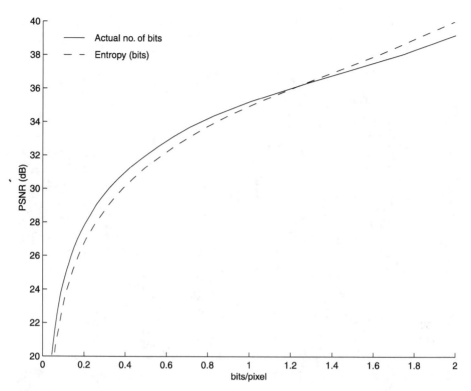

FIGURE 1.13
Plot of distortion (PSNR) versus bit rate showing both the entropy and actual coding rates.

matrix were to be calculated from randomly chosen blocks from arbitrary locations on the image, the data for generating the KLT would be different from the data used in encoding the image. Fig. 1.16 shows the results for both the KLT generated from the sequential set of blocks and a set of 4096 randomly chosen blocks. While the transform generated from the same data to be coded performs better, the improvement is not significant.

What happens if the KLT is generated based on an image completely different from the one being encoded? A second test image, "Goldhill," is shown in Fig. 1.17. This image was encoded using the KLT generated from the image and the KLT originally generated from the "Lena" image. The rate-distortion curves are shown for both cases in Fig. 1.18. As expected, using the same data for generating the transform as for encoding results in better performance than using different data to generate the transform. However, as the figure shows, this increase is only minor. In this case, the transformation based on the "Lena" image generalizes well to the other image.

FIGURE 1.14
Image after compression of 10:1, MSE = 24.8, PSNR = 34.2 dB. Reproduced by
Special Permission of *Playboy* magazine. Copyright ©1972, 2000 by Playboy.

1.4.4 Markov-1 Solution

To compare the usefulness of the Markov-1 solution to the KLT, we first look at
the autocorrelation of the image. As shown in Table 1.3, the autocorrelation does
appear to follow the Markov-1 model of $E[x_i x_j] = E[x^2]\rho^{|i-j|}$ with $\rho_H = 0.9543$
for horizontally neighboring pixels. A similar relationship also holds for vertically
neighboring pixels with $\rho_V = 0.9768$. For simplicity we will assume a separable,
isotropic distribution and choose $\rho = 0.9543$ for both directions. The resulting KLT
bases are shown in Fig. 1.19. Note the strong sinusoidal nature of the basis images.
The rate-distortion results for using this set of KLT bases are shown in Fig. 1.20 along
with the original results for the KLT generated from the image itself. Since the two

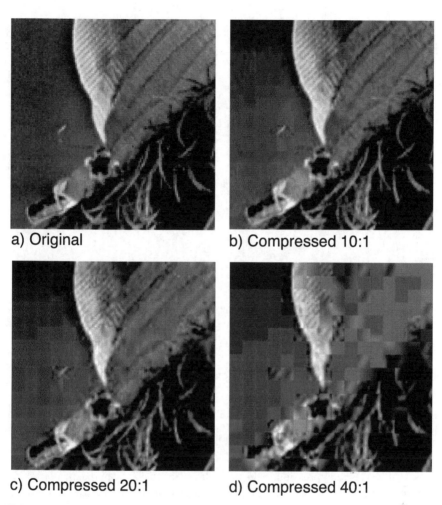

a) Original   b) Compressed 10:1

c) Compressed 20:1   d) Compressed 40:1

FIGURE 1.15
Details of image before and after 10:1, 20:1, and 40:1 compression. (a) Original, (b) Compressed 10:1, (c) Compressed 20:1, (d) Compressed 40:1. Reproduced by Special Permission of *Playboy* magazine. Copyright ©1972, 2000 by Playboy.

curves are almost identical, the savings in computational resources from having a closed form solution for the Markov-1 case incurs little if any cost in performance.

1.4.5 Medical Imaging

One of the most demanding application areas for the use of image compression is the compression of medical images. The implications of introducing any sort of distortion in this class of images are grave. There are numerous legal and regulatory issues which consequently are of concern [37]. As a result, there is an argument for

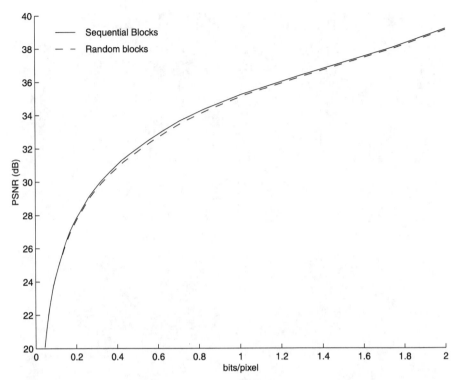

FIGURE 1.16
Plot of distortion versus bit rate for KLT calculated from both randomly chosen blocks and sequential blocks.

the use of lossless compression in this field; however, such an approach is of limited usefulness due to the theoretical limits on the maximum allowable compression.

The question, of course, is how much compression can be achieved? For lossy image compression methods, this is the same as asking how much distortion can be introduced in the reconstructed image. To answer this question, the end-use of the images must properly be defined. For the following example, as originally presented in Dony et al. [9], the application is for educational use. Currently, radiology residents acquire their diagnostic skills through examining actual clinical images of normal patients as well as those with various pathologies. With the growth in digital imaging, it is now possible to store such a library of images digitally in a computer database. The residents would be free to call up any of the images and examine them at their convenience. The evaluation criteria for this environment are quite different from, say, a diagnostic environment. In the educational environment, the diagnosis or pathology is given beforehand. It is sufficient that an image show clearly the pathology in question or the characteristics of a normal image. So, it is the overall quality of the image and the visibility of the pathology as judged by an experienced radiologist which must be measured.

FIGURE 1.17
Second test image, "Goldhill."

Nine digital chest radiographs (X-rays) obtained for clinical reasons were selected for evaluation as being representative of both normal anatomy and pathology. A sample image is shown in Fig. 1.21. Each of the nine images was compressed using an adaptive variation of the KLT at 10:1, 20:1, 30:1, and 40:1, and the five versions of each image were presented simultaneously to each of seven radiologists, in random order and without the evaluator knowing the degree of compression. The radiologists were asked to rank image quality and visibility of pathology in the context of their suitability for educational use. Possible ratings varied from excellent, good, and fair — acceptable — and poor or bad — unacceptable. A mean opinion score (MOS) was calculated by assigning a numeric value to each rating, e.g., excellent scored 5 points and bad 1 point [24].

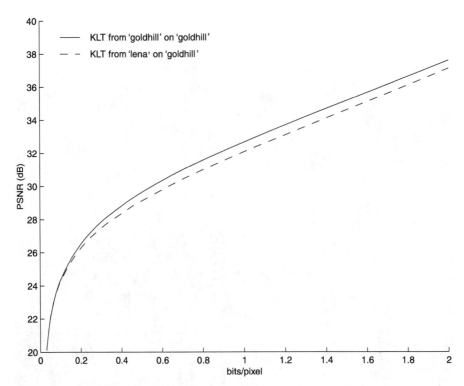

FIGURE 1.18
Distortion versus bit rate for "Goldhill" image using KLT from both "Goldhill"
image and "Lena" image.

The results of evaluation are summarized in Fig. 1.22 which shows the plot of the mean opinion score for both scoring criteria. The figure shows that the MOS at the various degrees of compression remains quite close to that of the original. For image quality, the MOS for the original is 4.28 and drops only to 4.01 at 40:1. The MOS for the pathology visibility is 4.33 for the original and 4.10 for the 40:1 compression ratio. Therefore the use of a compression method based on the KLT results in usable images at even relatively high compression.

1.4.6 Color Images

Another application of the decorrelation abilities of the KLT is the compression of color images. Color images can be represented by three color components per pixel. Typically these are the three primary colors, red, green, and blue (RGB), corresponding to the responses of the three color receptors in the retina of the human eye. Similarly, in most color vision systems, three color filters of red, green, and blue are used to produce, respectively, the three color components per pixel. From the original RGB data, there are numerous transformations that can represent color values

Table 1.3 Correlation Between First 8
Neighboring Pixels on the Rows

	$E[x_i x_j]$	$E[x_i x_j]/E[x_{i-1}x_j]$
$\|i - j\| = 0$	2657	-
$\|i - j\| = 1$	2589	0.9744
$\|i - j\| = 2$	2472	0.9546
$\|i - j\| = 3$	2338	0.9460
$\|i - j\| = 4$	2223	0.9510
$\|i - j\| = 5$	2111	0.9492
$\|i - j\| = 6$	2010	0.9524
$\|i - j\| = 7$	1914	0.9523

in different coordinate spaces [18]. Some, for example HSI, express the components in a form that follows more closely the human perceptions of color qualities such as hue, saturation, and intensity. Others, for example YIQ, attempt to decorrelate the chromatic and intensity information. For the following example, we will explore the use of the decorrelation property of the KLT on the raw RGB data.

A simple approach to compression would be to treat each of the three RGB components as separate images. However, this method does not exploit the correlation between the three color values at each pixel. An alternative is to include all three component pixel values within a block. For example, an 8×8 block will contain 192 individual values. The KLT can then decorrelate the component values allowing improved coding.

To show the difference in coding performance between combining and not combining the three component values, the image shown in Fig. 1.23 is used as a test image. The image is 512×768 pixels in size and each pixel has 3 RGB values of 8 bits each for a total of 24 bits per pixel. For the separate encoding, three transforms were calculated and applied, one for each component. The resulting rate-distortion relationship is shown as the dashed curve in Fig. 1.24. The bit rate combines the file sizes of all three components and the distortion is the mean across the components. For the combined method, the image was divided into blocks of 8×8 pixels $\times 3$ components for a total input dimension of 192. The performance of the KLT generated from this data is shown by the solid curve of Fig. 1.24. The figure shows that the difference in performance is substantial. For example, at a compression of 12:1 (2 bits per pixel), allowing the transform to decorrelate the RGB components results in a 4 dB increase in fidelity. Again, this example shows that the greater the decorrelation, the better the performance of the transform.

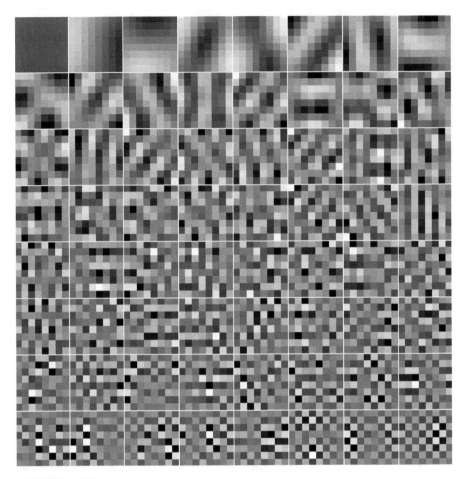

FIGURE 1.19
KLT basis images for Markov-1 model, $\rho = 0.9543$.

1.5 Summary

The Karhunen-Loève transform (KLT) is defined as the linear transformation whose basis vectors are the eigenvectors of the covariance matrix of the data. As it diagonalizes the covariance matrix, it decorrelates the data. The resulting set of coefficients can be encoded with fewer bits for a given distortion than the raw data.

The KLT is the optimal transformation in terms of minimizing the bit rate. The use of eigenvectors as the basis vectors ensures that the variance of the first coefficient is maximized, and, subject to the orthogonality of basis vectors, all subsequent coefficient variances are maximized in order. Maximizing each variance means that

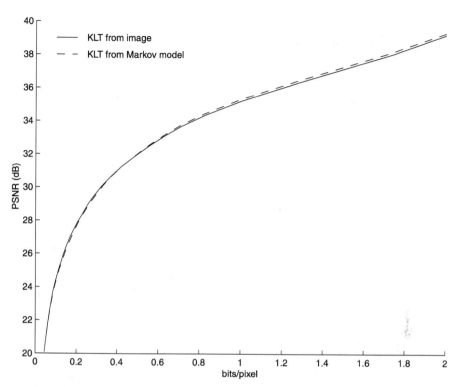

FIGURE 1.20
Plot of distortion (PSNR) versus bit rate for the KLT from the image covariance matrix and the KLT generated from the Markov-1 model.

the product of all the variances is minimized due to the energy preserving nature of any orthonormal transformation. Since the total differential entropy for the blocks increases with the product of the variances, the KLT minimizes the entropy thereby minimizing the bound on the bit rate.

The transform has a number of important performance characteristics for image compression. At moderate compression ratios, very little distortion is visible. As the compression ratio increases, more distortion becomes evident. However, because the transform is based on data from the image, some areas remain faithfully reproduced at even relatively low bit rates. The most prominent feature of the distortion as the compression ratio increases is the blocking effects of using finite sized blocks. While the KLT is calculated from the covariance matrix of an image and the covariances of different images are rarely identical, the transform based on one image can still perform well on a different image since the second order statistics of many images are rather similar. Even the use of the quite general Markov-1 model for the covariance results in performance almost as effective as the strictly image-specific transformation. As well, the decorrelating property of the transform can be used successfully on pixel

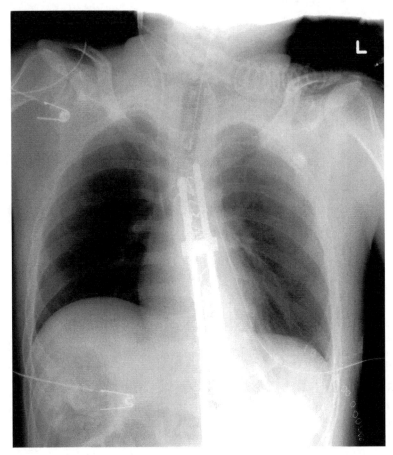

FIGURE 1.21
Sample chest radiograph for medical image compression evaluation.

data with more than one component, such as the three RGB components in color images.

While the KLT has the theoretically optimal decorrelation property, it has seldom been used in practice. While the transform can generalize well, the basis vectors must accompany an image or set of images for reconstruction if the Markov-1 model is not used. There are also the additional computational requirements of estimating the covariance and solving the eigensystem to extract the principal components. Further, the computation of the forward and inverse transform is considered "slow," requiring an order of $O(N^2)$ operations per block of N pixels or $O(N \times p)$ for an image of p pixels. Finally, while the transform may be optimal from an information-theoretic basis, the distortion criterion may not correspond well with our visual perception of distortion. For example, the block effect distortion is quite visible at high compression

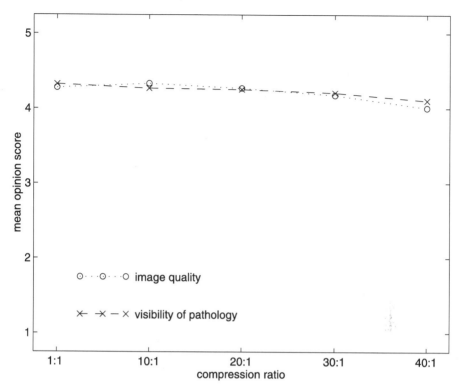

FIGURE 1.22
Mean opinion score across all images and evaluators.

FIGURE 1.23
Color test image, "Monarch."

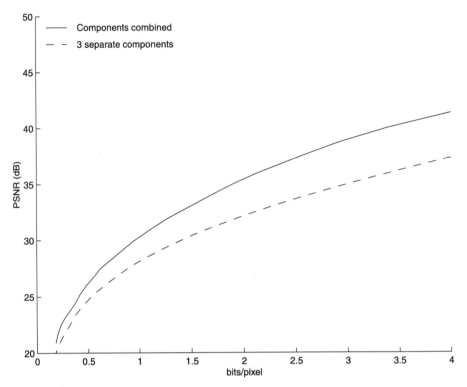

FIGURE 1.24
Distortion versus bit rate for "Monarch" image for encoding the RGB components separately and together.

ratios, yet it is not accounted for in the distortion criteria. A full frame KLT is theoretically possible, but it is only practical for sets of quite small images.

References

[1] Berger, T., *Rate Distortion Theory,* Prentice-Hall, Englewood Cliffs, NJ, 1971.

[2] Castleman, K.R., *Digital Image Processing,* Prentice-Hall, Englewood Cliffs, NJ, 1996.

[3] Chatterjee, C., Roychowdhury, V.P., and Chong, E.K.P., On relative convergence properties of principal component analysis algorithms, *IEEE Trans. Neural Networks,* 9(2):319–329, 1998.

[4] Chen, T., Hua, Y., and Yan, W.-Y., Global convergence of Oja's subspace algorithm for principal component extraction, *IEEE Trans. Neural Networks,* 9(1):58–67, 1998.

[5] Clarke, R.J., *Transform Coding of Images,* Academic Press, San Diego, CA, 1985.

[6] Clarke, R.J., *Digital Compression of Still Images and Video,* Academic Press, San Diego, CA, 1995.

[7] Diamantaras, K.I. and Kung, S.Y., *Principal Component Neural Networks: Theory and Applications,* John Wiley & Sons, New York, 1996.

[8] Dony, R.D. and Haykin, S., Optimally adaptive transform coding, *IEEE Trans. Image Processing,* 4(10):1358–1370, 1995.

[9] Dony, R.D., Haykin, S., Coblentz, C., and Nahmias, C., Compression of digital chest radiographs using a mixture of principal components neural network: an evaluation of performance, *RadioGraphics,* 16, 1996.

[10] Gersho, A. and Gray, R.M., *Vector Quantization and Signal Compression,* Kluwer Academic Publishers, Norwell, MA, 1992.

[11] GNU Octave, http://www.che.wisc.edu/octave.

[12] Gonzalez, R.C. and Woods, R.E., *Digital Image Processing,* Addison-Wesley, Reading, MA, 1993.

[13] Gray, R.M., *Source Coding Theory,* Kluwer Academic Publishers, Norwell, MA, 1990.

[14] Haykin, S., *Neural Networks: A Comprehensive Foundation,* Macmillan, New York, 1994.

[15] Hotelling, H., Analysis of a complex of statistical variables into principal components, *J. Educ. Psychol.,* 24:417–447, 498–520, 1933.

[16] Jolliffe, I., *Principal Component Analysis,* Springer-Verlag, New York, 1986.

[17] Karhunen, K., Über lineare methoden in der wahrscheinlich-keitsrechnung. *Ann. Acad. Sci. Fennicea,* Ser. A137, 1947. (Translated by Selin, I. in "On Linear Methods in Probability Theory," Doc. T-131, The RAND Corp., Santa Monica, CA, 1960.)

[18] Levkowitz, H., *Color Theory and Modeling for Computer Graphics, Visualization, and Multimedia Applications,* Kluwer Academic Publishers, Norwell, MA, 1997.

[19] Loève, M., Fonctions Aléatoires de second order, In Lévy, P., Ed., *Processus Stochastiques et Movement Brownien,* Hermann, Paris, 1948.

[20] MathWorks, http://www.mathworks.com.

[21] Netlib Repository, http://www.netlib.org/eispack.

[22] Netlib Repository, `http://www.netlib.org/lapack`.

[23] Netlib Repository, `http://www.netlib.org/linpack`.

[24] Netravali, A.N. and Haskell, B.G., *Digital Pictures: Representation and Compression,* Plenum Press, New York, 1988.

[25] Oja, E., A simplified neuron model as a principal component analyzer, *J. Math. Biology,* 15:267–273, 1982.

[26] Oja, E., Neural networks, principal components, and subspaces, *Int. J. Neural Systems,* 1(1):61–68, 1989.

[27] Oja, E. and Karhunen, J., On stochastic approximation of the eigenvectors and eigenvalues of the expectation of a random matrix, *J. Math. Analysis and Applications,* 106:69–84, 1985.

[28] Press, W.H., Flannery, B.P., Teukolsky, S.A., and Vetterling, W.T., *Numerical Recipes in C: The Art of Scientific Computing,* Cambridge University Press, Cambridge, UK, 1988.

[29] Rao, K.R. and Yip, P., *Discrete Cosine Transform: Algorithms, Advantages, Applications,* Academic Press, New York, 1990.

[30] Ray, W. and Driver, R.M., Further decomposition of the Karhunen-Loève series representation of a stationary random process, *IEEE Trans. Information Theory,* IT-16:663–668, 1970.

[31] Research Systems, `http://www.rsinc.com`.

[32] Rosenfeld, A. and Kak, A.C., *Digital Picture Processing,* Vol. I & II, 2nd ed., Academic Press, San Diego, CA, 1982.

[33] Sanger, T.D., Optimal unsupervised learning in a single-layer linear feedforward neural network, *Neural Networks,* 2:459–473, 1989.

[34] Shannon, C.E., A mathematical theory of communication, *The Bell System Technical J.,* 27(3):379–423, 623–656, 1948.

[35] Solo, V. and Kong, X., Performance analysis of adaptive eigenanalysis algorithms, *IEEE Trans. Signal Processing,* 46(3):636–645, 1998.

[36] Wallace, G.K., The JPEG still image compression standard, *Communications of the ACM,* 34(4):30–44, 1991.

[37] Wong, S., Zaremba, L., Gooden, D., and Huang, H.K., Radiologic image compression — A review, *Proc. IEEE,* 83(2):194–219, 1995.

Chapter 2

The Discrete Fourier Transform

Ivan W. Selesnick
Polytechnic University

Gerald Schuller
Bell Labs

2.1   Introduction

The discrete Fourier transform (DFT) is a fundamental transform in digital signal processing, with applications in frequency analysis, fast convolution, image processing, etc. Moreover, fast algorithms exist that make it possible to compute the DFT very efficiently. The algorithms for the efficient computation of the DFT are collectively called fast Fourier transforms (FFTs). The historic paper by Cooley and Tukey [15] made well known an FFT of complexity $N \log_2 N$, where N is the length of the data vector. A sequence of early papers [3, 11, 13, 14, 15] still serves as a good reference for the DFT and FFT. In addition to texts on digital signal processing, a number of books devote special attention to the DFT and FFT [4, 7, 10, 20, 28, 33, 36, 39, 48].

The importance of Fourier analysis in general is put forth very well by Leon Cohen [12]:

> . . . Bunsen and Kirchhoff, observed (around 1865) that light spectra can be used for recognition, detection, and classification of substances because they are unique to each substance.

> This idea, along with its extension to other waveforms and the invention of the tools needed to carry out spectral decomposition, certainly ranks as one of the most important discoveries in the history of mankind.

The kth DFT coefficient of a length N sequence $\{x(n)\}$ is defined as

$$X(k) = \sum_{n=0}^{N-1} x(n)\, W_N^{kn}, \qquad k = 0, \ldots, N-1 \tag{2.1}$$

where

$$W_N = e^{-j2\pi/N} = \cos\left(\frac{2\pi}{N}\right) - j\,\sin\left(\frac{2\pi}{N}\right)$$

is the principal N-th root of unity. Because W_N^{nk} as a function of k has a period of N, the DFT coefficients $\{X(k)\}$ are periodic with period N when k is taken outside the range $k = 0, \ldots, N - 1$. The original sequence $\{x(n)\}$ can be retrieved by the inverse discrete Fourier transform (IDFT)

$$x(n) = \frac{1}{N} \sum_{k=0}^{N-1} X(k)\, W_N^{-kn}, \qquad n = 0, \ldots, N - 1 \,.$$

The inverse DFT can be verified by using a simple observation regarding the principal N-th root of unity W_N. Namely,

$$\sum_{n=0}^{N-1} W_N^{nk} = N \cdot \delta(k), \qquad k = 0, \ldots, N - 1 \,,$$

where $\delta(k)$ is the Kronecker delta function defined as

$$\delta(n) = \begin{cases} 1 & n = 0 \\ 0 & n \neq 0 \,. \end{cases}$$

For example, with $N = 5$ and $k = 0$, the sum gives

$$1 + 1 + 1 + 1 + 1 = 5 \,.$$

For $k = 1$, the sum gives

$$1 + W_5 + W_5^2 + W_5^3 + W_5^4 = 0$$

which can be graphically illustrated as:

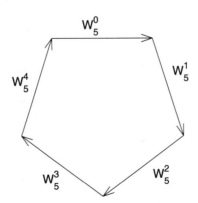

The sums can also be visualized by looking at the illustration of the DFT matrix in Fig. 2.1. Because W_N^{nk} as a function of k is periodic with period N, we can write

$$\sum_{n=0}^{N-1} W_N^{nk} = N \cdot \delta\left(\langle k \rangle_N\right)$$

where $\langle k \rangle_N$ denotes the remainder when k is divided by N, i.e., $\langle k \rangle_N$ is k modulo N.

To verify the inversion formula, we can substitute the DFT into the expression for the IDFT:

$$x(n) = \frac{1}{N} \sum_{k=0}^{N-1} \left(\sum_{l=0}^{N-1} x(l) \, W_N^{kl} \right) W_N^{-kn} , \tag{2.2}$$

$$= \frac{1}{N} \sum_{l=0}^{N-1} x(l) \sum_{k=0}^{N-1} W_N^{k(n-l)} , \tag{2.3}$$

$$= \frac{1}{N} \sum_{l=0}^{N-1} x(l) \, N \, \delta\left(\langle n - l \rangle_N\right) , \tag{2.4}$$

$$= x(n) . \tag{2.5}$$

2.2 The DFT Matrix

The DFT of a length N sequence $\{x(n)\}$ can be represented as a matrix-vector product. For example, a length 5 DFT can be represented as

$$\begin{bmatrix} X(0) \\ X(1) \\ X(2) \\ X(3) \\ X(4) \end{bmatrix} = \begin{bmatrix} 1 & 1 & 1 & 1 & 1 \\ 1 & W & W^2 & W^3 & W^4 \\ 1 & W^2 & W^4 & W^6 & W^8 \\ 1 & W^3 & W^6 & W^9 & W^{12} \\ 1 & W^4 & W^8 & W^{12} & W^{16} \end{bmatrix} \begin{bmatrix} x(0) \\ x(1) \\ x(2) \\ x(3) \\ x(4) \end{bmatrix}$$

where $W = W_5$, or as

$$\mathbf{X} = \mathbf{F}_N \cdot \mathbf{x} ,$$

where $\mathbf{F}_N$ is the $N \times N$ DFT matrix whose elements are given by

$$(\mathbf{F}_N)_{l,m} = W_N^{lm} \qquad 0 \le l, m \le N - 1 .$$

As the IDFT and DFT formulae are very similar, the IDFT represented as a matrix is closely related to $\mathbf{F}_N$,

$$\mathbf{F}_N^{-1} = \frac{1}{N} \mathbf{F}_N^* $$

where $\mathbf{F}_N^*$ represents the complex conjugate of $\mathbf{F}_N$.

It is very useful to illustrate the entries of the matrix $\mathbf{F}_N$ as in Fig. 2.1, where each complex value is shown as a vector. In Fig. 2.1, it can be seen that in the kth row of the matrix the elements consist of a vector rotating clockwise with a constant increment of $2\pi k/N$. In the first row $k = 0$ and the vector rotates in increments of 0. In the second row $k = 1$ and the vector rotates in increments of $2\pi/N$.

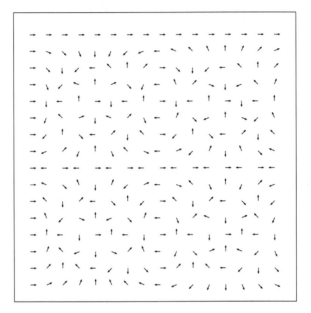

FIGURE 2.1
The 16-point DFT matrix.

2.3   An Example

The DFT is especially useful for efficiently representing signals that are comprised of a few frequency components. For example, the length 2048 signal shown in Fig. 2.2 is an electrocardiogram (ECG) recording from a dog[1]. The DFT of this real signal, shown in Fig. 2.2, is greatest at specific frequencies corresponding to the fundamental frequency and its harmonics. Clearly, the signal $\{x(n)\}$ can be represented well even when many of the small DFT $\{X(k)\}$ coefficients are set to zero. By discarding, or coarsely quantizing, the DFT coefficients that are small in absolute value, one obtains a

---

[1]The dog ECG data is available from the Signal Processing Information Base (SPIB) at URL http://spib.rice.edu/.

more efficient representation of $\{x(n)\}$. Fig. 2.3 illustrates the DFT coefficients when the 409 coefficients that are largest in absolute value are kept, and the remaining 1639 DFT coefficients are set to zero. Fig. 2.3 also shows the signal reconstructed from this truncated DFT. It can be seen that the reconstructed signal is a fairly accurate depiction of the original signal $\{x(n)\}$. For signals that are made up primarily of a few strong frequency components, the DFT is even more suitable for compression purposes.

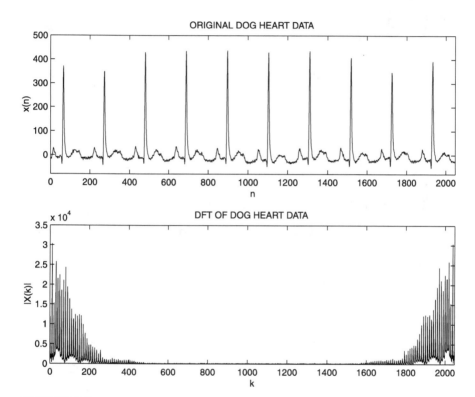

FIGURE 2.2
2048 samples recorded of a dog heart and its DFT coefficients. The magnitudes of the DFT coefficients are shown (see property 1 in Section 2.5.1).

2.4 DFT Frequency Analysis

To formalize the type of frequency analysis accomplished by the DFT, it is useful to view each DFT value $\{X(k)\}$ as the output of a length N FIR filter $h_k(n)$. The

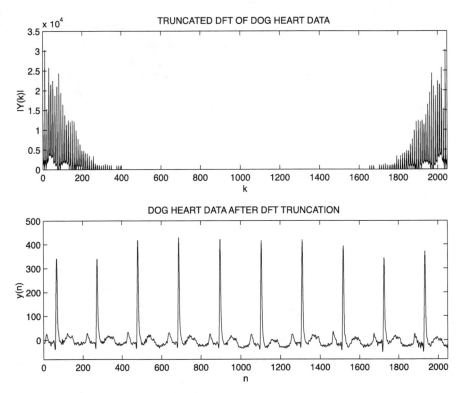

FIGURE 2.3
The truncated DFT coefficients and the time signal reconstructed from the truncated DFT.

output of the filter is given by the convolution sum

$$y_k(l) = \sum_{n=0}^{l} x(n) \, h_k(l - n) \, .$$

When the output $y_k(l)$ is evaluated at time $l = N - 1$, one has

$$y_k(N - 1) = \sum_{n=0}^{N-1} x(n) \, h_k(N - 1 - n) \, .$$

If the filter coefficients $h_k(n)$ are defined as

$$h_k(n) = \begin{cases} W_N^{k(N-1-n)} & 0 \le n \le N - 1 \\ 0 & \text{otherwise} \end{cases} \tag{2.6}$$

then one has

$$y_k(N-1) = \sum_{n=0}^{N-1} x(n) \, W_N^{kn} \,, \tag{2.7}$$

$$= X(k) \,. \tag{2.8}$$

Note that $h_k(n) = W_N^{k(N-1-n)}$ represents a reversal of the values W_N^{kn} for $n = 0, \ldots, N-1$, which in turn is the k-th row of the DFT matrix. Therefore, the DFT of a length N sequence $\{x(n)\}$ can be interpreted as the output of a bank of N FIR filters each of length N sampled at time $l = N - 1$.

Moreover, the impulse responses $h_k(n)$ are directly related to each other through DFT-modulation:

$$h_k(n) = W_N^{k(N-1-n)} \cdot p(n)$$

where the filter $h_0(n) = p(n)$ is given by

$$p(n) = \begin{cases} 1 & 0 \le n \le N-1 \\ 0 & \text{otherwise}\,. \end{cases} \tag{2.9}$$

This filter is called a *rectangular window* as it is not tapered at its ends. It follows that the Z-transforms of the filters are also simply related:

$$H_k(z) = \sum_{n=0}^{N-1} h_k(n) \, z^{-n} \tag{2.10}$$

$$= \sum_{n=0}^{N-1} W_N^{k(N-1-n)} \, p(n) \, z^{-n} \tag{2.11}$$

$$= W_N^{-k} \sum_{n=0}^{N-1} W_N^{-kn} \, p(n) \, z^{-n} \tag{2.12}$$

$$= W_N^{-k} \sum_{n=0}^{N-1} p(n) \left(W_N^k z \right)^{-n} \tag{2.13}$$

$$= W_N^{-k} \, P\left(W_N^k z \right) \tag{2.14}$$

where $P(z) = \sum_{n=0}^{N-1} p(n) \, z^{-n}$. That is, if each filter $h_k(n)$ in an N-channel filter bank is taken to be the time-flip of the k-th row of the DFT matrix, then their Z-transforms are given by $H_k(z) = W_N^{-k} P(W_N^k z)$. $H_0(z) = P(z)$, $H_1(z) = W_N^{-1} P(W_N z)$, etc. It is instructive to view the frequency responses of the N filters $h_k(n)$, as the frequency responses of the filters $H_k(z)$ indicate the effect of the DFT on a sequence. The magnitude of the frequency response of $H_k(z)$ and the zero plot in the z-plane are given in Fig. 2.4. Note that the zeros of $H_k(z)$ in the z-plane are simply rotated by $2\pi/N$, and that the frequency responses are shifted by the same

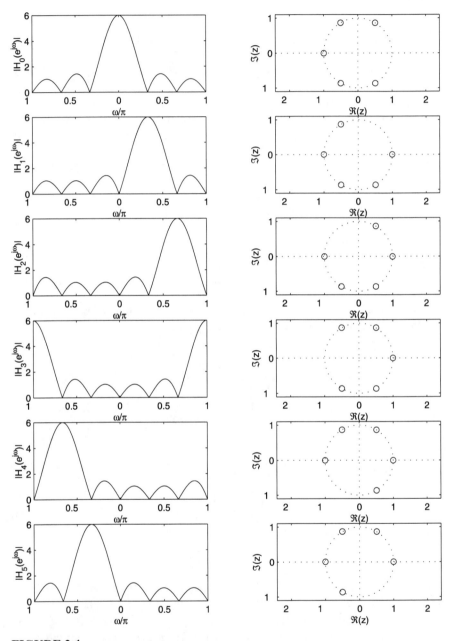

FIGURE 2.4
The magnitude of the frequency response of the filters $h_k(n)$ for $k = 0, \ldots, 5$, corresponding to a 6-point DFT. Shown on the right are the zeros of $H_k(z)$.

amount. The figure makes clear the way in which the DFT performs a frequency decomposition of a signal.

The frequency response of the filter h_k is given by $H_k(e^{j\omega})$, the discrete-time Fourier transform (DTFT) of the impulse response:

$$H_k(e^{j\omega}) = \sum_{n=0}^{N-1} h_k(n) e^{-j\omega n} . \qquad (2.15)$$

The frequency response of the rectangular window $p(n)$ is given by

$$P\left(e^{j\omega}\right) = \sum_{n=0}^{N-1} 1 \cdot e^{-j\omega n} \qquad (2.16)$$

$$= \frac{1 - e^{-jN\omega}}{1 - e^{-j\omega}} \qquad (2.17)$$

$$= \frac{e^{-j\omega N/2} \left(e^{j\omega N/2} - e^{-j\omega N/2}\right)}{e^{-j\omega/2} \left(e^{j\omega/2} - e^{-j\omega/2}\right)} \qquad (2.18)$$

$$= e^{-j\omega(N-1)/2} \cdot \frac{\sin \frac{N}{2}\omega}{\sin \frac{1}{2}\omega} . \qquad (2.19)$$

The function $\sin\left(\frac{N}{2}\omega\right)/\sin\left(\frac{1}{2}\omega\right)$ is called the *digital sinc* function, for its resemblance to the usual sinc function.

2.5 Selected Properties of the DFT

Of the many properties the DFT possesses, the symmetry properties are some of the most useful when using the DFT for compression.

Because the DFT operates on finite-length data sequences, it is useful to define two types of symmetries as follows. When $\{x(n)\}$ is periodically extended outside the range $n = 0, \ldots, N - 1$, the following definitions for symmetric and anti-symmetric sequences are consistent with their usual definitions for sequences that are not finite in length.

Symmetry: Let $\{x(n)\}$ be a real-valued length N data sequence, for $n = 0, \ldots, N - 1$, then $\{x(n)\}$ is *symmetric* if

$$x(N - n) = x(n), \qquad n = 1, \ldots, N - 1 .$$

Note that an *even*-length N symmetric sequence $\{x(n)\}$ is fully described by its first $N/2 + 1$ values. For example, a length 6 symmetric sequence is fully determined by its first 4 values as illustrated in Fig. 2.5. On the other hand, an *odd*-length N symmetric sequence $\{x(n)\}$ is fully described by its first $(N + 1)/2$ values. For

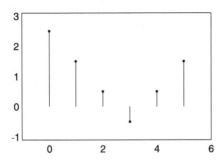

FIGURE 2.5
Illustration of even-length symmetric sequence.

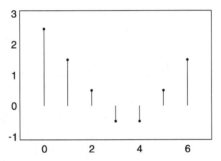

FIGURE 2.6
Illustration of odd-length symmetric sequence.

example, a length 7 symmetric sequence is fully determined by its first 4 values as illustrated in Fig. 2.6. For both even- and odd-length sequences, the number of values that determine a length N symmetric sequence is $\lfloor N/2 + 1 \rfloor$ where $\lfloor k \rfloor$ denotes the greatest integer smaller than or equal to k.

Anti-symmetry: A real-valued length N data sequence is *anti*-symmetric if

$$x(0) = 0 \quad \text{and} \quad x(N - n) = -x(n), \quad n = 1, \ldots, N - 1 .$$

Note that an *even*-length N anti-symmetric sequence $\{x(n)\}$ is fully described by $N/2 - 1$ values. For example, a length 6 anti-symmetric sequence is fully determined by 2 values (see Fig. 2.7). On the other hand, an *odd*-length N anti-symmetric sequence $\{x(n)\}$ is fully described by $(N - 1)/2$ values. For example, a length 7 anti-symmetric sequence is fully determined by 3 values (see Fig. 2.8). For both even- and odd-length sequences, the number of values that determine a length N anti-symmetric sequence is $\lceil N/2 - 1 \rceil$ where $\lceil k \rceil$ denotes the smallest integer greater than or equal to k.

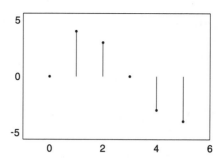

FIGURE 2.7
Illustration of even-length anti-symmetric sequence.

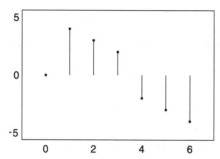

FIGURE 2.8
Illustration of odd-length anti-symmetric sequence.

2.5.1 Symmetry Properties

To state the symmetry properties of the DFT, it is useful to introduce the notation $\{X_r(k)\}$ and $\{X_i(k)\}$ for the real and imaginary parts of $\{X(k)\}$. Similarly, $\{x_r(n)\}$ and $\{x_i(n)\}$ are used to denote the real and imaginary parts of $\{x(n)\}$.

If $\{x(n)\}$ is a length N data vector and ...

1. if $\{x(n)\}$ is *real-valued,* then

$$X(k) = X^*(N - k), \qquad k = 1, \ldots, N - 1,$$

i.e., the real part of $\{X(k)\}$ is symmetric, and the imaginary part of $\{X(k)\}$ is anti-symmetric.

2. if $\{x(n)\}$ is *real-valued and symmetric,* then

$$X(k) = X_r(k) = X_r(N - k), \quad k = 1, \ldots, N - 1,$$

i.e., $\{X(k)\}$ is purely real and symmetric.

3. if $\{x(n)\}$ is *real-valued and anti-symmetric,* then

$$X(k) = j\, X_i(k) = -j\, X_i(N - k), \quad k = 1, \ldots, N - 1,$$

i.e., $\{X(k)\}$ is purely imaginary and anti-symmetric.

4. if $\{x(n)\}$ is *purely imaginary*, then

$$X(k) = -X^*(N-k), \qquad k = 1, \ldots, N-1,$$

i.e., the real part of $\{X(k)\}$ is anti-symmetric, and the imaginary part of $\{X(k)\}$ is symmetric.

5. if $\{x(n)\}$ is *purely imaginary and $\{x_i(n)\}$ is symmetric*, then

$$X(k) = j\,X_i(k) = j\,X_i(N-k), \quad k = 1, \ldots, N-1,$$

i.e., $\{X(k)\}$ is purely imaginary and symmetric.

6. if $\{x(n)\}$ is *purely imaginary and $\{x_i(n)\}$ is anti-symmetric*, then

$$X(k) = X_r(k) = -X_r(N-k), \quad k = 1, \ldots, N-1,$$

i.e., $\{X(k)\}$ is purely real and anti-symmetric.

These properties are summarized in Table 2.1.

These properties explain why the total number of parameters needed to describe the original data sequence $\{x(n)\}$ is the same after the DFT is performed. For example, consider a real-valued length 6 sequence $\{x(n)\}$ and its DFT:

$$\mathbf{x} = \begin{bmatrix} 1 \\ 3 \\ 5 \\ 6 \\ 7 \\ 2 \end{bmatrix} \qquad \mathbf{X} = \begin{bmatrix} 24.0000 \\ -8.5000 \\ -1.5000 \\ 2.0000 \\ -1.5000 \\ -8.5000 \end{bmatrix} + j \begin{bmatrix} 0 \\ 0.8660 \\ -2.5981 \\ 0 \\ 2.5981 \\ -0.8660 \end{bmatrix}.$$

It is clear that there are a total of 6 distinct values in the DFT coefficients $\{X(k)\}$ for this example.

In general, for a length N real-valued sequence $\{x(n)\}$, the symmetric $\{X_r(k)\}$ is determined by $\lfloor N/2 + 1 \rfloor$ values, and the anti-symmetric $\{X_i(k)\}$ is determined by $\lceil N/2 - 1 \rceil$ values. Therefore, even though the DFT $\{X(k)\}$ of a length N real-valued sequence $\{x(n)\}$ is complex-valued, it is fully determined by exactly N values. The number of parameters is the same in both $\{x(n)\}$ and $\{X(k)\}$.

Recall that an even-length real-valued symmetric sequence $\{x(n)\}$ is determined by its first $N/2 + 1$ values. By the symmetry property above, the same is true for the DFT $\{X(k)\}$. An odd-length real-valued symmetric sequence $\{x(n)\}$ is determined by its first $(N + 1)/2$ values. By the symmetry property above, the same is true for the DFT $\{X(k)\}$. The symmetry properties for real-valued symmetric sequences are especially useful because they can be used to develop useful DFT-based transforms that yield real-valued coefficients.

Table 2.1 DFT Symmetry Properties

x is purely real		$\mathbf{X}_r$ is symmetric,	$\mathbf{X}_i$ is anti-symmetric
x is purely real,	$\mathbf{x}_r$ is symmetric	$\mathbf{X}_r$ is symmetric,	**X** is purely real
x is purely real,	$\mathbf{x}_r$ is anti-symmetric	**X** is purely imaginary,	$\mathbf{X}_i$ is anti-symmetric
x is purely imaginary		$\mathbf{X}_r$ is anti-symmetric,	$\mathbf{X}_i$ is symmetric
x is purely imaginary,	$\mathbf{x}_i$ is symmetric	**X** is purely imaginary,	$\mathbf{X}_i$ is symmetric
x is purely imaginary,	$\mathbf{x}_i$ is anti-symmetric	$\mathbf{X}_r$ is anti-symmetric,	**X** is purely real

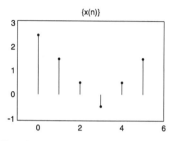
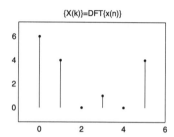

FIGURE 2.9

Illustration of DFT symmetry property for an even-length sequence.

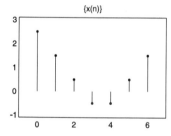
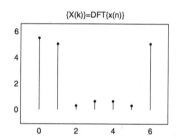

FIGURE 2.10

Illustration of DFT symmetry property for an odd-length sequence.

2.6 Real-Valued DFT-Based Transforms

In most applications the data are real-valued. For this reason, it can be beneficial to use the DFT in a specialized way so that it gives real values. This can be accomplished by suitably extending the given data sequence $\{x(n)\}$ so that it exhibits the necessary symmetry that makes the DFT $\{X(k)\}$ real-valued.

For example, given a length N real-valued sequence $\{x(n)\}$, which does not necessarily possess any symmetries, one can construct a symmetric sequence by symmetrically extending $\{x(n)\}$. There is more than one way to symmetrically extend a given sequence, depending on how the end points are treated. Different symmetric extensions give rise to the different types of DFT-based signal transforms that map real-valued sequences to real-valued sequences. One class of DFT-based real transforms is the *discrete cosine and sine transforms*. In fact, 16 different cosine and sine transforms are described in [32].

One way to symmetrically extend a finite length N sequence is illustrated in Fig. 2.11. The result is a symmetric sequence $\{x_1(n)\}$ of even length $2N-2$. $\{X_1(k)\}$,

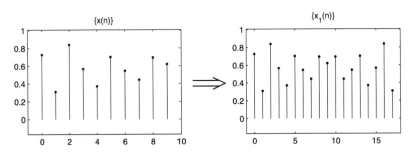

FIGURE 2.11
Illustration of symmetric extension.

the DFT of $\{x_1(n)\}$, is therefore real-valued symmetric and is determined by its first N values (see Fig. 2.12). Because $\{X_1(k)\}$ is determined by its first N values, this procedure gives an N-point real transform. The inverse of this transform is obtained by performing the same steps in reverse sequence. Given the first N values of $\{X_1(k)\}$, construct a symmetric extension as above to obtain a length $2N-2$ sequence $\{X_1(k)\}$, take the inverse DFT of the resulting sequence to obtain the length $2N-2$ sequence $\{x_1(n)\}$, from which $\{x(n)\}$ can be extracted.

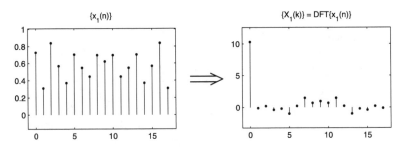

FIGURE 2.12
Illustration of symmetric extension.

The transform formulae can be found explicitly using the DFT formulae together with the symmetric extension.

$$X_1(k) = \text{DFT}\{x_1(n)\} \tag{2.20}$$

$$= \sum_{n=0}^{2N-3} x_1(n)\, W_{2N-2}^{kn} \tag{2.21}$$

$$= x(0) + \sum_{n=1}^{N-2} x(n)\left[W_{2N-2}^{nk} + W_{2N-2}^{2N-2-n} \right] + x(N-1)\, W_{2N-2}^{(N-1)k} \tag{2.22}$$

$$= x(0) + 2\sum_{n=1}^{N-2} x(n) \cos\left(\frac{nk\pi}{N-1} \right) + (-1)^k x(N-1) \tag{2.23}$$

where we have used the simplification $W_{2N-2}^{k(2N-2-n)} = W_{2N-2}^{-nk}$. Often the first and last values, $x(0)$ and $x(N-1)$, are scaled by $\sqrt{2}$ so that the transform is orthogonal. The inverse can also be derived in a similar way.

It is very interesting to look at the type of frequency analysis this type of discrete cosine transform (DCT) [1, 42, 61] performs, as was done for the DFT in Fig. 2.4. In Fig. 2.13, the frequency responses corresponding to this DCT are shown. Note that the plots of zeros in the z-plane are especially simple.

Another way to symmetrically extend a finite length N sequence is illustrated in Fig. 2.14. The result is a symmetric sequence $\{x_2(n)\}$ of odd length $2N - 1$. $\{X_2(k)\}$, the DFT of $\{x_2(n)\}$, is therefore real-valued symmetric and is determined by its first N values (see Fig. 2.15). Because $\{X_2(k)\}$ is determined by its first N values, this procedure gives an N-point real transform. The inverse of this transform is obtained by performing the same steps in reverse sequence. Given the first N values of $\{X_2(k)\}$, construct a symmetric extension as above to obtain a length $2N - 1$ sequence $\{X_2(k)\}$, and take the inverse DFT of the resulting sequence to obtain the length $2N-1$ sequence $\{x_2(n)\}$, from which $\{x(n)\}$ can be extracted.

Now consider a symmetric extension by simply mirroring the entire length N sequence,

$$\{x_1(n)\} = [x(0), \ldots, x(N-1), x(N-1), \ldots, x(0)]$$

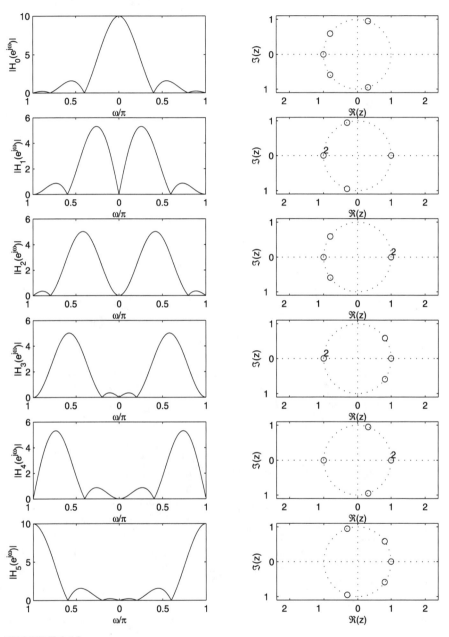

FIGURE 2.13
The discrete cosine transform (I) basis vectors illustrated in the frequency domain and in the z-plane. $N = 6$.

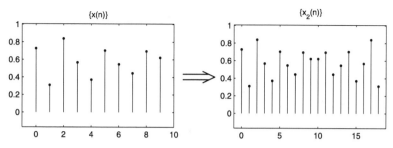

FIGURE 2.14
Illustration of DFT symmetry property.

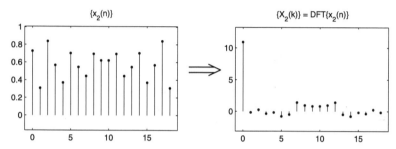

FIGURE 2.15
Illustration of DFT symmetry property.

for $0 \leq n \leq 2N - 1$ (a length $2N$ sequence). The DFT of this sequence becomes

$$X_1(k) \quad = \quad \text{DFT}\{x_1(n)\} \tag{2.24}$$

$$= \quad \sum_{n=0}^{2N-1} x_1(n) W_{2N}^{kn} \tag{2.25}$$

$$= \quad \sum_{n=0}^{N-1} x(n) \left[W_{2N}^{kn} + W_{2N}^{k(2N-1-n)} \right] \tag{2.26}$$

$$= \quad \sum_{n=0}^{N-1} x(n) W_{2N}^{-0.5k} \left[W_{2N}^{k(n+0.5)} + W_{2N}^{k(2N-0.5-n)} \right] \tag{2.27}$$

$$= \quad W_{2N}^{-0.5k} \sum_{n=0}^{N-1} x(n) \cos\left(\frac{\pi}{N} \cdot k \cdot (n+0.5) \right) \tag{2.28}$$

The phase factor $W_{2N}^{-0.5k}$ can be neglected in applications since it carries no information about the signal. Since the transform length is N, the frequency index has the range $k = 0, \ldots, N-1$, so that a quadratic cosine transform matrix is obtained. The

DCT thus obtained is a so-called DCT type II. Its transform matrix is

$$\mathbf{D}_{II}(k, n) := \sqrt{2/N} \cos\left(\frac{\pi}{N}k(n - 0.5)\right)$$

for $n, k = 0, \ldots, N - 1$. To make this transform matrix orthogonal, its first row is usually scaled to

$$\mathbf{D}_{II}(0, n) := \sqrt{1/N}$$

for $k = 0$. This transform divides the frequency axis as illustrated in Fig. 2.16. It can be seen that the width of the resulting frequency bins or bands is π/N, except for the lowest band for $k = 0$, as it is centered around DC. This results in a lowpass filter bandwidth of $1/(2N)$. The highest band for $k = N - 1$ is centered at $\pi(1 - 1/N)$, which means the required bandwidth to cover the entire frequency axis up to π is $2/N$. This means that for the design of filter banks with uniform frequency width for all bands, a shift of the frequency grid by $1/2$ would be suitable, so that the lowest band covers more bandwidth, and the highest band needs to cover less, as illustrated in Fig. 2.17. This results in a DCT type IV; its orthogonal transform matrix is

$$\mathbf{D}_{IV}(k, n) := \sqrt{2/N} \cos\left(\frac{\pi}{N}(k + 0.5)(n + 0.5)\right) . \tag{2.29}$$

Similarly a discrete sine transform of types II and IV are obtained by applying a DFT to the sequence

$$\{x_1\} = [x(0), \ldots, x(N - 1), -x(N - 1), \ldots, -x(0)]$$

for $0 \leq n \leq 2N - 1$. The resulting transform matrix for a DST type IV is

$$\mathbf{S}_{IV}(k, n) := \sqrt{2/N} \sin\left(\frac{\pi}{N}(k + 0.5)(n + 0.5)\right) . \tag{2.30}$$

Efficient ways to obtain DCTs with the help of FFTs can be found, for example, in Malvar [31].

FIGURE 2.16
The distribution of bands with a DCT II. Horizontally is the normalized frequency Ω/π. The band edges are marked with long vertical lines, and the band centers with short lines.

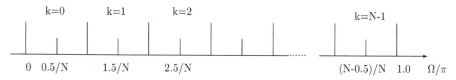

FIGURE 2.17
The distribution of bands with a DCT IV. The band edges are again marked with long vertical lines, and the band centers with short lines.

2.7   The Fast Fourier Transform

A fast Fourier transform (FFT) is any fast algorithm for computing the DFT. As stated earlier, FFT algorithms have a tremendous impact on computational aspects of signal processing. To introduce the FFT, recall the definition of the DFT in Eq. (2.1) and suppose the data vector $\{x(n)\}$ is of even length N. The basic derivation of the FFT begins by splitting the sum into two parts — one part for the even-indexed values $\{x(2n)\}$ and one part for the odd-indexed values $\{x(2n+1)\}$

$$X(k) = \sum_{\substack{n=0 \\ n \text{ even}}}^{N-1} x(n)\, W_N^{nk} + \sum_{\substack{n=0 \\ n \text{ odd}}}^{N-1} x(n)\, W_N^{nk}$$

which can be written as

$$X(k) = \sum_{n=0}^{N/2-1} x(2n)\, W_N^{2nk} + \sum_{n=0}^{N/2-1} x(2n+1)\, W_N^{(2n+1)k}$$

or as

$$X(k) = \sum_{n=0}^{N/2-1} x(2n)\, W_N^{2nk} + W_N^k \sum_{n=0}^{N/2-1} x(2n+1)\, W_N^{2nk} .$$

Note that W_N^{2nk} can be rewritten as follows:

$$W_N^{2nk} = e^{-j2\pi(2nk)/N} \tag{2.31}$$
$$= e^{-j2\pi(nk)/(N/2)} \tag{2.32}$$
$$= W_{N/2}^{nk} . \tag{2.33}$$

Hence the DFT values $\{X(k)\}$ can be written as

$$X(k) = \sum_{n=0}^{N/2-1} x(2n)\, W_{N/2}^{nk} + W_N^k \sum_{n=0}^{N/2-1} x(2n+1)\, W_{N/2}^{nk} .$$

Note that the first sum is the length $N/2$ DFT of the sequence $\{x(2n)\}$ and the second sum is the length $N/2$ DFT of the sequence $\{x(2n+1)\}$. Defining these sequences as $\{x_0(n)\} = \{x(2n)\}$ and $\{x_1(n)\} = \{x(2n+1)\}$ for $n = 0, \ldots, N-1$ makes them both sequences of length $N/2$. Then one has

$$X(k) = X_0(k) + W_N^k X_1(k), \qquad k = 0, \ldots, N-1,$$

where $\{X_0(k)\}$ and $\{X_1(k)\}$ are the DFTs of $\{x_0(n)\}$ and $\{x_1(n)\}$, respectively. It should be noted that in the definition of the length N DFT, $\{X(k)\}$ was defined for $k = 0, \ldots N-1$. As $\{x_0(n)\}$ is a sequence of length $N/2$, its DFT is also of length $N/2$, and therefore $\{X_0(k)\}$ would be defined for $k = 0, \ldots, N/2 - 1$. However, as noted in Section 2.1, when k is taken outside this range, the DFT coefficients are periodic — so $X_0(k) = X_0(k - N/2)$ for values of k from $N/2$ to $N-1$. Likewise for $X_1(k)$.

This expression shows how a length N DFT can be computed using two length $N/2$ DFTs. After taking the two length $N/2$ DFTs it remains only to multiply the result of the second DFT with the terms W_N^k and to add the results. The multipliers W_N^k are known as *twiddle factors*.

If $N/2$ can be further divided by 2, then the same procedure can be used to calculate the length $N/2$ DFTs. To determine the arithmetic complexity of this algorithm for computing the DFT, let $A(N)$ denote the number of complex additions for computing the DFT of a length N complex sequence $\{x(n)\}$. Let N be a power of 2, $N = 2^K$. Then, according to the above procedure, one has

$$A(N) = 2\,A(N/2) + N$$

as N complex additions are required to put the two length $N/2$ DFTs back together. Note that a length 2 DFT is simply a sum and difference:

$$X(0) = x(0) + x(1)$$
$$X(1) = x(0) - x(1).$$

Hence, the starting condition is $A(2) = 2$. [Or one can use $A(1) = 0$.] Then solving the recursive equation yields

$$A(N) = N \log_2 N \quad \text{complex additions.}$$

Similarly, one has a recursive formula for complex multiplications:

$$M(N) = 2\,M(N/2) + N/2$$

which gives

$$M(N) = \frac{N}{2} \log_2 N \quad \text{complex multiplications.}$$

In fact, this number can be reduced by a more careful examination of the multipliers W_N^k (the twiddle factors). In particular, the numbers $1, -1, j,$ and $-j$ will be among the twiddle factors W_N^k, when k is a multiple of $N/4$ — and so these multiplications need not be performed. Taking this into account, one has the following formulae for the number of real additions and real multiplications of the DFT of a sequence whose length is a power of 2:

$$A_r(N) = 7\frac{N}{2} \log_2 N - 5N + 8 \tag{2.34}$$

$$M_r(N) = 3\frac{N}{2} \log_2 N - 5N + 8 \tag{2.35}$$

where a complex addition counts as two real additions, and a complex multiplication counts as three real additions and three real multiplications.

The advantage of the efficient algorithm for computing the DFT is a reduction from an arithmetic complexity of N^2 for direct calculation to a complexity of $N \log_2 N$. This is a fundamental improvement in the complexity, and historically it led to many new developments in signal processing that would not otherwise have been possible or practical. Due to its fundamental quickening in calculating the DFT, the efficient algorithm for its computation is called the *fast Fourier transform* or FFT.

Many variations and enhancements of this basic algorithm have been developed in the literature and used in practice, and they are collectively called FFTs. Of particular note is the *split radix FFT* [16, 50, 56], which is a refinement of the algorithm that attains the lowest computational complexity of practical FFT variants for lengths that are powers of 2. FFT algorithms can be developed for lengths that are not powers of 2. Some types of FFTs, called *prime factor FFTs,* do not require the use of twiddle factors [9, 52, 53] and therefore have a reduced computational complexity (this is possible when the length N is factored into relatively prime integers. It is not applicable for lengths that are powers of 2). Implementations of the FFT for real-valued data are described in Sorenson et al. [51]. Most FFT algorithms depend on the ability to factor N, the length of the data vector $\{x(n)\}$; for prime-length DFTs a separate approach is needed to combine shorter FFTs. The algorithms for prime-length FFTs are based on work by Rader and Winograd [40, 60, 59]. FFT programs for prime lengths are discussed in several publications [25, 29, 46]. Descriptions of the different types of FFTs are available in several books [4, 7, 20, 10, 33, 35, 36, 54, 28] and book chapters [8, 18, 19, 49]. The complexity theory associated with the FFT is described in Winograd [60] and Heideman [22]. A comparison of different FFT implementations on DSP chips is described in Meyer and Schwarz [34].

A relevant issue in practice is the trade-off between computational complexity and implementation complexity. The right balance must be obtained for the best results and some FFT algorithms with improved computational complexity are more complex to implement than others. Moreover, for the fastest results, the variant of the FFT chosen should be matched to the hardware on which it will run. Methods for choosing the best variant of the FFT from among a family of FFTs have been the subject of recent research [23, 24, 21].

2.8   The DFT in Coding Applications

In coding applications the DFT is used in two broad classes — in power spectrum estimation and in subband coding, where it is used in the implementation of complex-, cosine- or sine-modulated filter banks. As an illustration, audio coding will be considered in the following.

In audio coding, the real-valued audio signal is decomposed into a number of subbands with a filter bank. The subband signals are then adaptively quantized and encoded [47, 6]. The subband decomposition has the purpose of obtaining a more efficient description of the signal (redundancy reduction) and applying a psycho-acoustic model to control the quantization noise such that it will be inaudible (irrelevance reduction); see Fig. 2.18.

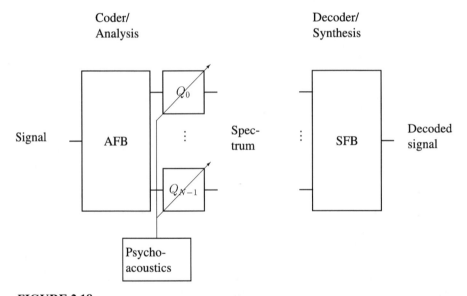

FIGURE 2.18
Audio coding based on filter banks, AFB: analysis filter bank, SFB: synthesis filter bank.

In audio coding, the subband decomposition is usually obtained with a filter bank called *modified discrete cosine transform* MDCT. It can often be switched between differing numbers of bands, for example, between 128 and 1024 bands. The MDCT is used, for example, in ASPEC, MPEG, MUSICAM, and PAC audio coders [30]. ASPEC and MUSICAM were later combined into MPEG-1 layer III, also known as MP3.

One way in which the DFT is used in subband coding is for the implementation of filter banks. Since the filters $h_k(n) = e^{j2\pi kn/N}, n = 0, \ldots, N-1, k = 0, \ldots, N-1$, can be seen as a rectangular window of length N multiplied with the exponential, the frequency localization is not very good, as can be seen in Fig. 2.4. Since this frequency localization is very important in coding applications, the DFT is used only indirectly in coding applications, e.g., for implementing the MDCT. The output of the MDCT is real valued for real-valued inputs, and its subband filter impulse responses $h_k(n)$ are longer and have a nonrectangular shape, such that the frequency localization is better than for the DFT. The MDCT filter bank can be implemented using a DCT of length N, which in turn can be implemented using FFTs of length $N/2$ [31].

In audio coding the DFT is also used as a complex filter bank. The psycho-acoustic model, used to control the quantization step size, needs to detect and estimate signals (sinusoids) in the subbands, i.e., it needs a reliable estimate of the time-varying power spectrum, with a time and frequency resolution as similar to the MDCT as possible. This is most reliably done with a complex valued spectral decomposition because it provides the phase and magnitude of signals in the subbands at every time step. To estimate the spectrum, only the magnitude of the subband signal is needed.

This would not be possible with a real-valued filter bank because in such a filter bank a sinusoid in a subband is still a sinusoid after filtering, which will pass through zero at certain times — so it may not be detected. That is, the estimated power of the signal at that frequency and time would be lower than it should be. That is why some audio coders [e.g., MPEG-AAC (Advanced Audio Coder) [30]] possess an FFT parallel to the MDCT as input to the psycho-acoustic model. But a problem is the insufficient frequency localization of the FFT, which reduces the accuracy of the psycho-acoustic model.

The so called Balian-Low theorem states that the rectangular window of the DFT gives rise to the only orthogonal FIR filter bank with complex Fourier modulation and critical sampling [57] (every N input samples produce N output samples). However, for the time-varying spectral estimation required for the psycho-acoustic model, critical sampling is not a constraint. That is why, for example, in the perceptual audio coder (PAC) [30] the input of the psycho-acoustic model is a complex signal, which is taken from two filter banks. The real part of the signal is the output of the real-valued MDCT filter bank with a cosine modulation function. Hence, only an appropriate imaginary part corresponding to this signal is needed to obtain a complete complex subband signal — which will have improved frequency localization and therefore a more accurate psycho-acoustic model. This imaginary part of the subband signal can be obtained by using a second filter bank which is based on the same window function as the MDCT, but with a sine modulation function instead of a cosine modulation function. Interestingly, this sine-modulated filter bank alone is again a perfect reconstruction (PR) filter bank, as is the cosine-modulated MDCT filter bank. These two filter banks, in parallel, can be seen as one complex filter bank which is twice oversampled. Hence the limitation the Balian-Low theorem no longer applies, as the filter bank system is not critically sampled.

2.9   The DFT and Filter Banks

Because the frequency content of many signals changes with time, it is often more desirable to first partition a signal into blocks and then apply the DFT to each block individually. This *block-wise* DFT leads to a point of view based on filter banks. If the independent variable of the input signal is time (e.g., an audio signal), then this results in a time-frequency representation. If the input data is arranged in a matrix

$$\mathbf{x} = \begin{bmatrix} x(0) & x(N) & x(2N) & \cdots \\ x(1) & x(N+1) & x(2N+1) & \cdots \\ \vdots & & & \\ x(N-1) & x(2N-1) & x(3N-1) & \cdots \end{bmatrix}$$

and $\mathbf{F}_N$ is the DFT matrix, then the block-wise DFT can be written as

$$\mathbf{X} = \mathbf{F}_N \cdot \mathbf{x} \tag{2.36}$$

where each column of the matrix $\mathbf{X}$ is a DFT spectrum. Clearly this operation is easily inverted with

$$\mathbf{x} = (\mathbf{F}_N)^{-1} \cdot \mathbf{X}. \tag{2.37}$$

Depending on the amount of data, the matrices for $\mathbf{X}$ and $\mathbf{x}$ can be quite large. To simplify the mathematical description and to obtain a more general formulation, the Z-transform can be used. Then each block, or time frame, of $\mathbf{X}$ and $\mathbf{x}$ is associated with a power of z^{-1}, and the data becomes a vector of polynomials in z^{-1},

$$\mathbf{x}(z) = \begin{bmatrix} x(0) + x(N)\, z^{-1} + x(2N)\, z^{-2} + \cdots \\ x(1) + x(N+1)\, z^{-1} + x(2N+1)\, z^{-2} + \cdots \\ \vdots \\ x(N-1) + x(N+N-1)\, z^{-1} + x(2N+N-1)\, z^{-2} + \cdots \end{bmatrix}.$$

This leads to

$$\mathbf{X}(z) = \mathbf{F}_N \cdot \mathbf{x}(z) \tag{2.38}$$

and

$$\mathbf{x}(z) = (\mathbf{F}_N)^{-1} \cdot \mathbf{X}(z). \tag{2.39}$$

These equations are quite similar to Eqs. (2.36) and (2.37), but now the data $\mathbf{x}$ and $\mathbf{X}$ are in the form of a simple vector instead of a possibly infinite matrix. The operation of applying the DFT to blocks of the signal can now also be viewed as a filter bank, as seen in Fig. 2.19. The symbol $\downarrow N$ means a downsampling operation, i.e., only every N-th sample is let through. This figure shows an analysis filter bank on the left which

corresponds to Eq. (2.36), and a synthesis filter bank on the right which corresponds
to Eq. (2.37). Since the DFT is invertible, the signal $\{x(n)\}$ can be directly obtained
from the block-wise DFT coefficients using the inverse DFT on each block. This
inverse can also be interpreted in terms of filter banks as illustrated by the synthesis
filter bank in Fig. 2.19.

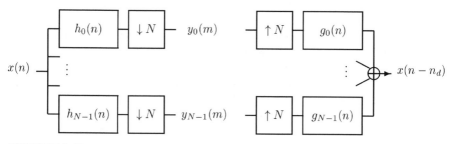

FIGURE 2.19

**An N-channel filter bank with critical downsampling, perfect reconstruction,
and a system delay of n_d samples.**

Because the analysis filter bank is invertible, it is said to have the perfect reconstruction (PR) property. Because the total number of samples in the input signal
$\{x(n)\}$ equals the total number of samples in the subbands (the N channels), it is said
to be critically sampled. In coding applications, critical downsampling is important
because it leads to an accurate and complete description of a signal with the least possible number of samples, and it leads to computationally efficient implementations.
The analysis filter bank is used in the encoder, and the synthesis filter bank in the
decoder.

To see that the matrix formulation can also be represented by a filter bank structure
(see also Vaidyanathan [55]), consider the following. For simplicity, we assume a
time-shifted sequence $\{x(n+N-1)\}$. The filtering (convolution) and downsampling
operation can be written as

$$y_k(m) = \sum_n h_k(n) \cdot x(mN + N - 1 - n), \qquad 0 \le k \le N - 1 . \qquad (2.40)$$

On the other hand, Eq. (2.36) can also be written as

$$\mathbf{X}_{k,m} = \sum_{n=0}^{N-1} W_N^{kn} \cdot x(mN + n), \qquad 0 \le k \le N - 1 . \qquad (2.41)$$

If this equation is compared to Eq. (2.40), it can be seen by a substitution of the index
variable that they are identical if the filters $\{h_k(n)\}$ are defined as

$$h_k(n) = W_N^{k(N-1-n)} = W_N^{-k} W_N^{-kn} \quad \text{for } n = 0, \ldots N - 1$$

and

$$h_k(n) = 0 \quad \text{otherwise}$$

[see also Eq. (2.6)]. It was noted in Section 2.4 that these filters are complex-modulated versions of the rectangular window function. The resulting frequency responses of $\{h_k(n)\}$ are frequency-shifted versions of the frequency response of the rectangular window function $p(n)$, as can also be seen in Fig. 2.4. The block-wise interpretation of this DFT-modulated filter bank leads to an efficient algorithm for its implementation using an FFT.

The rectangular window does not have a good frequency localization because of its limited length and its rectangular shape. Fig. 2.4 shows that the main lobe of the frequency response (its passband) is quite wide, and the side lobes are not very low — the stopband attenuation is not very high. A solution is to increase the window length and to give it a different shape, such that the passband becomes more narrow and the stopband attenuation is improved (see Bellanger [2]). To this end, first consider a general window function $\{p(n)\}$ of length N, the shape of which is not necessarily rectangular. ($\{p(n)\}$ denotes the analysis *prototype* filter or window function; the synthesis prototype filter will be denoted by $\{q(n)\}$.) The filters $\{h_k(n)\}$ in this case are given by

$$h_k(n) = W_N^{-k} \cdot W_N^{-kn} \cdot p(n) \tag{2.42}$$

or in terms of Z-transforms, as

$$H_k(z) = W_N^{-k} \, H_a\left(W_N^k z\right) \, .$$

The analysis equation can then be written using a diagonal matrix as

$$\mathbf{X}(z) = \mathbf{F}_N \cdot \begin{bmatrix} p(N-1) & 0 & \cdots & 0 \\ 0 & p(N-2) & & \vdots \\ \vdots & & \ddots & \\ 0 & \cdots & & p(0) \end{bmatrix} \cdot \mathbf{x}(z) \, . \tag{2.43}$$

The diagonal matrix is also called a *filter matrix,* denoted by $\mathbf{F}_a$ for the analysis. The inverse gives the equation for the synthesis stage

$$\mathbf{x}(z) = \begin{bmatrix} 1/p(N-1) & 0 & \cdots & 0 \\ 0 & 1/p(N-2) & & \vdots \\ \vdots & & \ddots & \\ 0 & \cdots & & 1/p(0) \end{bmatrix} \cdot \mathbf{F}_N^{-1} \cdot \mathbf{X}(z) \, . \tag{2.44}$$

The analysis window function $p(n)$ leads to a synthesis window function of $1/p(n)$, e.g., the synthesis window is the point-wise inverse of the analysis window. Consequently, a window with improved frequency localization properties in the analysis stage can lead to worse frequency localization in the synthesis stage, which is often not desired. Also, the limited length of N of the window still is an important limiting factor in the design of better window functions.

When the filter $p(n)$ is longer than N, say LN, then Eq. (2.41) becomes

$$\mathbf{X}_{k,m} = \sum_{n=0}^{LN-1} W_N^{kn} \cdot p(LN - 1 - n)\, x(mN + n)\,.$$

Since $W_N^{k(n+N)} = W_N^{kn}$, we can replace n by $lN + n$ to obtain

$$\mathbf{X}_{k,m} = \sum_{n=0}^{N-1} W_N^{kn} \cdot \sum_{l=0}^{L} p(LN - 1 - n - lN)\, x(mN + n + lN)\,.$$

The inner sum can be interpreted as a convolution, which is written as a product in the z-domain, with

$$P_n(z) = \sum_{l=0}^{L-1} p(n + lN) \cdot z^{-l}$$

$$X_n(z) = \sum_{l=0}^{\infty} x(n + lN) \cdot z^{-l}\,.$$

This leads to

$$\mathbf{X}_k(z) = \sum_{n=0}^{N-1} W_N^{kn} \cdot P_{N-1-n}(z) \cdot X_n(z)$$

so that Eq. (2.43) becomes

$$\mathbf{X}(z) = \mathbf{F}_N \cdot \begin{bmatrix} P_{N-1}(z) & 0 & 0 & \cdots \\ 0 & P_{N-2}(z) & 0 & \cdots \\ \vdots & & & \\ 0 & & \cdots & 0 & P_0(z) \end{bmatrix} \cdot \mathbf{x}(z)\,. \qquad (2.45)$$

At this point it becomes clear that the synthesis requires the inverse functions $1/P_n(z)$, which represent IIR filters, whose stability is difficult to control. Consequently, a critically sampled filter bank based on filters $\{h_k(n)\}$ that are related through DFT modulation, as in Eq. (2.42), can have the perfect reconstruction property with FIR filters in both the analysis stage *and* the synthesis stage only if the filters are not longer than the downsampling rate N and have no overlap in time with neighboring blocks. To obtain FIR synthesis for longer filters, the filter bank must have a different structure.

2.9.1 Cosine-Modulated Filter Banks

We saw that a discrete cosine transform is obtained by applying a DFT to a symmetrically extended real valued signal. This suggests that a DCT would lead to a different filter matrix $\mathbf{F}_a$, with elements off the diagonal. In many applications, as in video, audio, or speech coding, the signal is indeed represented as real values. Now it would be interesting to see the shape of the resulting filter matrix for a filter bank based on a DCT IV modulation [compare to Eq. (2.29)]. In this case, the filters $\{h_k(n)\}$ are modulated with cosine functions (the factor $\sqrt{2/N}$ is neglected for simplicity),

$$h_k(n) = \cos\left(\frac{\pi}{N}(k+0.5)(n+0.5)\right) \cdot p(LN - 1 - n) , \qquad (2.46)$$

and the transform (the subband signals) can be written as

$$\mathbf{X}_{k,m} = \sum_{n=0}^{LN-1} \cos\left(\frac{\pi}{N}(k+0.5)(n+0.5)\right) \cdot p(LN - 1 - n)\, x(mN+n) . \quad (2.47)$$

We will exploit the symmetries embodied in the identities

$$\cos\left(\frac{\pi}{N}(k+0.5)((n+N)+0.5)\right)$$
$$= -\cos\left(\frac{\pi}{N}(k+0.5)((N-1-n)+0.5)\right) \qquad (2.48)$$

and

$$\cos\left(\frac{\pi}{N}(k+0.5)((n+2N)+0.5)\right) = -\cos\left(\frac{\pi}{N}(k+0.5)(n+0.5)\right) \qquad (2.49)$$

This means that every second block of N input samples "reverses the direction" of the cosine transform. A close examination of these symmetries and replacing n by $n + 2lN$ and $n + N + 2lN$ shows that the analysis equation (2.47) can be written as a type of *folding* operation followed by a cosine transform, as can be seen in the following.

Again the filtering can be written more easily in the z-domain, with

$$P_n(z) = \sum_{l=0}^{L-1} p(n+2lN) \cdot z^{-l}$$

with $n = 0, \ldots, N-1$,

$$X_n(z) = \sum_{l=0}^{\infty} x(n+lN) \cdot z^{-l} .$$

with $n = 0, \ldots, 2N-1$. Using $\mathbf{D}_{IV}$ as the DCT IV matrix leads to

$$\mathbf{X}(z) = \mathbf{D}_{IV} \cdot \mathbf{F}_a(z) \cdot \mathbf{x}(z)$$

where

$$\mathbf{F}_a(z) =$$

$$
\begin{bmatrix}
z^{-1}p_{2N-1}\left(-z^2\right) & & 0 & & & & p_0\left(-z^2\right) \\
& \ddots & & & & \iddots & \\
& & z^{-1}p_{N+N/2}\left(-z^2\right) & p_{N/2-1}\left(-z^2\right) & & & \\
0 & & p_{N/2}\left(-z^2\right) & z^{-1}p_{N+N/2-1}\left(-z^2\right) & & 0 & \\
& \iddots & & & \ddots & & \\
p_{N-1}\left(-z^2\right) & & 0 & & & z^{-1}p_N\left(-z^2\right) &
\end{bmatrix}
$$

$$(2.50)$$

This form of $\mathbf{F}_a(z)$ assumes that the window length factor L is even, which can always be obtained by appending zeros. The filter matrix $\mathbf{F}_a(z)$ has a bi-diagonal structure, i.e., it has nonzero elements not only on the diagonal but also on the antidiagonal. This means a window function can be designed such that the inverse of the filter matrix leads to FIR filters. An example is the classical MDCT or TDAC filter bank [37]. It results from inserting an additional phase shift of $N/2$ in the modulating cosine function:

$$\mathbf{X}_{k,m} = \sum_{n=0}^{2N-1} \cos\left(\frac{\pi}{N}(k+0.5)(n+0.5+N/2)\right) \cdot p(LN-1-n)\,x(mN+n).$$

This phase shift leads to a shift of the structure of the filter matrix downwards by $N/2$. For example, for a window function $p(n)$ for $n = 0, \dots, 2N-1$, the filter matrix has the following form,

$$
\mathbf{F}_a(z) =
\begin{bmatrix}
0 & & z^{-1}p(1.5N) & z^{-1}p(1.5N-1) & & 0 \\
& \iddots & & & \ddots & \\
z^{-1}p(2N-1) & & 0 & & & z^{-1}p(N) \\
p(N-1) & & & & & -p(0) \\
& \ddots & & & \iddots & \\
0 & & p(N/2) & -p(N/2-1) & & 0
\end{bmatrix}
$$

The inverse for the synthesis matrix is

$$
z^{-1}\mathbf{F}_a^{-1}(z) =
\begin{bmatrix}
0 & & q(0) & z^{-1}q(N) & & 0 \\
& \iddots & & & \ddots & \\
q(N/2-1) & & 0 & & & z^{-1}q(1.5N-1) \\
q(N/2) & & & & & -z^{-1}q(1.5N) \\
& \ddots & & & \iddots & \\
0 & & q(N-1) & -z^{-1}q(2N-1) & & 0
\end{bmatrix}
$$

with

$$q(n) = \frac{p(n)}{p(2N - 1 - n)p(n) + p(N - 1 - n)p(N + n)}$$

$$q(N + n) = \frac{p(N + n)}{p(2N - 1 - n)p(n) + p(N - 1 - n)p(N + n)}$$

where $n = 0, \ldots, N-1$. This inverse is used in the synthesis filter bank to reconstruct the signal, e.g., in a decoder. The synthesis side has a filter matrix with the same shape as the analysis side, so the synthesis filter bank is again a cosine-modulated filter bank, with $q(n)$ as its window function. Observe that $q(n) = p(n)$ if the denominator for the computation of the inverse becomes one.

The DCT leads to a filter matrix which has a form enabling us to design filter banks with critical sampling and FIR filters for analysis as well as for the synthesis. Therefore filter banks based on DCTs are the predominant tools for time-frequency decomposition in audio coding.

To design filter banks with longer filters and more freedom in the design process, the filter matrix Eq. (2.50) can be written as a product of simpler matrices. These simpler matrices can be unitary, such that the product is a unitary matrix, whose inverse is then obtained by simply transposing it and replacing z by z^{-1} [31]. Or these simpler matrices can be bi-orthogonal, so that the resulting filter bank is bi-orthogonal [45]. The latter is a more general solution, which enables us to design, for example, filter banks with a lower end-to-end delay than unitary or orthogonal filter banks [43, 44].

2.9.2   Complex DFT-Based Filter Banks

A disadvantage of the DCT is that it delivers no phase or magnitude information, as the DFT does. For example, in audio coding the magnitudes of the subband signals are needed as inputs to psycho-acoustic models which control the quantization process, as seen in Fig. 2.18. Such is the basic structure of, for example, the PAC audio coder. The DCT can be seen as the real part of a DFT of a real valued signal, so what is needed is the imaginary part to obtain complex subband signals and hence their magnitudes. The imaginary part can be obtained by using a filter bank based on a DST. For a cosine-modulated filter bank with a DCT IV, the corresponding sine-modulated filter bank uses a DST IV (2.30). The equality

$$\sin\left(\frac{\pi}{N}(k + 0.5)(n + 0.5)\right) = \cos\left(\frac{\pi}{N}(k + 0.5)(n - N + 0.5)\right)$$

shows, that the sine modulation function has the same symmetries in time n as the cosine modulation function [Eqs. (2.48) and (2.49)] but is shifted by N samples. This leads to the same conditions on the window function for perfect reconstruction, so the same window function can be used for the cosine- and the sine-modulated filter banks, hence for the real and imaginary parts of the resulting complex valued filter bank. This is important for obtaining the precise magnitude and phase information of a signal.

In audio coding, the signal consists of real values. If the input signal to the complex filter bank consists of complex values, as in applications such as synthetic aperture radar (SAR) [30], the filter bank needs to cover positive as well as negative frequencies to obtain perfect reconstruction. If AFB_C is the output of the cosine-modulated analysis filter bank, and AFB_S is the output of the sine-modulated filter bank, then the positive frequencies are obtained by taking $AFB_C - jAFB_S$ and the negative frequencies by $AFB_C + jAFB_S$, similar to the DFT. This means the analysis filter bank consists of $2N$ bands

$$[AFB_C - jAFB_S, AFB_C + jAFB_S] .$$

The synthesis filter bank for perfect reconstruction has an analogous structure,

$$[SFB_C + jSFB_S, SFB_C - jSFB_S] ,$$

where SFB_C, SFB_S are the outputs of the synthesis filter banks.

It is easy to see that this synthesis filter bank leads to perfect reconstruction if the cosine and sine filter banks have the perfect reconstruction property of their own. Observe that this is not the only solution for perfect reconstruction since the filter bank is, in effect, oversampled at twice the rate. But this solution for the synthesis has an advantage because it has an analogous structure, hence similar properties, as the analysis part, which is often desirable in coding applications.

Figs. 2.21–2.23 show a comparison of the frequency responses of the window functions of a direct FFT approach, as used in the MPEG-AAC audio coder as input for the psycho-acoustic model, and the complex filter bank. Fig. 2.22 shows the frequency response of a 1024 band FFT filter bank, and Fig. 2.21 shows the frequency response of a complex low-delay filter bank with 1024 bands, an analysis/synthesis delay of 2047 samples, and filter length of 4096 taps. Figs. 2.23 and 2.24 show an enlargement with the passband on the left. The passband of the complex filter bank is narrower, and the stopband attenuation is much higher than with the direct FFT application.

Figs. 2.25–2.27 show an application example for a stereo audio signal that is encoded and decoded at two different bit rates. Fig. 2.25 shows a piece of the original audio signal (jazz music), the left channel, sampled at 32000 samples/s. In this uncompressed representation, each sample is represented with a 16 bit integer number, which leads to a bit rate of $16 \cdot 2 \cdot 32000 = 1024$ kb/s. Fig. 2.26 shows that signal, but coded and decoded with a bit rate of 67 kb/s for the stereo signal (i.e., 35 kb/s per channel, or a compression ratio of over 14). The resulting audio quality is comparable to FM radio. It can be seen that there are slight differences to the original, but most of the differences are still inaudible because of the application of the psycho-acoustic model. Fig. 2.27 shows the signal at 30 kb/s stereo (a compression ratio of over 34). The resulting quality is comparable to AM radio. There are now more pronounced differences to the original; it is much smoother, which means it contains fewer high frequencies. Here the difference to the original is easy to hear, but the psycho-acoustic model is used such that the audible distortions are minimized.

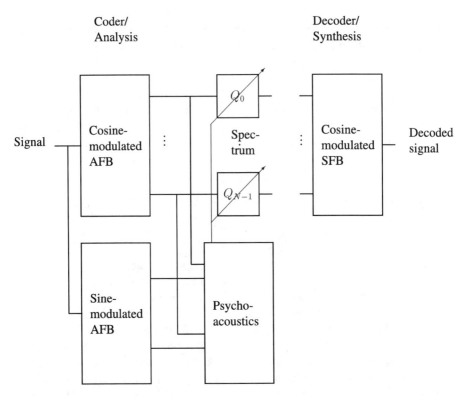

FIGURE 2.20
Audio coding based on filter banks, AFB: analysis filter bank, SFB: synthesis filter bank.

2.10   Conclusion

This chapter introduced the DFT and some of its basic properties. Even though it is a complex-valued transform, because of its symmetry properties, the DFT of a real-valued N-point signal can be represented again by N real values. A set of real-valued discrete cosine transforms can be derived using the DFT. The derivation of a fast algorithm for computing the DFT (the FFT) was also described here.

The DFT has many applications in coding. For example, the FFT is used for the efficient implementation of DCTs, the MDCT, and low delay filter banks. Furthermore, the complex output is used for power spectrum estimation, in particular, to drive psycho-acoustic models in audio coding, and it can be used to implement complex-valued filter banks for improved power spectrum estimation.

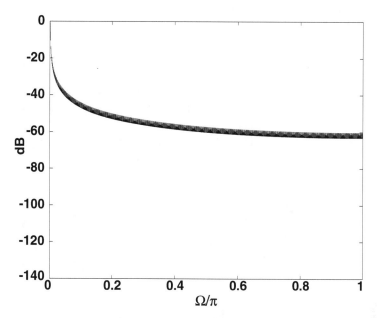

FIGURE 2.21
Magnitude of the frequency response of the rectangular window of a DFT of length 1024.

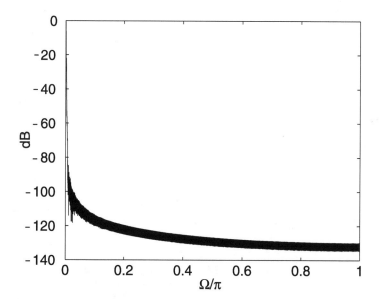

FIGURE 2.22
Magnitude of the frequency response of the window of a low delay filter bank with 1024 bands and filter length 4096.

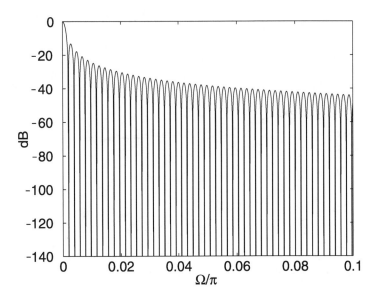

FIGURE 2.23
Enlargement of the first part of the magnitude of the frequency response of the
rectangular window of the DFT.

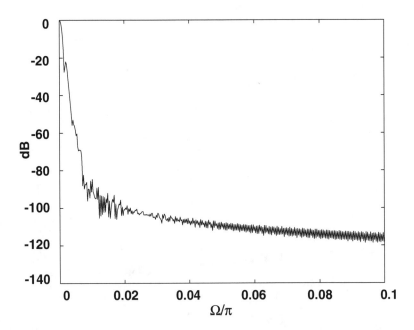

FIGURE 2.24
Enlargement of the first part of the magnitude of the frequency response of the
window of the low delay filter bank.

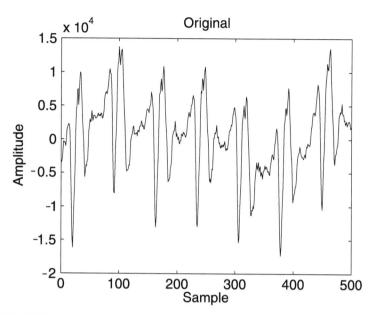

FIGURE 2.25
A piece of an example audio signal, sampled at 32 khz. Shown is the left channel
of the stereo signal.

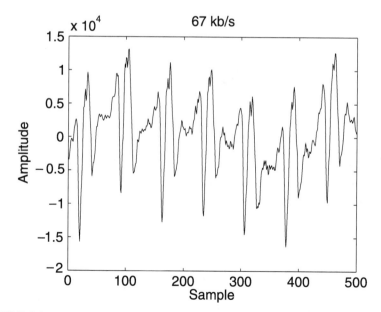

FIGURE 2.26
The stereo audio signal, coded and decoded with 67 kb/s. The left channel is
shown.

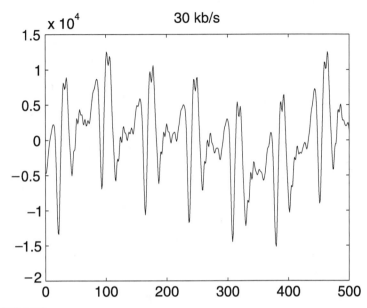

FIGURE 2.27
The left channel of the stereo audio signal, coded and decoded, but with 30 kb/s.

2.11   FFT Web sites

The following list reflects some of the available software and information on Web sites devoted to the FFT (September 1999).

- FFTW
  http://www.fftw.org/index.html
  http://www.fftw.org/benchfft/doc/ffts.html

- FFTPACK
  http://www.netlib.org/fftpack/

- FFT for Pentium (Bernstein)
  ftp://koobera.math.uic.edu/www/djbfft.html

- FFT software (comp.speech FAQ Q2.4)
  http://svr-www.eng.cam.ac.uk/comp.speech/
  Section2/Q2.4.html

- One-dimensional real fast Fourier transforms
  http://www.hr/josip/DSP/fft.html

- FXT package FFT code (Arndt)
  `http://www.jjj.de/fxt/`

- FFT (Don Cross)
  `http://www.intersrv.com/~dcross/fft.html`

- Public domain FFT code
  `http://risc1.numis.nwu.edu/ftp/pub/transforms/`
  `http://risc1.numis.nwu.edu/fft/`

- DFT (Paul Bourke)
  `http://www.swin.edu.au/astronomy/pbourke/`
  `sigproc/dft/`

- FFT code for TMS320 processors
  `ftp://ftp.ti.com/mirrors/tms320bbs/`

- Fast Fourier Transforms (Kifowit)
  `http://ourworld.compuserve.com/homepages/`
  `steve_kifowit/fft.htm`

- Nielsen's MIXFFT page
  `http://home.get2net.dk/jjn/fft.htm`

- Parallel FFT homepage
  `http://www.arc.unm.edu/Workshop/FFT/fft/fft.html`

- FFT public domain algorithms
  `http://www.arc.unm.edu/Workshop/FFT/fft/fft.html`

- Numerical recipes
  `http://www.nr.com/`

- General purpose FFT package
  `http://momonga.t.u-tokyo.ac.jp/õoura/fft.html`

- FFT links
  `http://momonga.t.u-tokyo.ac.jp/õoura/fftlinks.html`

- FFT, performance, accuracy, and code (Mayer)
  `http://www.geocities.com/ResearchTriangle/8869/`
  `fft_summary.html`

- Prime-length FFT
  `http://www.dsp.rice.edu/software/RU-FFT/`
  `pfft/pfft.html`

- Notes on the FFT (Burrus)
  `http://www.dsp.rice.edu/research/fft/fftnote.asc`

- Yahoo FFT Web site list
  `http://dir.yahoo.com/Science/Mathematics/Software/`
  `Fast_Fourier_Transform__FFT_/`

References

[1] Ahmed, N., Natarajan, T., and Rao, K.R., Discrete cosine transform, *IEEE Trans. Comput.,* 23:90–93, 1974, also in [41].

[2] Bellanger, M., *Digital Processing of Signals, Theory and Practice,* John Wiley & Sons, Chichester, NY, 1989.

[3] Bergland, G.D., A guided tour of the fast Fourier transform, *IEEE Spectrum,* 6:41–52, 1969, also in [39].

[4] Blahut, R.E., *Fast Algorithms for Digital Signal Processing.* Addison-Wesley, Reading, MA, 1985.

[5] Bracewell, R.N., *The Fourier Transform and its Applications,* McGraw Hill, Reading, MA, 1986.

[6] Brandenburg, K. and Bosi, M., Overview of MPEG audio: current and future standards for low bit rate audio coding, *J. Audio Eng. Soc.,* 45(1/2):4–21, 1997.

[7] Brigham, E.O., *The Fast Fourier Transform and its Applications,* Prentice-Hall, Englewood Cliffs, NJ, 1988.

[8] Burrus, C.S., Efficient Fourier transform and convolution algorithms, in Lim, J.S. and Oppenheim, A.V., Eds., *Advanced Topics in Signal Processing,* Prentice-Hall, Englewood Cliffs, NJ, 1988.

[9] Burrus, C.S. and Eschenbacher, P.W., An in-place, in-order prime factor FFT algorithm, *IEEE Trans. on Acoust., Speech, Signal Proc.,* 29(4):806–817, 1981.

[10] Burrus, C.S. and Parks, T.W., *DFT/FFT and Convolution Algorithms,* John Wiley & Sons, Chichester, NY, 1985.

[11] Cochran, J.W., Favin, D.L., Helms, H.D., Kaenel, R.A., Lang, W.W., Maling, G.C., Nelson, D.E., Rader, C.M., and Welch, P.D., What is the fast Fourier transform?, *IEEE Trans. Audio Electroacoust.,* 15:45–55, 1967, also in [39].

[12] Cohen, L., *Time-Frequency Analysis.* Prentice-Hall, Englewood Cliffs, NJ, 1995.

[13] Cooley, J.W., Lewis, P.A.W., and Welch, P.D., Historical notes on the fast Fourier transform, *IEEE Trans. Audio Electroacoust.,* 15:76–79, 1967, also in [39].

[14] Cooley, J.W., Lewis, P.A.W., and Welch, P.D., The finite Fourier transform, *IEEE Trans. Audio Electroacoust.,* 17:77–85, 1969, also in [39].

[15] Cooley, J.W. and Tukey, J.W., An algorithm for the machine calculation of complex Fourier series, *Mathematics of Computation,* 19(90):297–301, 1965, also in [39].

[16] Duhamel, P., Implementation of "split radix" FFT algorithm, *IEEE Trans. on Acoust., Speech, Signal Proc.,* 34(2):285–295, 1986.

[17] Duhamel, P. and Vetterli, M., Fast Fourier transforms: a tutorial review and a state of the art, *Signal Processing,* 19:259–299, 1990, also in [18].

[18] Duhamel, P. and Vetterli, M., Fast Fourier transforms: a tutorial review and a state of the art, in Madisetti, V.K. and Williams, D.B., Eds., *The Digital Signal Processing Handbook,* chapter 7, CRC Press, 1998, also appears as [17].

[19] Elliott, D.F., Fast Fourier transforms, in Elliott, D.F., Ed., *Handbook of Digital Signal Processing,* Chapter 7, pages 527–631, Academic Press, New York, 1987.

[20] Elliott, D.F. and Rao, K.R., *Fast Transforms: Algorithms, Analyses, Applications,* Academic Press, New York, 1982.

[21] Frigo, M. and Johnson, S.G., FFTW, FFT software developed at MIT, http://www.fftw.org/index.html.

[22] Heideman, M.T., *Multiplicative Complexity, Convolution, and the DFT,* Springer-Verlag, New York, Berlin, 1988.

[23] Johnson, H.W. and Burrus, C.S., The design of optimal DFT algorithms using dynamic programming, *IEEE Trans. on Acoust., Speech, Signal Proc.,* 31(2):378–387, 1983.

[24] Johnson, J., Automatic implementation and generation of FFT algorithms, SIAM parallel processing FFT session, March 1999, see SPIRAL webpage http://www.ece.cmu.edu/~spiral/.

[25] Jones, K.J., Prime number DFT computation via parallel circular convolvers, *IEE Proceedings, Part F,* 137(3):205–212, 1990.

[26] Karp, T. and Fliege, N.J., MDFT filter banks with perfect reconstruction, in *IEEE International Symposium on Circuits and Systems,* Seattle, WA, 1995.

[27] Lim, J.S. and Oppenheim, A.V., *Advanced Topics in Signal Processing,* Prentice-Hall, Englewood Cliffs, NJ, 1988.

[28] Van Loan, C., *Computational Frameworks for the Fast Fourier Transform,* SIAM, Philadelphia, PA, 1992.

[29] Lu, C., Cooley, J.W., and Tolimieri, R., FFT algorithms for prime transform sizes and their implementations on VAX, IBM3090VF, and IBM RS/6000, *IEEE Trans. on Acoust., Speech, Signal Proc.,* 41(2):638–648, 1993.

[30] Madisetti, V.K. and Williams, D.B., *The Digital Signal Processing Handbook,* CRC Press and IEEE Press, Boca Raton, FL, 1997.

[31] Malvar, H., *Signal Processing with Lapped Transforms,* Artech House, Boston, MA, London, 1992.

[32] Martucci, S., Symmetric convolution and the discrete sine and cosine transforms, *IEEE Trans. on Signal Processing,* 42:1038–1051, 1994.

[33] McClellan, J.H. and Rader, C.M., *Number Theory in Digital Signal Processing,* Prentice-Hall, Englewood Cliffs, NJ, 1979.

[34] Meyer, R. and Schwarz, K., FFT implementation on DSP-chips, theory and practice, in *Proc. IEEE Int. Conf. Acoust., Speech, Signal Processing (ICASSP),* 1503–1506, April 1990.

[35] Myers, D.G., *Digital Signal Processing: Efficient Convolution and Fourier Transform Techniques,* Prentice-Hall, Englewood Cliffs, NJ, 1990.

[36] Nussbaumer, H.J., *Fast Fourier Transform and Convolution Algorithms,* Springer-Verlag, New York, Berlin, 1982.

[37] Princen, J.P. and Bradley, A.B., Analysis/synthesis filter bank design based on time domain aliasing cancellation, *IEEE Trans. on Signal Processing,* 34(10):1153–1161, 1986.

[38] Proakis, J.G., Rader, C.M., Ling, F., and Nikias, C.L., *Advanced Digital Signal Processing,* Macmillan, New York, 1992.

[39] Rabiner, L.R. and Rader, C.M., Eds., *Digital Signal Processing,* IEEE Press, Piscataway, NJ, 1972.

[40] Rader, C.M., Discrete Fourier transform when the number of data samples is prime, *Proc. IEEE,* 56(6):1107–1108, 1968.

[41] Rao, K.R., Ed., *Discrete Transforms and Their Applications,* Krieger, Malabar, FL, 1990.

[42] Rao, K.R. and Yip, P., *Discrete Cosine Transform: Algorithms, Advantages, Applications,* Academic Press, New York, 1990.

[43] Schuller, G., Time-varying filter banks with variable system delay, in *Proc. IEEE ICASSP,* Vol. 3, 2469–2472, Munich, Germany, 1997.

[44] Schuller, G. and Karp, T., Modulated filter banks with arbitrary system delay: efficient implementations and the time-varying case, *IEEE Trans. on Signal Processing,* 48(3), 2000.

[45] Schuller, G.D.T. and Smith, M.J.T., New framework for modulated perfect reconstruction filter banks, *IEEE Trans. on Signal Processing,* 44(8):1942–1954, 1996.

[46] Selesnick, I.W. and Burrus, C.S., Automatic generation of prime length FFT programs, *IEEE Trans. on Signal Processing*, 44(1):14–24, 1996.

[47] Sinha, D., Johnston, J.D., Dorward, S., and Quackenbush, S., The perceptual audio coder (PAC), in Madisetti, V. and Williams, D.B., Eds., *The Digital Signal Processing Handbook*, chapter 42, CRC Press and IEEE Press, Boca Raton, FL, 1997.

[48] Smith, W.W. and Smith, J.M., *Handbook of Real-Time Fast Fourier Transforms*, IEEE Press, Piscataway, NJ, 1995.

[49] Sorensen, H.V. and Burrus, C.S., Fast DFT and convolution algorithms, in Mitra, S.K. and Kaiser, J.F., Eds., *Handbook For Digital Signal Processing*, chapter 8, 491–610, John Wiley & Sons, New York, 1993.

[50] Sorenson, H.V., Heideman, M.T., and Burrus, C.S., On computing the split-radix FFT, *IEEE Trans. on Acoust., Speech, Signal Proc.*, 34(1):152–156, 1986.

[51] Sorenson, H.V., Jones, D.L., Heideman, M.T., and Burrus, C.S., Real-valued fast Fourier transform algorithms, *IEEE Trans. on Acoust., Speech, Signal Proc.*, 35(6):849–863, 1987.

[52] Temperton, C., Implementation of a self-sorting in-place prime factor FFT algorithm, *J. of Computational Physics*, 58:283–299, 1985.

[53] Temperton, C., Self-sorting in-place fast Fourier transforms, *SIAM J. on Scientific and Statistical Computing*, 12(4):808–823, 1991.

[54] Tolimieri, R., An, M., and Lu, C., *Algorithms for Discrete Fourier Transform and Convolution*, Springer-Verlag, New York, Berlin, 1989.

[55] Vaidyanathan, P.P., *Multirate Systems and Filter Banks*, Prentice-Hall, Englewood Cliffs, NJ, 1992.

[56] Vetterli, M. and Duhamel, P., Split-radix algorithms for length-p^m DFTs, *IEEE Trans. on Acoust., Speech, Signal Proc.*, 37(1):57–64, 1989.

[57] Vetterli, M. and Kovačević, J., *Wavelets and Subband Coding*, Prentice-Hall, Englewood Cliffs, NJ, 1995.

[58] Wickershauser, M.L., *Adapted Wavelet Analysis from Theory to Software*, A.K. Peters, Wellesley, MA, 1994.

[59] Winograd, S., Some bilinear forms whose multiplicative complexity depends on the field of constants, *Mathematical Systems Theory*, 10:169–180, 1977.

[60] Winograd, S., *Arithmetic Complexity of Computations*, SIAM, Philadelphia, PA, 1980.

[61] Yip, P. and Rao, K.R., Fast discrete transforms, in Elliott, D.F., Ed., *Handbook of Digital Signal Processing*, chapter 6, 481–525, Academic Press, New York, 1987.

Chapter 3

Comparametric Transforms for Transmitting Eye Tap Video with Picture Transfer Protocol (PTP)

W. Steve G. Mann

University of Toronto

Eye Tap video is a new genre of video imaging facilitated by and for the apparatus of the author's eyeglass-based "wearable computer" invention [1]. This invention gives rise to a new genre of video that is best processed and compressed by way of comparametric equations, and comparametric image processing. These new methods are based on an Edgertonian philosophy, in sharp departure from the traditional Nyquist philosophy of signal processing. A new technique is given for estimating the comparameters (relative parameters between successive frames of an image sequence) taken with a camera (or Eye Tap device) that is free to pan, tilt, rotate about its optical axis, and zoom. This technique solves the problem for two cases of static scenes: images taken from the same location of an arbitrary 3-D scene and images taken from arbitrary locations of a flat scene, where it is assumed that the gaze pattern of the eye sweeps on a much faster time scale than the movement of the body (e.g., an assumption that image flow across the retina induced by change in eye location is small compared to that induced by gaze pattern).

3.1 Introduction: Wearable Cybernetics

Wearable cybernetics is based on the WearComp invention of the 1970s, originally intended as a wearable electronic photographer's assistant [2].

3.1.1   Historical Overview of WearComp

A goal of the author's WearComp/WearCam (wearable computer and personal imaging) inventions of the 1970s and early 1980s (Fig. 3.1) was to make the metaphor of technology as an extension of the mind and body into a reality. In some sense, these inventions transformed the body into not just a camera, but also a networked cybernetic entity. The body thus became part of a system always seeking the best picture, in all facets of ordinary day-to-day living. These systems served to illustrate the concept of the camera as a true extension of the mind and body of the wearer.

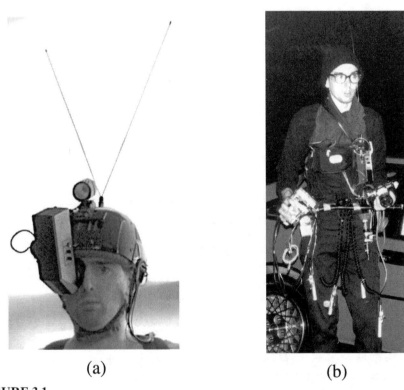

(a)   (b)

FIGURE 3.1
Personal Imaging in the 1970s and 1980s: **Early embodiments of the author's WearComp invention that functioned as a "photographer's assistant" for use in the field of personal imaging. (a) Author's early headgear. (b) Author's early "smart clothing" including cybernetic jacket and cybernetic pants** *(continued).*

3.1.2   Eye Tap Video

Eye Tap video [3] is video captured from the pencil of rays that would otherwise pass through the center of the lens of the eye. The Eye Tap device is typically worn like eyeglasses.

(c)

FIGURE 3.1
(Cont.) **(c) Author's 1970s chording keyboard comprising switches mounted to a light source, similar to the mid 1980s version depicted in author's right hand in (b).**

3.2 The Edgertonian Image Sequence

Traditional image sequence compression, such as MPEG [4, 5] (see, for example, the Moving Picture Expert Group FAQ), is based on processing frames of video as a continuum. The integrity of motion is often regarded as being more important than, or at least as important as, the integrity of each individual frame of the image sequence. However, it can be argued that temporal integrity is not always of the utmost importance and can, in fact, often be sacrificed with good reason.

3.2.1 Edgertonian versus Nyquist Thinking

Consider the very typical situation in which the frame rate of a picture acquisition process vastly exceeds the frame rate at which it is possible to send pictures of satisfactory quality over a given bandwidth-limited communications channel. This situation arises, for example, with Web-based cameras, including the Wearable Wireless Webcam [6].

Suppose that the camera provides 30 pictures per second, but the channel allows us to send only one picture per second (ignore for the moment the fact that we can trade spatial resolution, temporal resolution, and compression quality to adjust the frame rate). In order to downsample our 30 pictures per second to one picture per second, the "Nyquist school of thought" would suggest that we temporally lowpass filter the image sequence in order to remove any temporal frequencies that would exceed the Nyquist frequency. To apply this standard "lowpass filter then downsample" approach, we might average each 30 successive pictures to obtain one output picture. Thus, fast moving objects would be blurred to prevent temporal aliasing.

We might be tempted to think that this blurring is desirable, given temporal aliasing that would otherwise result. However, cinematographers and others who produce motion pictures often disregard concepts' temporal aliasing. Most notably, Harold E. Edgerton [7], inventor of the electronic flash and known for his movies of high speed events in which objects are "frozen" in time, has produced movies and other artifacts that defy any avoidance of temporal aliasing. Edgerton's movies provide us with a temporal sampling that is more like a Dirac comb (downsampling of reality) than a lowpass-filtered and then downsampled version of reality. For the example of downsampling from 30 frames per second to one frame per second, an Edgertonian thinker would likely advocate simply taking every 30th frame from the original sequence and throwing all the others away.

The Edgertonian downsampling philosophy gives rise to image sequences in which propeller blades or wagon wheel spokes appear to spin backwards or stand still. The Nyquist philosophy, on the other hand, gives rise to image sequences in which the propeller blades or wagon wheel spokes visually disappear. The author believes that it is preferable that the propeller blades and wagon wheel spokes appear to spin backwards, or stand still, rather than visually disappear. More generally, an important assumption upon which the thesis of this chapter rests is that it is preferable to have a series of crisp well-defined "snapshots" of reality, rather than the blur of images that one would get by following the antialiasing approach of traditional signal processing.

The author's personal experience with his wearable Eye Tap video camera invention, wearing the camera often 8 to 16 hours a day, led to an understanding of how the world looks through Web-based video. On this system, it was possible to choose from among various combinations of Edgertonian and Nyquist sampling strategies. It was found that experiencing the world through "Edgertonian eyes" was generally preferable to the Nyquist approach.

3.2.2   Frames versus Rows, Columns, and Pixels

There is a trend now toward processing sequences of images as spatio-temporal volumes, e.g., as a function $f(x, y, t)$. Within this conceptual framework, motion pictures are treated as static three-dimensional volumes of data. So-called *spatio-temporal filters* $h(x, y, t)$ are applied to these spatio-temporal volumes $f(x, y, t)$.

However, this unified treatment of the three dimensions (discretized to row, column, and frame number) ignores the fact that the time dimension has a much different intuitive meaning than the other two dimensions. Apart from the progressive (forward-only) direction of time, there is the more important fact (even for stored image sequences) that a snapshot in time (a still picture selected from the sequence) often has immediate meaning to the human observer. A single row of pixels across a picture or a single column of pixels down a picture do not generally have similar significance to the human observer. Likewise, a single pixel means little to the human observer in the absence of surrounding pixels.

Notwithstanding their utility, slices of the form $f(x, y_0, t)$ or of the form $f(x_0, y, t)$ are often confusing at best, compared to the still picture $f(x, y, t_0)$ that remains as an extraction from a picture sequence which is far more meaningful to a typical human observer. Thus the author believes that downsampling across rows or downsampling down columns of an image should be preceded by lowpass filtering, whereas temporal downsampling should not.

There is, therefore, a special significance to the notion of a "snapshot in time" and the processing, storage, transmission, etc. of a motion picture as a sequence of such snapshots. The object of this chapter is to better understand the relationship between individual sharply defined frames of an Edgertonian sequence of pictures.

3.3 Picture Transfer Protocol (PTP)

When applying data compression to a stream of individual pictures that will be viewed in real-time (for example, in videoconferencing, such as the first-person-perspective videoconferencing of the wearable Eye Tap device), it is helpful to consider the manner in which the data will be sent. Most notably, pictures are typically sent over a packet-based communications channel. For example, Wearable Wireless Webcam used the AX25 Amateur Radio [8] protocol. Accordingly, packets typically arrive either intact or corrupted. Packets that are corrupt traditionally would be resent. An interesting approach is to provide data compression on a per-image basis, and to vary the degree of compression so that the size of each picture in the image sequence is exactly equal to the length of one packet.

Together with the prior assumption (that images are acquired at a rate that exceeds the channel capacity), it will generally be true that by the time we know that a packet (which is a complete picture) is corrupt at the receiver, a newer picture will have already been acquired. For example, if the round trip time (RTT) were 100 ms (which is equal to the time it takes to generate three pictures), there would be little sense in resending a picture that was taken three pictures ago. The commonly arising situation in which pictures are captured at a rate that exceeds the RTT suggests that there will always be newer picture information at the transmit site than what would be resent in the event of a lost packet.

This approach forms the basis for the Picture Transfer Protocol (PTP) proposed by the author. In particular, PTP is based on the idea of treating each snapshot in time as a single entity, in isolation, and compressing it into a single packet, so it will have either arrived in its entirety or not arrived at all (and therefore can be discarded). It should be clear that the philosophical underpinnings of PTP are closely related to those of Edgertonian downsampling.

3.4   Best Case Imaging and Fear of Functionality

A direct result of Edgertonian sampling is that a single picture from a picture sequence has a high degree of relevance and meaning even when it is taken in isolation. Similarly, a direct result of PTP is that a single packet from a packet sequence has a high degree of relevance and meaning even when it is taken in isolation (for example, when the packets before and after it have been corrupted). It is therefore apparent that if a system were highly unreliable, to the extent that pictures could be transmitted only occasionally and unpredictably, then the Edgertonian sampling combined with PTP would provide a system that would degrade gracefully.

Indeed, if we were to randomly select just a few frames from one of Edgerton's motion pictures, we would likely have a good summary of the motion picture, since any given frame would provide us with a sharp picture in which subject matter of interest could be clearly discerned. Likewise, if we were to randomly select a few packets from a stream of thousands of packets of PTP, we would have data that would provide a much more meaningful interpretation to the human observer than if all we had were randomly selected packets from an MPEG sequence.

Personal imaging systems are characterized by a wearable incidentalist "always ready" mode of operation in which the system need not always be functioning to be of benefit. It is the *potential* functionality, rather than the actual functionality, of such a system that makes it so different from other imaging systems such as hand-held cameras and the like. Accordingly, an object of the personal imaging project is to provide a system that transmits pictures in harsh or hostile environments. One application of such a system is the personal safety device (PSD) [9]. The PSD differs from other wireless data transmission systems in the sense that it was designed for "best case" operation. Ordinarily, wireless transmissions are designed for worst case scenarios, such as might guarantee a certain minimum level of performance throughout a large metropolitan area. The PSD, however, is designed to make it hard for an adversary to guarantee total nonperformance.

It is not a goal of the PSD to guarantee connectivity in the presence of hostile jamming of the radio spectrum but, rather, to make it difficult for the adversary to guarantee the absence of connectivity. Therefore, an otherwise potential perpetrator of a crime would never be able to be certain that the wearer's device was nonoperational and would therefore need to be on his or her best behavior at all times.

Traditional surveillance networks, based on so-called public safety camera systems, have been proposed to reduce the allegedly rising levels of crime. However, building such surveillance superhighways may do little to prevent, for example, crime by representatives of the surveillance state, or those who maintain the database of images. Human rights violations can continue, or even increase, in a police state of total state surveillance. The same can be true of owners of an establishment where surveillance systems are installed and maintained by these establishment owners. An example is the famous Latasha Harlins case, in which a shopper was falsely accused of shoplifting by a shopkeeper and was then shot dead by the shopkeeper. Therefore, what is needed is a PSD to function as a crime deterrent, particularly with regard to crimes perpetrated by those further up the organizational hierarchy.

Since there is the possibility that only one packet, which contains just one picture, would provide incriminating evidence of wrongdoing, individuals can wear a PSD to protect themselves from criminals, assailants, and attackers, notwithstanding any public or corporate video surveillance system already in place.

An important aspect of this paradigm is the fear of functionality (FoF) model. The balance is usually tipped in favor of the state or large organization in the sense that state- or corporate-owned surveillance cameras are typically mounted on fixed mount points and networked by way of high bandwidth land lines. The PSD, on the other hand, would be connected by way of wireless communication channels of limited bandwidth and limited reliability. For example, in the basement of a department store, the individual has a lesser chance of getting a reliable data connection than does the store-owned surveillance cameras. Just as many department stores use a mixture of fake, nonfunctional cameras and real ones, so the customer never knows whether or not a given camera is operational, what is needed is a similar means of best case video transmission. Not knowing whether or not one is being held accountable for his actions, one must be on his best behavior at all times. Thus, a new philosophy, based on FoF, can become the basis of design for image compression, transmission, and representation.

Fig. 3.2(a) illustrates an example of a comparison between two systems, SYSTEM A, and SYSTEM B. These systems are depicted as two plots, in a hypothetical parameter space. The parameter space could be time, position, or the like. For example, SYSTEM A might work acceptably (e.g., meet a certain guaranteed degree of functionality F_{GUAR}) everywhere at all times, whereas SYSTEM B might work very well sometimes and poorly at others. Much engineering is motivated by an *articulability* model, i.e., that one can make an articulable basis for choosing SYSTEM A because it gives the higher worst case degree of functionality.

A new approach, however, reverses this argument by regarding functionality as a bad thing — bad for the perpetrator of a crime — rather than a good thing. Thus we turn the whole graph on its head, and, looking at the problem in this reversed light, come to a new solution, namely that SYSTEM B is better because there are times when it works really well.

Imagine, for example, a user in the sub-basement of a building, inside an elevator. Suppose SYSTEM A would have no hope of connecting to the outside world. SYS-

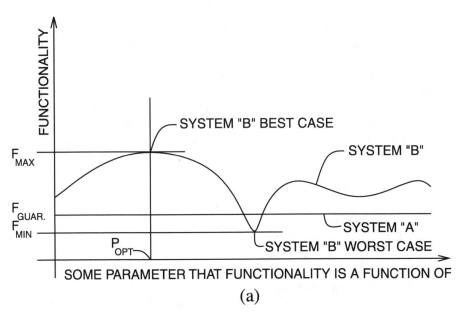

FIGURE 3.2

Fear of Functionality (FoF): (a) Given two different systems, SYSTEM A having a guaranteed minimum level of functionality F_{GUAR} that exceeds that of SYSTEM B, an articulable basis for selecting SYSTEM A can be made. Such an articulable basis might appeal to lawyers, insurance agents, and others who are in the business of guaranteeing easily defined articulable boundaries. However, a thesis of this chapter is that SYSTEM B might be a better choice. Moreover, given that we are designing and building a system like SYSTEM B, traditional worst case engineering would suggest focusing on the lowest point of functionality of SYSTEM B (*continued*).

TEM B, however, through some strange quirk of luck, might actually work, but we don't know in advance one way or the other.

The fact of the matter, however, is that one who was hoping that the system would not function, would be more afraid of SYSTEM B than SYSTEM A because it would take more effort to ensure that SYSTEM B would be nonfunctional.

The FoF model means that if the possibility exists that the system might function part of the time, a would-be perpetrator of a crime against the wearer of the PSD must be on his or her best behavior at all times.

Fig. 3.2(b) depicts what we might do to further improve the "fear factor" of SYSTEM B, to arrive at a new SYSTEM $\tilde{B}$. The new SYSTEM $\tilde{B}$ is characterized by being even more idiosyncratic; the occasional times that SYSTEM $\tilde{B}$ works, it works very well, but most of the time it either doesn't work at all or works very poorly.

Other technologies, such as the Internet, have been constructed to be robust enough to resist the hegemony of central authority (or an attack of war). However, an impor-

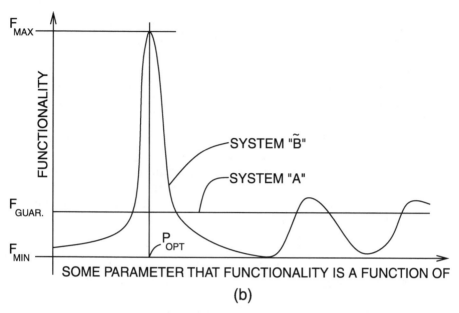

FIGURE 3.2

(Cont.) **(b) Instead, it is proposed that one might focus one's efforts on the highest point of functionality of SYSTEM B, to make it even higher, at the expense of further degrading the SYSTEM B worst case, and even at the expense of decreasing the overall average performance. The new SYSTEM $\tilde{B}$ is thus sharply serendipitous (peaked in its space of various system parameters).**

tant difference here is that the FoF paradigm is not suggesting the design of *robust* data compression and transmission networks.

Quite the opposite is true!

The FoF paradigm suggests the opposite of robustness in that SYSTEM $\tilde{B}$ is even more sensitive to mild perturbations in the parameter space about the optimal operating point, P_{OPT}, than is SYSTEM B. In this sense, our preferred SYSTEM $\tilde{B}$ is actually much less robust than SYSTEM B. Clearly it is not robustness, in and of itself, that the author is proposing here. The PSD doesn't need to work constantly but rather must simply present criminals with the possibility that it could work sometimes or even just occasionally. This scenario forms the basis for best-case design as an alternative to the usual worst-case design paradigm.

The personal imaging system therefore transmits video, but the design of the system is such that it will, at the very least, occasionally transmit a meaningful still image. Likewise, the philosophy for data compression and transforms needs to be completely rethought for this FoF model.

This rethinking extends from the transforms and compression approach right down to the physical hardware. For example, typically the wearer's jacket functions as a large low frequency antenna, providing transmission capability in a frequency band that is very hard to stop. For example, the 10-meter band is a good choice because of its unpredictable performance (owing to various "skip" phenomena, etc.). However, other frequencies are also used in parallel. For example, a peer-to-peer form of infrared communication is also included to "infect" other participants with the possibility of having received an image. In this way, it becomes nearly impossible for a police state to suppress the signal because of the *possibility* that an image may have escaped an iron-fisted regime.

It is not necessary to have a large aggregate bandwidth to support an FoF network. In fact, quite the opposite. Since it is not necessary that everyone transmit everything they see, at all times, very little bandwidth is needed. It is only necessary that anyone *could* transmit a picture at any time. This potential transmission (e.g., fear of transmission) does not even need to be done on the Internet; for example, it could simply be from one person to another.

3.5   Comparametric Image Sequence Analysis

Video sequences from the PSD are generally collected and assembled into a small number of still images, each still image being robust to the presence or absence of individual constituent frames of the video sequence from which it is composed.

Processing video sequences from the apparatus of the author's Eye Tap camera requires finding the coordinate transformation between two images of the same scene or object. Whether to recover gaze motion between video frames, stabilize retinal images, relate or recognize Eye Tap images taken from two different eyes, compute depth within a 3-D scene, or align images for lookpainting (high-resolution enhancement resulting from looking around), it is desired to have both a precise description of the coordinate transformation between a pair of Eye Tap video frames, and some indication as to its accuracy.

Traditional *block matching* [10] (such as used in *motion estimation*) is really a special case of a more general *coordinate transformation*. This chapter proposes a solution to the *motion estimation* problem using this more general estimation of a coordinate transformation, together with a technique for automatically finding the comparametric projective coordinate transformation that relates two frames taken of the same static scene. The technique takes two frames as input and automatically outputs the comparameters of the exact model to align the frames. It does not require the tracking or correspondence of explicit features, yet it is computationally practical. Although the theory presented makes the typical assumptions of static scene and no parallax, the estimation technique is robust to deviations from these assumptions. In particular, the technique is applied to image resolution enhancement and lookpainting [11], illustrating its success on a variety of practical and difficult cases, including some that violate the nonparallax and static scene assumptions.

A coordinate transformation maps the image coordinates, $\mathbf{x} = [x, y]^T$, to a new set of coordinates, $\tilde{\mathbf{x}} = [\tilde{x}, \tilde{y}]^T$. Generally, the approach to finding the coordinate transformation relies on assuming that it will take one of the models in Table 3.1, and then estimating the two to twelve scalar parameters of the chosen model. An illustration showing the effects possible with each of these models is given in Fig. 3.3.

Table 3.1 Image Coordinate Transformations Discussed in this Chapter: The Translation, Affine, and Projective Models Are Expressed in Vector Form; e.g., $\mathbf{x} = [x, y]^T$ is a Vector of dimension 2, and $\mathbf{A} \in \mathbb{R}^{2\times 2}$ is a Matrix of Dimension 2 by 2, etc.

Model	Coordinate transformation from x to x̃	Parameters
Translation	$\tilde{\mathbf{x}} = \mathbf{x} + \mathbf{b}$	$\mathbf{b} \in \mathbb{R}^2$
Affine	$\tilde{\mathbf{x}} = \mathbf{A}\mathbf{x} + \mathbf{b}$	$\mathbf{A} \in \mathbb{R}^{2\times 2}, \mathbf{b} \in \mathbb{R}^2$
Bilinear	$\tilde{x} = q_{\tilde{x}xy}xy + q_{\tilde{x}x}x + q_{\tilde{x}y}y + q_{\tilde{x}}$ $\tilde{y} = q_{\tilde{y}xy}xy + q_{\tilde{y}x}x + q_{\tilde{y}y}y + q_{\tilde{y}}$	$q_* \in \mathbb{R}$
Projective	$\tilde{\mathbf{x}} = \frac{\mathbf{A}\mathbf{x}+\mathbf{b}}{\mathbf{c}^T\mathbf{x}+1}$	$\mathbf{A} \in \mathbb{R}^{2\times 2}, \mathbf{b}, \mathbf{c} \in \mathbb{R}^2$
Pseudoperspective	$\tilde{x} = q_{\tilde{x}x}x + q_{\tilde{x}y}y + q_{\tilde{x}} + q_\alpha x^2 + q_\beta xy$ $\tilde{y} = q_{\tilde{y}x}x + q_{\tilde{y}y}y + q_{\tilde{y}} + q_\alpha xy + q_\beta y^2$	$q_* \in \mathbb{R}$
Biquadratic	$\tilde{x} = q_{\tilde{x}x^2}x^2 + q_{\tilde{x}xy}xy + q_{\tilde{x}y^2}y^2 + q_{\tilde{x}x}x + q_{\tilde{x}y}y + q_{\tilde{x}}$ $\tilde{y} = q_{\tilde{y}x^2}x^2 + q_{\tilde{y}xy}xy + q_{\tilde{y}y^2}y^2 + q_{\tilde{y}x}x + q_{\tilde{y}y}y + q_{\tilde{y}}$	$q_* \in \mathbb{R}$

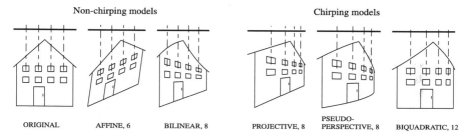

Non-chirping models Chirping models

ORIGINAL   AFFINE, 6   BILINEAR, 8   PROJECTIVE, 8   PSEUDO-PERSPECTIVE, 8   BIQUADRATIC, 12

FIGURE 3.3

Pictorial effects of the six coordinate transformations of Table 3.1, arranged left to right by number of parameters. Note that translation leaves the original house unchanged, except in its location. Most importantly, only the three coordinate transformations at the right affect the periodicity of the window spacing (e.g., induce the desired "chirping" which corresponds to what we see in the real world). Of these, only the projective coordinate transformation preserves straight lines. The 8-parameter projective coordinate transformation "exactly" describes the possible camera motions.

The most common assumption (especially in motion estimation for coding and optical flow for computer vision) is that the coordinate transformation between frames is a translation. Tekalp, Ozkan, and Sezan [12] have applied this assumption to high-resolution image reconstruction. Although translation is the least constraining and simplest to implement of the six coordinate transformations in Table 3.1, it is poor at handling large changes due to camera zoom, rotation, pan, and tilt.

Zheng and Chellappa [13] considered a subset of the affine model — translation, rotation, and scale — in image registration. Other researchers [14, 15] have assumed affine motion (six parameters) between frames. For the assumptions of static scene and no parallax, the affine model exactly describes rotation about the optical axis of the camera, zoom of the camera, and pure shear, which the camera does not do except in the limit as the lens focal length approaches infinity. The affine model cannot capture camera pan and tilt and, therefore, cannot accurately express the "chirping" and "keystoning" seen in the real world (see Fig. 3.3). Consequently, the affine model tries to fit the wrong parameters to these effects. When the parameter estimation is not done properly to align the images, a greater burden is placed on designing post-processing to enhance the poorly aligned images.

The 8-parameter *projective* model gives the exact eight desired parameters to account for all the possible camera motions. However, its parameters have traditionally been mathematically and computationally too hard to find. Consequently, a variety of approximations have been proposed. Before the solution to estimating the projective parameters is presented, it will be helpful to better understand these approximate models.

Going from first order (affine) to second order gives the 12-parameter biquadratic model. This model properly captures both the chirping (change in spatial frequency with position) and converging lines (keystoning) effects associated with projective coordinate transformations, although, despite its larger number of parameters, there is still considerable discrepancy between a projective coordinate transformation and the best-fit biquadratic coordinate transformation. Why stop at second order? Why not use a 20-parameter bicubic model? While an increase in the number of model parameters will result in a better fit, there is a tradeoff where the model begins to fit noise. The physical camera model fits exactly in the 8-parameter projective group; therefore, we know that "eight is enough." Hence, it is appealing to find an approximate model with only eight parameters.

The 8-parameter bilinear model is perhaps the most widely used [16] in the fields of image processing, medical imaging, remote sensing, and computer graphics. This model is easily obtained from the biquadratic model by removing the four x^2 and y^2 terms. Although the resulting bilinear model captures the effect of converging lines, it completely fails to capture the effect of chirping.

The 8-parameter *pseudo-perspective* model [17] does, in fact, capture both the converging lines and the chirping of a projective coordinate transformation. This model may first be thought of as the removal of two of the quadratic terms ($q_{\tilde{x}y^2} = q_{\tilde{y}x^2} = 0$), which results in a 10-parameter model (the *q-chirp* of Navab and Mann [18]) and then the constraining of the four remaining quadratic parameters to have two degrees of freedom. These constraints force the chirping effect (captured by $q_{\tilde{x}x^2}$ and $q_{\tilde{y}y^2}$) and the converging effect (captured by $q_{\tilde{x}xy}$ and $q_{\tilde{y}xy}$) to work together in the "right" way to match, as closely as possible, the effect of a projective coordinate transformation. By setting $\mathbf{q}_\alpha = \mathbf{q}_{\tilde{x}x^2} = \mathbf{q}_{\tilde{y}xy}$, the chirping in the x-direction is forced to correspond with the converging of parallel lines in the x-direction (and likewise for the y-direction). Therefore, of the 8-parameter approximations to the true projective, we would expect the *pseudo-perspective* model to perform the best.

Of course, the desired "exact" eight parameters come from the projective model, but they have been notoriously difficult to estimate. The parameters for this model have been solved by Tsai and Huang [19], but their solution assumed that features had been identified in the two frames, along with their correspondences. In this chapter, a simple featureless means of registering images by estimating their comparameters is presented.

Other researchers have looked at projective estimation in the context of obtaining 3-D models. Faugeras and Lustman [20], Shashua and Navab [21], and Sawhney [22] have considered the problem of estimating the projective parameters while computing the motion of a rigid planar patch, as part of a larger problem of finding 3-D motion and structure using parallax relative to an arbitrary plane in the scene. Kumar, Anandan, and Hanna [23] have also euggested registering frames of video by computing the flow along the *epipolar* lines, for which there is also an initial step of calculating the gross camera movement assuming no parallax. However, these methods have relied on feature correspondences and were aimed at 3-D scene modeling. Our focus is not on recovering the 3-D scene model, but on aligning 2-D images of 3-D scenes. Feature correspondences greatly simplify the problem; however, they also have many problems which are reviewed below. The focus of this chapter is a simple featureless approach to estimating the projective coordinate transformation between image frames.

Two similar efforts exist to the new work presented here. Mann [24] and Szeliski and Coughlan [25] independently proposed featureless registration and compositing of either pictures of a nearly flat object or pictures taken from approximately the same location. Both used a 2-D projective model and searched over its 8-parameter space to minimize the mean square error (or maximize the inner product) between one frame and a 2-D projective coordinate transformation of the next frame. However, in both these earlier works, the algorithm relies on nonlinear optimization techniques which we are able to avoid with the new technique presented here.

3.5.1 Camera, Eye, or Head Motion: Common Assumptions and Terminology

Two assumptions are relevant to this work. The first is that the scene is relatively constant — changes of scene content and lighting are small between frames, relative to changes that are induced by camera, eye, or head motion (e.g., a person can turn his or her head, hence turning an Eye Tap camera, and induce a much greater image flowfield than that induced by movement of objects in the scene). The second assumption is that of an ideal pinhole camera — implying unlimited depth of field with everything in focus (infinite resolution) and implying that straight lines map to straight lines.[1] This assumption is particularly valid for laser Eye Tap cameras which actually do have infinite depth of focus. Consequently, the camera, eye, or head has three degrees of freedom in 2-D space and eight degrees of freedom in 3-D space: translation (X, Y, Z), zoom (scale in each of the image coordinates x and y), and rotation (rotation about the optical axis, pan, and tilt).

In this chapter, an "uncalibrated camera" refers to one in which the principal point[2] is not necessarily at the center (origin) of the image and the scale is not necessarily isotropic. It is assumed that the film, sensor, retina, or the like is flat (although we know in fact that the retina is curved).

It is assumed that the zoom is continually adjustable by the camera user, and that we do not know the zoom setting or if it changed between recording frames of the image sequence. We also assume that each element in the camera sensor array returns a quantity that is linearly proportional to the quantity of light received.[3]

3.5.2   VideoOrbits

Tsai and Huang [19] noted that the elements of the projective *group* give the true camera motions with respect to a planar surface. They explored the group structure associated with images of a 3-D rigid planar patch, as well as the associated *Lie algebra,* although they assume that the correspondence problem has been solved. The solution presented in this chapter (which does not require prior solution of correspondence) also relies on projective group theory. We briefly review the basics of this theory, before presenting the new solution in the next section.

Projective Group in 1-D

For simplicity, the theory is first reviewed for the projective coordinate transformation in one dimension:[4] $\tilde{x} = (ax + b)/(cx + 1)$, where the images are functions of one variable, x. The set of all projective coordinate transformations for which $a \neq 0$ forms a group, **P**, the *projective group.* When $a \neq 0$ and $c = 0$, it is the affine group. When $a = 1$ and $c = 0$, it becomes the translation group.

Of the six coordinate transformations in the previous section, only the projective, affine, and translation operations form groups. A group of operators together with the set of 1-D images (operands) form a *group operation.*[5] The new set of images

---

[1]When using low cost wide-angle lenses, there is usually some barrel distortion which we correct using the method of Campbell and Bobick [26].

[2]The principal point is where the optical axis intersects the film, retina, sensor, or the like, as the case may be.

[3]This condition can be enforced over a wide range of light intensity levels, by using the Wyckoff principle [27, 28].

[4]In a 2-D world, the "camera" consists of a center of projection (pinhole lens) and a line (1-D sensor array or 1-D "film").

[5]Also known as a *group action* or *G-set* [29].

that results from applying all possible operators from the group to a particular image from the original set is called the *orbit* of that image under the group operation [29].

A camera at a fixed location, and free to zoom and pan, gives rise to a resulting pair of 1-D frames taken by the camera, which are related by the coordinate transformation from x_1 to x_2, given by [30]:

$$
\begin{aligned}
x_2 &= z_2 \tan\left(\arctan\left(x_1/z_1\right) - \theta\right), \quad \forall x_1 \neq o_1 \\
&= \left(ax_1 + b\right)/\left(cx_1 + 1\right), \quad \forall x_1 \neq o_1
\end{aligned}
\tag{3.1}
$$

where $a = z_2/z_1$, $b = -z_2 \tan(\theta)$, $c = \tan(\theta)/z_1$, and $o_1 = z_1 \tan(\pi/2 + \theta) = -1/c$ is the location of the singularity in the domain. We should emphasize that c, the degree of perspective, has been given the interpretation of a chirp-rate [30]. The coordinate transformations of Eq. (3.1) form a group operation. This result and the proof of this group's isomorphism to the group corresponding to nonsingular projections of a flat object are given in Mann and Picard [31].

Projective Group in 2-D

The theory for the projective, affine, and translation groups also holds for the familiar 2-D images taken of the 3-D world. The video orbit of a given 2-D frame is defined to be the set of all images that can be produced by applying operators from the 2-D projective group to the given image. Hence, we restate the coordinate transformation problem: given a set of images that lie in the same orbit of the group, we wish to find for each image pair that operator in the group which takes one image to the other image.

If two frames, say f_1 and f_2, are in the same orbit, then there is a group operation $\mathbf{p}$ such that the mean squared error (MSE) between f_1 and $f_2' = \mathbf{p} \circ f_2$ is zero, where the symbol $\circ$ denotes the operation of $\mathbf{p}$ acting on frame f_2. In practice, however, we find which element of the group takes one image "nearest" the other, for there will be a certain amount of parallax, noise, interpolation error, edge effects, changes in lighting, depth of focus, etc. Fig. 3.4 illustrates the operator $\mathbf{p}$ acting on frame f_2 to move it nearest to frame f_1. (This figure does not, however, reveal the precise shape of the orbit, which occupies an 8-D space.)

The primary assumptions in these cases are that of no parallax and of a static scene. Because the 8-parameter projective model is "exact," it is theoretically the right model to use for estimating the coordinate transformation. The examples that follow demonstrate that it also performs better in practice than the other proposed models. In the next section, a new technique for estimating its eight parameters is shown.

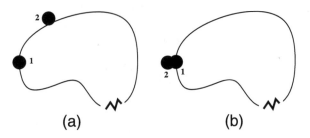

FIGURE 3.4

Video orbits. (a) The orbit of frame 1 is the set of all images that can be produced by acting on frame 1 with any element of the operator group. Assuming that frames 1 and 2 are from the same scene, frame 2 will be close to one of the possible projective coordinate transformations of frame 1. In other words, frame 2 lies near the orbit of frame 1. (b) By bringing frame 2 along its orbit (which is nearly the same orbit as the orbit of frame 1), we can determine how closely the two orbits come together at frame 1.

3.6   Framework: Comparameter Estimation and Optical Flow

Before the new results are presented, existing methods of comparameter estimation for coordinate transformations are reviewed. Comparameters refer to the relative parameters that transform one image into another, between a pair of images from an image sequence. Estimation of comparameters in a pairwise fashion can be dealt with globally based on the group properties, assuming the parameters in question trace an orbit of a group.

We classify existing methods into two categories: feature-based and featureless. Of the featureless methods, consider two subcategories: methods based on minimizing MSE (generalized correlation, direct nonlinear optimization) and methods based on spatio-temporal derivatives and optical flow. Note that variations such as *multiscale* have been omitted from these categories; multiscale analysis can be applied to any of them. The new algorithm developed in this chapter (with final form given in Section 3.7) is featureless and is based on multiscale spatio-temporal derivatives.

Some of the descriptions below are presented for hypothetical 1-D images taken in a 2-D space. This simplification yields a clearer comparison of the estimation methods. The new theory and applications will be presented subsequently for 2-D images taken in a 3-D space.

3.6.1   Feature-Based Methods

Feature-based methods [32, 33] assume that point correspondences in both images are available. In the projective case, given at least three correspondences between point pairs in the two 1-D images, we find the element $\mathbf{p} = \{a, b, c\} \in \mathbf{P}$ that maps the second image into the first. Let x_k, $k = 1, 2, 3, \ldots$ be the points in one image, and let $\tilde{x}_k$ be the corresponding points in the other image. Then, $\tilde{x}_k = (ax_k + b)/(cx_k + 1)$. Rearranging yields $ax_k + b - x_k\tilde{x}_k c = \tilde{x}_k$, so that a, b, and c can be found by solving $k \geq 3$ linear equations in three unknowns:

$$\begin{bmatrix} x_k & 1 & -\tilde{x}_k x_k \end{bmatrix}\begin{bmatrix} a & b & c \end{bmatrix}^T = \begin{bmatrix} \tilde{x}_k \end{bmatrix} \tag{3.2}$$

using least squares if there are more than three correspondence points. The extension from 1-D images to 2-D images is conceptually identical; for the affine and projective models, the minimum number of correspondence points needed in 2-D is three and four, respectively.

A major difficulty with feature-based methods is finding the features. Good features are often hand-selected or computed, possibly with some degree of human intervention [34]. A second problem with features is their sensitivity to noise and occlusion. Even if reliable features exist between frames, these features may be subject to signal noise and occlusion. The emphasis in the rest of this chapter is on robust featureless methods.

3.6.2 Featureless Methods Based on Generalized Cross-Correlation

Cross-correlation of two frames is a featureless method of recovering translation model comparameters. Affine and projective comparameters can also be recovered using generalized forms of cross-correlation between two images (e.g., comparing two images using cross correlation and related methods).

Generalized cross-correlation is based on an inner-product formulation which establishes a similarity metric between two functions, such as g and h, where $h \approx \mathbf{p} \circ g$ is an approximately coordinate-transformed version of g but the comparameters of the coordinate transformation $\mathbf{p}$ are unknown.[6] We can find, by exhaustive search (applying all possible operators, $\mathbf{p}$, to h), the "best" $\mathbf{p}$ as the one that maximizes the inner product:

$$\int_{-\infty}^{\infty} g(x)\frac{\mathbf{p}^{-1} \circ h(x)}{\int_{-\infty}^{\infty} \mathbf{p}^{-1} \circ h(x)dx}dx \tag{3.3}$$

where we have normalized the energy of each coordinate-transformed h before making the comparison. Equivalently, instead of maximizing a similarity metric, we can minimize an anti-similarity metric, such as MSE, given by $\int_{-\infty}^{\infty} \left(g(x) - \mathbf{p}^{-1} \circ h(x)\right)^2 dx$. Solving Eq. (3.3) has an advantage over finding MSE when one image is not only a coordinate-transformed version of the other but is also an amplitude-scaled version, as generally happens when there is an automatic gain control or an automatic iris in the camera.

---

[6]In the presence of additive white Gaussian noise, this method, also known as "matched filtering," leads to a maximum likelihood estimate of the parameters [35].

In 1-D, the affine model permits only dilation and translation. Given h, an affine coordinate-transformed version of g, generalized correlation amounts to estimating the parameters for dilation a and translation b by exhaustive search. The collection of all possible coordinate transformations, when applied to one of the images (say, h) serves to produce a family of templates to which the other image, g, can be compared. If we normalize each template so that all have the same energy

$$h_{a,b}(x) = \frac{1}{\sqrt{a}} h(ax + b)$$

then the maximum likelihood estimate corresponds to selecting the member of the family that gives the largest inner product:

$$\langle g(x), h_{a,b}(x) \rangle = \int_{-\infty}^{\infty} g(x) h_{a,b}(x) dx$$

This result is known as a *cross-wavelet transform.* A computationally efficient algorithm for the cross-wavelet transform has recently been presented [36]. (See Weiss [37] for a good review on wavelet-based estimation of affine coordinate transformations.)

Just like the cross-correlation for the translation group and the cross-wavelet for the affine group, the *cross-chirplet* can be used to find the comparameters of a projective coordinate transformation in 1-D, searching over a 3-parameter space. The chirplet transform [38] is a generalization of the wavelet transform. The *projective-chirplet* has the form

$$h_{a,b,c} = h\left(\frac{ax + b}{cx + 1}\right) \tag{3.4}$$

where h is the *mother chirplet,* analogous to the *mother wavelet* of wavelet theory. Members of this family of functions are related to one another by projective coordinate transformations.

With 2-D images, the search is over an 8-parameter space. A dense sampling of this volume is computationally prohibitive. Consequently, combinations of coarse-to-fine and iterative or repetitive gradient-based search procedures are required. Adaptive variants of the chirplet transform have been previously reported in the literature [39]. However, there are still many problems with the adaptive chirplet approach; thus, featureless methods based on spatio-temporal derivatives are now considered.

3.6.3   Featureless Methods Based on Spatio-Temporal Derivatives

Optical Flow — Translation Flow

When the change from one image to another is small, optical flow [40] may be used. In 1-D, the traditional optical flow formulation assumes each point x in frame t is a translated version of the corresponding point in frame $t + \Delta t$, and that Δx and Δt

are chosen in the ratio $\Delta x / \Delta t = u_f$, the translational flow velocity of the point in question. The image brightness $E(x, t)$ is described by

$$E(x, t) = E(x + \Delta x, t + \Delta t), \quad \forall(x, t). \tag{3.5}$$

In the case of pure translation, u_f is constant across the entire image. More generally though, a pair of 1-D images are related by a quantity, $u_f(x)$ at each point in one of the images.

Expanding the right side of Eq. (3.5) in a Taylor series and cancelling 0th order terms give the well-known optical flow equation $u_f E_x + E_t + h.o.t. = 0$, where E_x and E_t are the spatial and temporal derivatives, respectively, and $h.o.t.$ denotes higher order terms. Typically, the higher order terms are neglected, giving the expression for the optical flow at each point in one of the two images:

$$u_f E_x + E_t \approx 0. \tag{3.6}$$

Affine Fit and Affine Flow: a New Relationship

Given the optical flow between two images, g and h, we wish to find the coordinate transformation to apply to h to make it look most like g. We now describe two approaches based on the affine model: (1) finding the optical flow at every point and then fitting this flow with an affine model (*affine fit*), and (2) rewriting the optical flow equation in terms of an affine (not translation) motion model (*affine flow*).

Wang and Adelson have proposed fitting an affine model to an optical flow field [41] of 2-D images. We briefly examine their approach with 1-D images (1-D images simplify analysis and comparison to other methods). Denote coordinates in the original image, g, by x, and in the new image, h, by $\tilde{x}$. Suppose that h is a dilated and translated version of g, so $\tilde{x} = ax + b$ for every corresponding pair $(\tilde{x}, x)$. Equivalently, the affine model of velocity (normalizing $\Delta t = 1$), $u_m = \tilde{x} - x$, is given by $u_m = (a - 1)x + b$. We can expect a discrepancy between the flow velocity, u_f, and the model velocity, u_m, due to either errors in the flow calculation or errors in the affine model assumption. Accordingly, we apply linear regression to obtain the best least-squares fit by minimizing:

$$\varepsilon_{fit} = \sum_x \left(u_m - u_f\right)^2 = \sum (u_m + E_t/E_x)^2. \tag{3.7}$$

The constants a and b that minimize ε_{fit} over the entire patch are found by differentiating Eq. (3.7), and setting the derivatives to zero. This results in the *affine fit* equations [42]:

$$\begin{bmatrix} \sum_x x^2, \sum_x x \\ \sum_x x, \sum_x 1 \end{bmatrix} \begin{bmatrix} a - 1 \\ b \end{bmatrix} = - \begin{bmatrix} \sum_x x E_t/E_x \\ \sum_x E_t/E_x \end{bmatrix}. \tag{3.8}$$

Alternatively, the affine coordinate transformation may be directly incorporated into the brightness change constraint equation (3.5). Bergen et al. [43] have proposed this method, which has been called *affine flow* to distinguish it from the affine fit model of Wang and Adelson Eq. (3.8). Let us show how affine flow and affine fit are related. Substituting $u_m = (ax + b) - x$ directly into Eq. (3.6) in place of u_f and summing the squared error

$$\varepsilon_{\text{flow}} = \sum_x (u_m E_x + E_t)^2 \tag{3.9}$$

over the whole image, differentiating, and equating the result to zero gives a linear solution for both a and b:

$$\begin{bmatrix} \sum_x x^2 E_x^2, \sum_x x E_x^2 \\ \sum_x x E_x^2, \sum_x E_x^2 \end{bmatrix} \begin{bmatrix} a - 1 \\ b \end{bmatrix} = - \begin{bmatrix} \sum_x x E_x E_t \\ \sum_x E_x E_t \end{bmatrix} . \tag{3.10}$$

To see how this result compares to the affine fit we rewrite Eq. (3.7)

$$\varepsilon_{fit} = \sum_x \left(\frac{u_m E_x + E_t}{E_x} \right)^2 \tag{3.11}$$

and observe, comparing Eqs. (3.9) and (3.11), that affine flow is equivalent to a weighted least-squares fit, where the weighting is given by E_x^2. Thus the affine flow method tends to put more emphasis on areas of the image that are spatially varying than does the affine fit method. Of course, one is free to separately choose the weighting for each method in such a way that affine fit and affine flow methods both give the same result. Practical experience tends to favor the affine flow weighting, but, more generally, perhaps we should ask, "what is the best weighting?" For example, maybe there is an even better answer than the choice among these two. Lucas and Kanade [44], among others, have considered weighting issues.

Another approach to the affine fit involves computation of the optical flow field using the multiscale iterative method of Lucas and Kanade, and *then* fitting to the affine model. An analogous variant of the affine flow method involves multiscale iteration as well, but in this case the iteration and multiscale hierarchy are incorporated directly into the affine estimator [43]. With the addition of multiscale analysis, the fit and flow methods differ in additional respects beyond just the weighting. Experience indicates that the direct multiscale affine flow performs better than the affine fit to the multiscale flow. Multiscale optical flow makes the assumption that blocks of the image are moving with pure translational motion, and then, paradoxically, the affine fit refutes this pure-translation assumption. However, fit provides some utility over flow when it is desired to segment the image into regions undergoing different motions [45], or to gain robustness by rejecting portions of the image not obeying the assumed model.

Projective Fit and Projective Flow: New Techniques

Analogous to the affine fit and affine flow of the previous section, two new methods are proposed: *projective fit* and *projective flow*. For the 1-D affine coordinate transformation, the graph of the range coordinate as a function of the domain coordinate is a straight line; for the projective coordinate transformation, the graph of the range coordinate as a function of the domain coordinate is a rectangular hyperbola [31]. The affine fit case used linear regression; however, in the projective case hyperbolic regression is used. Consider the flow velocity given by Eq. (3.6) and the model velocity:

$$u_m = \tilde{x} - x = \frac{ax + b}{cx + 1} - x \qquad (3.12)$$

and minimize the sum of the squared difference paralleling Eq. (3.9):

$$\varepsilon = \sum_x \left(\frac{ax + b}{cx + 1} - x + \frac{E_t}{E_x} \right)^2 . \qquad (3.13)$$

For projective-flow (p-flow) we use, as for affine flow, the Taylor series of u_m:

$$u_m + x = b + (a - bc)x + (bc - a)cx^2 + (a - bc)c^2 x^3 + \cdots \qquad (3.14)$$

and again use the first three terms, obtaining enough degrees of freedom to account for the 3 comparameters being estimated. Letting $\epsilon = \sum (-h.o.t.)^2 = \sum((b + (a - bc - 1)x + (bc - a)cx^2)E_x + E_t)^2$, $q_2 = (bc - a)c$, $q_1 = a - bc - 1$, and $q_0 = b$, and differentiating with respect to each of the 3 comparameters of $\mathbf{q}$, setting the derivatives equal to zero, and verifying with the second derivatives, gives the linear system of equations for projective flow:

$$\begin{bmatrix} \sum x^4 E_x^2 & \sum x^3 E_x^2 & \sum x^2 E_x^2 \\ \sum x^3 E_x^2 & \sum x^2 E_x^2 & \sum x E_x^2 \\ \sum x^2 E_x^2 & \sum x E_x^2 & \sum E_x^2 \end{bmatrix} \begin{bmatrix} q_2 \\ q_1 \\ q_0 \end{bmatrix} = - \begin{bmatrix} \sum x^2 E_x E_t \\ \sum x E_x E_t \\ \sum E_x E_t \end{bmatrix} \qquad (3.15)$$

In Section 3.7 we extend this derivation to 2-D images and show how a repetitive approach may be used to compute the parameters, $\mathbf{p}$, of the exact model. A feedback system is used where the feedforward loop involves computation of the approximate parameters, $\mathbf{q}$, in the extension of Eq. (3.15) to 2-D.

As with the affine case, projective fit and projective flow Eq. (3.15) differ only in the weighting assumed, although projective fit provides the added advantage of enabling the motion within an arbitrary subregion of the image to be easily found. In this chapter only global image motion is considered, for which the projective flow model has been found to be best [42].

3.7   Multiscale Projective Flow Comparameter Estimation

In the previous section, two new techniques, p-fit and p-flow, were proposed. Now we describe our algorithm for estimating the projective coordinate transformation for 2-D images using p-flow. We begin with the brightness constancy constraint equation for 2-D images [40] which gives the flow velocity components in the x and y directions, analogous to Eq. (3.6):

$$u_f E_x + v_f E_y + E_t \approx 0 \,. \tag{3.16}$$

As is well known [40], the optical flow field in 2-D is underconstrained.[7] The model of *pure translation* at every point has two comparameters, but there is only one equation (3.16) to solve. Thus it is common practice to compute the optical flow over some neighborhood, which must be at least two pixels but is generally taken over a small block, 3×3, 5×5, or sometimes larger (e.g., the entire image, as in this chapter).

Our task is not to deal with the 2-D translation flow but with the 2-D projective flow, estimating the eight comparameters in the coordinate transformation:

$$\tilde{\mathbf{x}} = \begin{bmatrix} \tilde{x} \\ \tilde{y} \end{bmatrix} = \frac{\mathbf{A}[x, y]^T + \mathbf{b}}{\mathbf{c}^T [x, y]^T + 1} = \frac{\mathbf{A}\mathbf{x} + \mathbf{b}}{\mathbf{c}^T \mathbf{x} + 1} \,. \tag{3.17}$$

The desired eight scalar parameters are denoted by $\mathbf{p} = [\mathbf{A}, \mathbf{b}; \mathbf{c}, 1]$, $\mathbf{A} \in \mathbb{R}^{2 \times 2}$, $\mathbf{b} \in \mathbb{R}^{2 \times 1}$, and $\mathbf{c} \in \mathbb{R}^{2 \times 1}$.

As with the 1-D images, we make similar assumptions in expanding Eq. (3.17) in its own Taylor series, analogous to Eq. (3.14). If we take the Taylor series up to second order terms, we obtain the biquadratic model mentioned in Section 3.5. As mentioned there, by appropriately constraining the twelve parameters of the biquadratic model, we obtain a variety of 8-parameter approximate models. In our algorithm for estimating the exact projective group parameters, we will use one of these approximate models in an intermediate step.[8] We illustrate the algorithm below using the bilinear approximate model since it has the simplest notation.[9] First, we incorporate the approximate model directly into the generalized fit or generalized flow. The Taylor series for the bilinear case gives

$$
\begin{aligned}
u_m + x &= q_{\tilde{x}xy}xy + (q_{\tilde{x}x} + 1)\,x + q_{\tilde{x}y}y + q_{\tilde{x}} \\
v_m + y &= q_{\tilde{y}xy}xy + q_{\tilde{y}x}x + \left(q_{\tilde{y}y} + 1\right) y + q_{\tilde{y}}
\end{aligned}
\tag{3.18}
$$

Incorporating these into the flow criteria yields a simple set of eight scalar "linear"

---

[7] Optical flow in 1-D did not suffer from this problem.

[8] Use of an approximate model that does not capture chirping or preserve straight lines can still lead to the true projective parameters as long as the model captures at least eight degrees of freedom.

[9] The pseudo-perspective gives slightly better performance; its development is the same but with more notation.

(correctly speaking, affine) equations in eight scalar unknowns, for "bilinear flow":

$$
\begin{bmatrix}
\sum x^2y^2E_x^2, & \sum x^2yE_x^2, & \sum xy^2E_x^2, & \sum xyE_x, & \sum x^2y^2E_yE_x, & \sum x^2yE_yE_x, & \sum xy^2E_yE_x, & \sum E_yxyE_x \\
\sum x^2yE_x^2, & \sum x^2E_x^2, & \sum xyE_x^2, & \sum xE_x^2, & \sum x^2yE_yE_x, & \sum x^2E_yE_x, & \sum xyE_yE_x, & \sum E_yxE_x \\
\sum xy^2E_x^2, & \sum xyE_x^2, & \sum y^2E_x^2, & \sum yE_x^2, & \sum xy^2E_yE_x, & \sum xyE_yE_x, & \sum y^2E_yE_x, & \sum E_yyE_x \\
\sum xyE_x^2, & \sum xE_x^2, & \sum yE_x^2, & \sum E_x^2, & \sum xyE_yE_x, & \sum xE_yE_x, & \sum yE_yE_x, & \sum E_yE_x \\
\sum x^2y^2E_xE_y, & \sum x^2yE_xE_y, & \sum xy^2E_xE_y, & \sum E_xxyE_y, & \sum x^2y^2E_y^2, & \sum x^2yE_y^2, & \sum xy^2E_y^2, & \sum xyE_y^2 \\
\sum x^2yE_xE_y, & \sum x^2E_xE_y, & \sum xyE_xE_y, & \sum E_xxE_y, & \sum x^2yE_y^2, & \sum x^2E_y^2, & \sum xyE_y^2, & \sum xE_y^2 \\
\sum xy^2E_xE_y, & \sum xyE_xE_y, & \sum y^2E_xE_y, & \sum E_xyE_y, & \sum xy^2E_y^2, & \sum xyE_y^2, & \sum y^2E_y^2, & \sum yE_y^2 \\
\sum xyE_xE_y, & \sum xE_xE_y, & \sum yE_xE_y, & \sum E_xE_y, & \sum xyE_y^2, & \sum xE_y^2, & \sum yE_y^2, & \sum E_y^2
\end{bmatrix}
$$

$$
\begin{bmatrix}
q_{\tilde{x}xy} \\
q_{\tilde{x}x} \\
q_{\tilde{x}y} \\
q_{\tilde{x}} \\
q_{\tilde{y}xy} \\
q_{\tilde{y}x} \\
q_{\tilde{y}y} \\
q_{\tilde{y}}
\end{bmatrix}
= -\begin{bmatrix} \sum E_txyE_x, \sum E_txE_x, \sum E_tyE_x, \sum E_tE_x, \sum E_txyE_y, \sum E_txE_y, \sum E_tyE_y, \sum E_tE_y \end{bmatrix}^T
$$

$$(3.19)$$

The summations are over the entire image (all x and y) if computing global motion (as is done in this chapter), or over a windowed patch if computing local motion. This equation looks similar to the 6×6 matrix equation presented in Bergen et al. [43], except that it serves to address projective geometry rather than the affine geometry of Bergen et al. [43].

In order to see how well the model describes the coordinate transformation between 2 images, say g and h, one might *warp*[10] h to g, using the estimated motion model, and then compute some quantity that indicates how different the resampled version of h is from g. The MSE between the reference image and the warped image might serve as a good measure of similarity. However, since we are really interested in how the *exact model* describes the coordinate transformation, we assess the goodness of fit by first relating the parameters of the approximate model to the exact model, and then find the MSE between the reference image and the comparison image after applying the coordinate transformation of the exact model. A method of finding the parameters of the exact model, given the approximate model, is presented in Section 3.7.1.

3.7.1 Four Point Method for Relating Approximate Model to Exact Model

Any of the approximations above, after being related to the exact projective model, tend to behave well in the neighborhood of the identity, $\mathbf{A} = \mathbf{I}, \mathbf{b} = \mathbf{0}, \mathbf{c} = \mathbf{0}$. In 1-D, we explicitly expanded the Taylor series model about the identity; here, although we do not explicitly do this, we assume that the terms of the Taylor series of the model correspond to those taken about the identity. In the 1-D case, we solve the three linear equations in three unknowns to estimate the comparameters of the approximate motion model, and then we relate the terms in this Taylor series to the exact comparameters,

---

[10] The term *warp* is appropriate here, since the approximate model does not preserve straight lines.

a, *b*, and *c* (which involves solving another set of three equations in three unknowns, the second set being nonlinear, although very easy to solve).

In the extension to 2-D, the estimate step is straightforward, but the relate step is more difficult because we now have eight nonlinear equations in eight unknowns, relating the terms in the Taylor series of the approximate model to the desired exact model parameters. Instead of solving these equations directly, we now propose a simple procedure for relating the parameters of the approximate model to those of the exact model, which we call the *four point method:*

1. Select four ordered pairs (such as the four corners of the bounding box containing the region under analysis, or the four corners of the image if the whole image is under analysis). Here suppose, for simplicity, that these points are the corners of the unit square: $\mathbf{s} = [s_1, s_2, s_3, s_4] = [(0, 0)^T, (0, 1)^T, (1, 0)^T, (1, 1)^T]$.

2. Apply the coordinate transformation using the Taylor series for the approximate model [e.g., Eq. (3.18)] to these points: $\mathbf{r} = \mathbf{u}_m(\mathbf{s})$.

3. Finally, the correspondences between $\mathbf{r}$ and $\mathbf{s}$ are treated just like features. This results in four easy-to-solve linear equations:

$$\begin{bmatrix} \tilde{x}_k \\ \tilde{y}_k \end{bmatrix} = \begin{bmatrix} x_k, y_k, 1, 0, 0, 0, -x_k\tilde{x}_k, -y_k\tilde{x}_k \\ 0, 0, 0, x_k, y_k, 1, -x_k\tilde{y}_k, -y_k\tilde{y}_k \end{bmatrix}$$
$$\begin{bmatrix} a_{\tilde{x}x}, a_{\tilde{x}y}, b_{\tilde{x}}, a_{\tilde{y}x}, a_{\tilde{y}y}, b_{\tilde{y}}, c_x, c_y \end{bmatrix}^T \qquad (3.20)$$

where $1 \le k \le 4$ is resulting in the exact eight parameters, $\mathbf{p}$.

We remind the reader that the four corners are **not** feature correspondences as used in the feature-based methods of Section 3.6.1, but, rather, are used so that the two featureless models (approximate and exact) can be related to one another.

It is important to realize the full benefit of finding the exact parameters. While the approximate model is sufficient for small deviations from the identity, it is not adequate to describe large changes in perspective. However, if we use it to track small changes incrementally, and each time relate these small changes to the exact model Eq. (3.17), then we can accumulate these small changes using the *law of composition* afforded by the group structure. This is an especially favorable contribution of the group framework. For example, with a video sequence, we can accommodate very large accumulated changes in perspective in this manner. The problems with cumulative error can be eliminated, for the most part, by constantly propagating forward the true values, computing the residual using the approximate model, and each time relating this to the exact model to obtain a goodness-of-fit estimate.

3.7.2 Overview of the New Projective Flow Algorithm

Below is an outline of the new algorithm for estimation of *projective flow*. Details of each step are in subsequent sections.

Frames from an image sequence are compared pairwise to test whether or not they lie in the same orbit:

1. A Gaussian pyramid of three or four levels is constructed for each frame in the sequence.

2. The comparameters **p** are estimated at the top of the pyramid, between the two lowest-resolution images of a frame pair, g and h, using the repetitive method depicted in Fig. 3.5.

3. The estimated **p** is applied to the next higher-resolution (finer) image in the pyramid, **p** ∘ g, to make the two images at that level of the pyramid nearly congruent before estimating the **p** between them.

4. The process continues down the pyramid until the highest-resolution image in the pyramid is reached.

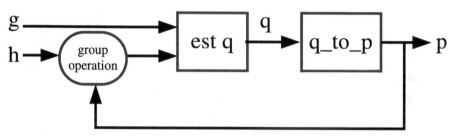

FIGURE 3.5
Method of computation of eight comparameters p between two images from the same pyramid level, g and h. The approximate model parameters q are related to the exact model parameters p in a feedback system.

3.7.3   Multiscale Repetitive Implementation

The Taylor-series formulations we have used implicitly assume smoothness; the performance is improved if the images are blurred before estimation. To accomplish this, we do not downsample critically after lowpass filtering in the pyramid. However, after estimation we use the original (unblurred) images when applying the final coordinate transformation.

The strategy we present differs from the multiscale iterative (affine) strategy of Bergen et al. in one important respect beyond simply an increase from six to eight parameters. The difference is the fact that we have two motion models, the "exact motion model" Eq. (3.17) and the "approximate motion model," namely the Taylor series approximation to the motion model itself. The approximate motion model is used to iteratively converge to the exact motion model, using the algebraic *law of composition* afforded by the exact projective group model. In this strategy, the exact parameters are determined at each level of the pyramid, and passed to the next level. The steps involved are summarized schematically in Fig. 3.5, and described below:

1. Initialize: set $h_0 = h$ and set $\mathbf{p}_{0,0}$ to the identity operator.

2. Iterate ($k = 1 \ldots K$):

(a) **Estimate:** estimate the 8 or more terms of the approximate model between two image frames, g and h_{k-1}. This results in approximate model parameters $\mathbf{q}_k$.

(b) **Relate:** relate the approximate parameters $\mathbf{q}_k$ to the exact parameters using the "four point method." The resulting exact parameters are $\mathbf{p}_k$.

(c) **Resample:** apply the *law of composition* to accumulate the effect of the $\mathbf{p}_k$'s. Denote these composite parameters by $\mathbf{p}_{0,k} = \mathbf{p}_k \circ \mathbf{p}_{0,k-1}$. Then set $h_k = \mathbf{p}_{0,k} \circ h$. (This should have nearly the same effect as applying $\mathbf{p}_k$ to h_{k-1}, except that it will avoid additional interpolation and anti-aliasing errors you would get by resampling an already resampled image [16].)

Repeat until either the error between h_k and g falls below a threshold, or until some maximum number of repetitions is achieved. After the first repetition, the parameters $\mathbf{q}_2$ tend to be near identity since they account for the residual between the "perspective-corrected" image h_1 and the "true" image g. We find that only two or three repetitions are usually needed for frames from nearly the same orbit.

A rectangular image assumes the shape of an arbitrary quadrilateral when it undergoes a projective coordinate transformation. In coding the algorithm, we pad the undefined portions with the quantity NaN, a standard IEEE arithmetic [46] value, so that any calculations involving these values automatically inherit NaN without slowing down the computations. The algorithm, running in Matlab on an HP 735, takes about six seconds per repetition for a pair of 320x240 images. A C language version, optimized, compiled, and running on the wearable computer portion of various PSDs built by the author, typically runs in a fraction of a second, in some cases on the order of 1/10th of a second or so. A Xilinx FPGA-based version of the PSD is currently being built by the author, together with Professor Jonathan Rose and others at the University of Toronto, and is expected to run the entire process in less than 1/60th of a second.

3.7.4 Exploiting Commutativity for Parameter Estimation

A fundamental uncertainty [47] is involved in the simultaneous estimation of parameters of a noncommutative group, akin to the Heisenberg uncertainty relation of quantum mechanics. In contrast, for a commutative[11] group (in the absence of noise), we can obtain the exact coordinate transformation.

Segman, Rubinstein, and Zeevi [48] considered the problem of estimating the parameters of a commutative group of coordinate transformations, in particular, the

---

[11] A commutative (or *Abelian*) group is one in which elements of the group commute. For example, translation along the x-axis commutes with translation along the y-axis, so the 2-D translation group is commutative.

parameters of the affine group [49]. Their work also deals with noncommutative groups, in particular, in the incorporation of scale in the Heisenberg group[12] [50].

Estimating the parameters of a commutative group is computationally efficient, e.g., through the use of Fourier cross-spectra [51]. We exploit this commutativity for estimating the parameters of the noncommutative 2-D projective group by first estimating the parameters that commute. For example, we improve performance if we first estimate the two parameters of translation, correct for the translation, and then proceed to estimate the eight projective parameters. We can also simultaneously estimate both the isotropic-zoom and the rotation about the optical axis by applying a log-polar coordinate transformation followed by a translation estimator. This process may also be achieved by a direct application of the Fourier-Mellin transform [52]. Similarly, if the only difference between g and h is a camera pan, then the pan may be estimated through a coordinate transformation to cylindrical coordinates, followed by a translation estimator.

In practice, we run through the following commutative initialization before estimating the parameters of the projective group of coordinate transformations:

1. Assume that h is merely a translated version of g.

    (a) Estimate this translation using the method of Girod and Kuo [51].

    (b) Shift h by the amount indicated by this estimate.

    (c) Compute the *MSE* between the shifted h and g and compare to the original MSE before shifting.

    (d) If an improvement has resulted, use the shifted h from now on.

2. Assume that h is merely a rotated and isotropically zoomed version of g.

    (a) Estimate the two parameters of this coordinate transformation.

    (b) Apply these parameters to h.

    (c) If an improvement has resulted, use the coordinate-transformed (rotated and scaled) h from now on.

3. Assume that h is merely an x-chirped (panned) version of g and similarly x-dechirped h. If an improvement results, use the x-dechirped h from now on. Repeat for y (tilt.)

Compensating for one step may cause a change in choice of an earlier step. Thus it might seem desirable to run through the commutative estimates repetitively. However, our experience on lots of real video indicates that a single pass usually suffices and, in particular, will catch frequent situations where there is a pure zoom, pure pan, pure tilt, etc. both saving the rest of the algorithm computational effort, as well as accounting for simple coordinate transformations such as when one image is an upside-down

---

[12]While the Heisenberg group deals with translation and frequency-translation (modulation), some of the concepts could be carried over to other more relevant group structures.

version of the other. (Any of these pure cases corresponds to a single parameter group, which is commutative.) Without the commutative initialization step, these parameter estimation algorithms are prone to getting caught in local optima and thus never converging to the global optimum.

3.8 Performance/Applications

3.8.1 A Paradigm Reversal in Resolution Enhancement

Much of the previous work on resolution enhancement [14, 53, 54] has been directed toward military applications, where one cannot get close to the subject matter; therefore, lenses of very long focal lengths were generally used. In this case, there was very little change in *perspective* and the motion could be adequately approximated as affine. Budgets also permitted lenses of exceptionally high quality, so the resolving power of the lens far exceeded the resolution of the sensor array.

Sensor arrays in earlier applications generally had a small number of pixels compared to today's sensors, leaving considerable "dead space" between pixels. Consequently, using multiple frames from the image sequence to fill in gaps between pixels was perhaps the single most important consideration in combining multiple frames of video.

We argue that in the current age of consumer video, the exact opposite is generally true: subject matter generally subtends a larger angle (e.g., is either closer, or more *panoramic* in content), and the desire for low cost has led to cheap plastic lenses that have very large distortion. Moreover, sensor arrays have improved dramatically. Accurate solution of the projective model is more important than ever in these new applications.

In addition to consumer video, there will be a large market in the future for small wearable wireless cameras. A prototype, the *wearable wireless webcam* (an eyeglass-based video production facility uplinked to the Internet [11]) has provided one of the most extreme testbeds for the algorithms explored in this research, as it captures noisy transmitted video frames, grabbed by a camera attached to a human head, free to move at the will of the individual. The projective model is especially well-suited to this new application, as people can turn their heads (camera rotation about an approximately fixed center of projection) much faster than they can undergo locomotion (camera translation). The new algorithm described in this chapter has consistently performed well on noisy data gathered from the headcam, even when the scene is not static and there is parallax.

Four Ways by which Resolution May be Enhanced:

1. **Sub-pixel** — "Filling in the gaps."

2. **Scene widening** — Increased spatial extent; stitching together images in a panorama.

3. **Saliency** — Suppose we have a wide shot of a scene, and then zoom into one person's face in the scene. In order to insert the face without downsampling it, we need to upsample the wide shot, increasing the meaningful pixel count of the whole image.

4. **Perspective** — In order to seamlessly mosaic images from panning with a wide angle lens, images need to be brought into a common system of coordinates resulting in a keystoning effect on the previously rectangular image boundary. Thus, we must hold the pixel resolution constant on the "squashed" side and upsample on the "stretched" side, resulting in increased *pixel resolution* of the entire mosaic.

The first of these four may arise from either microscopic camera movement (inducing image motion on the order of a pixel or less) or macroscopic camera movement (inducing motion on the order of many pixels). However, as movement increases, errors in registration will tend to increase, and enhancement due to sub-pixels will be reduced, while the enhancement due to scene widening, saliency, and perspective will increase.

Results of applying the proposed method to subpixel resolution enhancement are not presented in this chapter but may be found in Mann and Picard [31].

3.8.2 Increasing Resolution in the "Pixel Sense"

Fig. 3.6 shows some frames from a typical image sequence. Fig. 3.7 shows the same frames transformed into the coordinate system of frame (c); that is, the middle frame was chosen as the *reference frame.*

Given that we have established a means of estimating the projective coordinate transformation between any pair of images, there are two basic methods we use for finding the coordinate transformations between all pairs of a longer image sequence. Because of the group structure of the projective coordinate transformations, it suffices to arbitrarily select one frame and find the coordinate transformation between every other frame and this frame. The two basic methods are:

1. **Differential comparameter estimation**: the coordinate transformations between successive pairs of images, $p_{0,1}, p_{1,2}, p_{2,3}, \ldots$, estimated.

2. **Cumulative comparameter estimation**: the coordinate transformation between each image and the reference image is estimated directly. Without loss of generality, select frame zero (E_0) as the reference frame and denote these coordinate transformations as $p_{0,1}, p_{0,2}, p_{0,3}, \ldots$

Theoretically, the two methods are equivalent:

$$E_0 = p_{0,1} \circ p_{1,2} \circ \ldots \circ p_{n-1,n} E_n \text{ — differential method}$$
$$E_0 = p_{0,n} E_n \text{ — cumulative method} \tag{3.21}$$

FIGURE 3.6
Received frames of image sequence transformed by way of comparameters with respect to frame (c). Frames from original image orbit, sent from the apparatus of the author's WearComp ("wearable computer") invention [1], connected to eyeglass-based imaging apparatus. (Note the apparatus captures a sideways view so that it can "paint" out the image canvas with a wider "brush," when sweeping across for a panorama.) The entire sequence, consisting of all 20 color frames, is available (see note at end of the references section), together with examples of applying the proposed algorithm to this data.

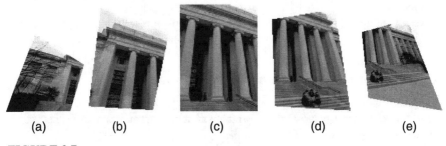

FIGURE 3.7
Received frames from image video orbit, transformed by way of comparameters with respect to frame (c). This transformed sequence involves moving them along the orbit to the reference frame (c). The coordinate-transformed images are alike except for the region over which they are defined. Note that the regions are not parallelograms; thus, methods based on the traditional affine model fail.

However, in practice the two methods differ for two reasons:

1. **Cumulative error**: in practice, the estimated coordinate transformations between pairs of images register them only approximately, due to violations of the assumptions (e.g., objects moving in the scene, center of projection not fixed, camera swings around to bright window and automatic iris closes, etc.). When a large number of estimated parameters are composed, cumulative error sets in.

2. **Finite spatial extent of image plane**: theoretically, the images extend infinitely in all directions, but, in practice, images are cropped to a rectangular bounding box. Therefore, a given pair of images (especially if they are far from adjacent in the orbit) may not overlap at all; hence, it is not possible to estimate the parameters of the coordinate transformation using those two frames.

The frames of Fig. 3.6 were brought into register using the differential parameter estimation and "cemented" together seamlessly on a common canvas. Cementing involves piecing the frames together, for example by median, mean, or trimmed mean, or combining on a subpixel grid [31]. (Trimmed mean was used here, but the particular method made little visible difference.) Fig. 3.8 shows this result (projective/projective), with a comparison to two nonprojective cases. The first comparison is to affine/affine where affine parameters were estimated (also multiscale) and used for the coordinate transformation. The second comparison, affine/projective, uses the six affine parameters found by estimating the eight projective parameters and ignoring the two chirp parameters c (which capture the essence of tilt and pan). These six parameters A, b are more accurate than those obtained using the affine estimation, as the affine estimation tries to fit its shear parameters to the camera pan and tilt. In other words, the affine estimation does worse than the six affine parameters within the projective estimation. The affine coordinate transform is finally applied, giving the image shown. Note that the coordinate-transformed frames in the affine case are parallelograms.

3.9 Summary

Some new connections between different motion estimation approaches, in particular a relation between affine fit and affine flow have been presented. This led to the proposal of two new techniques, projective fit and projective flow which estimate the projective comparameters (coordinate transformation) between pairs of images, taken with a camera that is free to pan, tilt, rotate about its optical axis and zoom.

A new multiscale repetitive algorithm for projective flow was presented and applied to comparametric transformations for sending images over a serendipitous communications channel. The algorithm solves for the 8 parameters of the "exact" model (the projective group of coordinate transformations), is fully automatic, and converges quickly.

The proposed method was found to work well on image data collected from both good-quality and poor-quality video under a wide variety of transmission conditions (noisy communications channels, etc.) as well as a wide variety of visual conditions (sunny, cloudy, day, night). It has been tested primarily with an eyeglass-mounted PSD, and performs successfully even in the presence of noise, interference, scene motion (such as people walking through the scene), and parallax (such as the author's head moving freely.)

projective/projective

affine/projective

affine/affine

FIGURE 3.8

Frames of Fig. 3.7 "cemented" together on single image "canvas," with comparison of affine and projective models. Note the good registration and nice appearance of the projective/projective image despite the noise in the serendipitous transmitter of the wearable Personal Safety Device, wind-blown trees, and the fact that the rotation of the camera was not actually about its center of projection. To see this image in color, see http://wearcam.org/orbits **where additional examples (e.g., some where the algorithm still worked despite "crowd noise" where many people were entering and leaving the building) also appear. Selecting just a few of the 20 frames produces approximately the same picture. In this way the methodology makes it difficult for a criminal to jam or prevent the operation of the Personal Safety Device. Note also that the affine model fails to properly estimate the motion parameters (affine/affine), and even if the "exact" projective model is used to estimate the affine parameters, there is no affine coordinate transformation that will properly register all of the image frames.**

By looking at image sequences as collections of still pictures related to one another by global comparameters, the images were expressed as part of the orbit of a group of coordinate transformations. This comparametric philosophy for transforms, image sequence coding, and transmission suggests that rather than sending every frame of a video sequence, we might send a reference frame, and the comparameters relating this reference frame to the other frames. More generally, we can send a photoquantigraphic image composite [1], along with a listing of the comparameters from which each image in the sequence may be drawn.

A new framework for constructing transforms, based on an Edgertonian rather than a Nyquist sampling philosophy, was proposed. Concomitant with Edgertonian sampling, was the principle of Fear of Functionality (FoF). By putting ourselves in the shoes of one who would regard functionality as undesirable, a new framework emerges in which unpredictability is a good thing. While the FoF framework seems at first paradoxical, it leads the way to new kinds of image transforms and image compression schemes. For example, the proposed comparametric image compression is based on a best case FoF model.

This model of comparametric compression is best suited to a wearable serendipitous personal imaging system, especially one that naturally taps the mind's eye, with the *possibility* that at any time what goes in the eye might also go into an indestructible (e.g., distributed on the World Wide Web) photographic/videographic memory recall system.

In the future, it is expected that many people will wear personal imaging devices, and that there will be a growing market for EyeTap (TM) video cameras once they are manufactured in mass production. The fundamental issue of limited bandwidth over wireless networks will make it desirable to further develop and refine this comparametric image compression and transmission approach. Moreover, a robust best-case wireless network may well supplant the current worst-case engineering approach used with many wireless networks.

PTP, a lossy, connectionless, serendipitously updated transmission protocol, will find new applications in the future world of ubiquitous Eye Tap video transmissions of first-person experiences.

3.10 Acknowledgements

This work was made possible by assistance from Kodak, Digital Equipment Corporation, Xybernaut Corp., CITO, NSERC, CLEARnet, and many others.

The author would also like to express thanks to many individuals for suggestions and encouragement. In particular, thanks goes to Roz Picard, Jonathan Rose, Will Waites, Robert Erlich, Lee Campbell, Shawn Becker, John Wang, Nassir Navab, Ujjaval Desai, Chris Graczyk, Walter Bender, Fang Liu, Constantine Sapuntzakis, Alex Drukarev, and Jeanne Wiseman. Some of the programs to implement the p-chirp models were developed in collaboration with Shawn Becker.

James Fung, Jordan Melzer, Eric Moncrieff, and Felix Tang are currently contributing further effort to this project.

Much of the success of this project can be attributed to the Free Source movement in general, of which the GNU project is one of the best examples. Richard Stallman, founder of the GNU effort, deserves acknowledgement for having set forth the general philosophy upon which many of these ideas are based.

Free computer programs distributed under the GNU General Public License (GPL) to implement the VideoOrbits work described in this article are available from
http: //wearcam.org/orbits/index.html or
http://wearcomp.org/ orbits/index.html.

This work was funded, in part, by the Canadian government, using taxpayer dollars. Accordingly, every attempt was made to ensure that the fruits of this labor made are freely available to any taxpayer, without the need to purchase any computer programs or use computer programs in which the principle of operation of the programs has been deliberately obfuscated (see http://wearcam.org/publicparks/index.html). Accordingly, the above computer programs were developed for use under the GNUX (GNU + Linux) operating system and environment which may be downloaded freely from various sites, such as http://gnux.org.

This manuscript was typeset using LaTeX running on a small wearable computer designed and built by the author. LaTeX is free and runs under GNUX. The computer programs to conduct this research and produce the results contained herein were also free and run under the GNUX system.

References

[1] Mann, S., Humanistic intelligence/humanistic computing: "wearcomp" as a new framework for intelligent signal processing, *Proceedings of the IEEE*, 86, 2123–2151, Nov. 1998, http://wearcam.org/procieee.htm.

[2] Mann, S., An historical account of the "WearComp" and "WearCam" projects developed for "personal imaging," in *International Symposium on Wearable Computing*, IEEE, Cambridge, MA, October 13–14, 1997.

[3] Mann, S., Eyeglass mounted wireless video: computer-supported collaboration for photojournalism and everyday use, *IEEE ComSoc*, 144–151, 1998, special issue on wireless video.

[4] Moving pictures expert group, mpeg standard, http://www.wearcam.org/mpeg/.

[5] Rosenberg, J., Kraut, R.E., Gomez, L., and Buzzard, C.A., Multimedia communications for users, *IEEE Communications Magazine*, 20-36, 1992.

[6] Mann, S., *Wearable Wireless Webcam*, 1994, http://wearcam.org.

[7] Edgerton, H.E., *Electronic flash, strobe*, MIT Press, Cambridge, MA, 1979.

[8] Terry Dawson, V., Ax.25 amateur packet-radio link-layer protocol, and ax25-howto, amateur radio, 1984, http://www.wearcam.org/ax25/.

[9] Mann, S., Smart clothing: the wearable computer and wearcam, *Personal Technologies*, 1(1), 21–27, 1997,

[10] Xu, J.B., Po, L.M., and Cheung, C.K., Adaptive motion tracking block matching algorithms for video coding, *IEEE Trans. Circ. Syst. and Video Technol.*, 97, 1025–1029, 1999.

[11] Mann, S., Personal imaging and lookpainting as tools for personal documentary and investigative photojournalism, *ACM Mobile Networking,* 4(1), 23–36, 1999, special issue on wearable computing.

[12] Tekalp, A., Ozkan, M., and Sezan, M., High-resolution image reconstruction from lower-resolution image sequences and space-varying image restoration, in *Proc. of the Int. Conf. on Acoust., Speech and Sig. Proc.,* III-169, IEEE, San Francisco, CA, Mar. 23–26, 1992.

[13] Zheng, Q. and Chellappa, R., A computational vision approach to image registration, *IEEE Transactions Image Processing,* 2(3), 311–325, 1993.

[14] Irani, M. and Peleg, S., Improving resolution by image registration, *CVGIP,* 53, 231–239, 1991.

[15] Teodosio, L. and Bender, W., Salient video stills: content and context preserved, *Proc. ACM Multimedia Conf.,* 39–46, August 1993.

[16] Wolberg, G., *Digital Image Warping,* IEEE Computer Society Press, Los Alamitos, CA, 1990, IEEE Computer Society Press Monograph.

[17] Adiv, G., Determining 3D motion and structure from optical flow generated by several moving objects, *IEEE Trans. Pattern Anal. Machine Intell.,* PAMI-7(4), 384–401, 1985.

[18] Navab, N. and Mann, S., Recovery of relative affine structure using the motion flow field of a rigid planar patch, *Mustererkennung 1994, Tagungsband.,* 1994.

[19] Tsai, R.Y., and Huang, T.S., Estimating three-dimensional motion parameters of a rigid planar patch I, *IEEE Trans. Accoust., Speech, and Sig. Proc.,* ASSP(29), 1147–1152, 1981.

[20] Faugeras, O.D. and Lustman, F., Motion and structure from motion in a piecewise planar environment, *International Journal of Pattern Recognition and Artificial Intelligence,* 2(3), 485–508, 1988.

[21] Shashua, A. and Navab, N., Relative affine: theory and application to 3D reconstruction from perspective views, *Proc. IEEE Conference on Computer Vision and Pattern Recognition,* 1994.

[22] Sawhney, H., Simplifying motion and structure analysis using planar parallax and image warping, *ICPR,* 1, 403–8, 1994, 12th IAPR.

[23] Kumar, R., Anandan, P., and Hanna, K., Shape recovery from multiple views: a parallax based approach, *ARPA Image Understanding Workshop,* Nov. 10, 1994.

[24] Mann, S. Compositing multiple pictures of the same scene, in *Proc. 46th Annual IS&T Conference,* 50–52, The Society of Imaging Science and Technology, Cambridge, MA, May 9–14, 1993.

[25] Szeliski, R. and Coughlan, J., Hierarchical spline-based image registration, *CVPR,* 194–201, 1994.

[26] Campbell, L. and Bobick, A., Correcting for radial lens distortion: a simple implementation, TR 322, M.I.T. Media Lab Perceptual Computing Section, Cambridge, MA, 1995.

[27] Wyckoff, C.W., An experimental extended response film, *S.P.I.E. Newsletter,* 16–20, 1962.

[28] Mann, S. and Picard, R., Being "undigital" with digital cameras: extending dynamic range by combining differently exposed pictures, Tech. Rep. 323, M.I.T. Media Lab Perceptual Computing Section, Cambridge, MA, 1994. (Also in IS&T's 48th annual conference, 422–428, May 7–11, 1995, `http://wearcam.org/ist95.htm`, Washington, DC.)

[29] Artin, M., *Algebra,* Prentice-Hall, Englewood Cliffs, NJ, 1991.

[30] Mann, S., Wavelets and chirplets: time-frequency perspectives, with applications, in *Advances in Machine Vision, Strategies and Applications,* Archibald, P., Ed., World Scientific Series in Computer Science, 32, World Scientific, Singapore, New Jersey, London, Hong Kong, 1992.

[31] Mann, S. and Picard, R.W., Virtual bellows: constructing high-quality images from video, in *Proc. IEEE First International Conference on Image Processing,* 363–367, Austin, TX, Nov. 13–16, 1994.

[32] Tsai, R.Y. and Huang, T.S., Multiframe image restoration and registration, in *Advances in Computer Vision and Image Processing,* JAI, 1, 317–339, 1984.

[33] Huang, T.S. and Netravali, A.N., Motion and structure from feature correspondences: a review, *Proc. IEEE,* 82(2), 252–268, 1984.

[34] Navab, N. and Shashua, A., Algebraic description of relative affine structure: connections to euclidean, affine and projective structure, *MIT Media Lab Memo No. 270,* 1994.

[35] Van Trees, H.L., *Detection, Estimation, and Modulation Theory (Part I),* John Wiley & Sons, New York, 1968.

[36] Young, R.K., *Wavelet Theory and Its Applications,* Kluwer Academic Publishers, Boston, 1993.

[37] Weiss, L.G., Wavelets and wideband correlation processing, *IEEE Signal Processing Magazine,* 13–32, 1994.

[38] Mann, S. and Haykin, S., The chirplet transform — a generalization of Gabor's logon transform, *Vision Interface '91,* June 3–7, 1991.

[39] Mann, S. and Haykin, S., Adaptive chirplet transform: an adaptive generalization of the wavelet transform, *Optical Engineering,* 31, 1243–1256, 1992.

[40] Horn, B. and Schunk, B., Determining optical flow, *Artificial Intelligence,* 17, 185–203, 1981.

[41] Wang, J.Y. and Adelson, E.H., Spatio-temporal segmentation of video data, in *SPIE Image and Video Processing II,* 120–128, San Jose, CA, February 7–9, 1994.

[42] Mann, S. and Picard, R.W., Video orbits of the projective group; a simple approach to featureless estimation of parameters, TR 338, MIT, Cambridge, MA, see `http://hi.eecg.toronto.edu/tip.html` 1995. (Also appears in IEEE Trans. Image Proc., Sept 1997, 6(9), 1281–1295.)

[43] Bergen, J., Burt, P.J., Hingorini, R., and Peleg, S., Computing two motions from three frames, in *Proc. Third Int'l Conf. Comput. Vision,* 27–32, Osaka, Japan, December 1990.

[44] Lucas, B.D. and Kanade, T., An iterative image-registration technique with an application to stereo vision, in *Image Understanding Workshop,* 121–130, 1981.

[45] Wang, J.Y.A. and Adelson, E.H., Representing moving images with layers, *Image Processing Spec. Iss: Image Seq. Compression,* 12, 625–638, 1994.

[46] Hennessy, J.L. and Patterson, D.A., *Computer Architecture: A Quantitative Approach.* Morgan Kauffman, 2nd ed., 1995.

[47] Wilson, R. and Granlund, G.H., The uncertainty principle in image processing, *IEEE Transactions on Pattern Analysis and Machine Intelligence,* 6, 758–767, 1984.

[48] Segman, J., Rubinstein, J., and Zeevi, Y.Y., The canonical coordinates method for pattern deformation: theoretical and computational considerations, *IEEE Trans. on Patt. Anal. and Mach. Intell.,* 14, 1171–1183, 1992.

[49] Segman, J., Fourier cross correlation and invariance transformations for an optimal recognition of functions deformed by affine groups, *Journal of the Optical Society of America, A,* 9, 895–902, 1992.

[50] Segman, J. and Schempp, W., *Two methods of incorporating scale in the Heisenberg group, JMIV* special issue on wavelets, 1993.

[51] Girod, B. and Kuo, D., Direct estimation of displacement histograms, *OSA Meeting on Image Understanding and Machine Vision,* 1989.

[52] Sheng, Y., Lejeune, C., and Arsenault, H.H., Frequency-domain Fourier-Mellin descriptors for invariant pattern recognition, *Optical Engineering,* 27, 354–7, 1988.

[53] Burt, P.J. and Anandan, P., Image stabilization by registration to a reference mosaic, *ARPA Image Understanding Workshop,* Nov. 10, 1994.

[54] Hansen, M., Anandan, P., Dana, K., van der Wal, G., and Burt, P.J., Real-time scene stabilization and mosaic construction, *ARPA Image Understanding Workshop,* Nov. 10, 1994.

Chapter 4

Discrete Cosine and Sine Transforms

Vladimir Britanak

Slovak Academy of Sciences

4.1 Introduction

The discrete cosine transform (DCT) and discrete sine transform (DST) are members of a family of sinusoidal unitary transforms. They have found applications in digital signal and image processing and particularly in transform coding systems for data compression/decompression. Among the various versions of DCT, types II and III have received much attention in digital signal processing. Besides being real, orthogonal, and separable, its properties are relevant to data compression and fast algorithms for its computation have proved to be of practical value. Recently, DCT has been employed as the main processing tool for data compression/decompression in international image and video coding standards [31]. An alternative transform used in transform coding systems is DST. In fact, the alternate use of modified forms of DST and DCT has been adopted in the international audio coding standards MPEG-1 and MPEG-2 (Moving Picture Experts Group) [31].

In this chapter, the definitions and basic mathematical properties of four even types of DCT and the DST are discussed. Then, the properties of DCT and DST relevant to data compression are briefly outlined. For each DCT and DST, a fast computational algorithm is described, and a corresponding regular generalized signal flow graph is shown, followed by its implementation in C. Finally, to illustrate the compression capability of DCT, a real DCT-based data compression application is considered. The simple and efficient JPEG (Joint Photographic Experts Group) DCT-based image compression and decompression system [31] and its implementation is described in detail. Generally, this chapter contains many implemented algorithms that can be useful not only in data compression applications but also in any other DCT- and DST-related applications.

4.2   The Family of DCTs and DSTs

DCTs and DSTs are members of the class of sinusoidal unitary transforms developed by Jain [1]. A sinusoidal unitary transform is an invertible linear transform whose kernel describes a set of complete, orthogonal discrete cosine and/or sine basis functions. The well-known Karhunen–Loève transform (KLT) [30], generalized discrete Fourier transform [2], generalized discrete Hartley transform [3] or equivalently generalized discrete W transform [4], and various types of the DCT and DST are members of this class of unitary transforms.

The set of DCTs and DSTs introduced by Jain [1] is not complete. The complete set of DCTs and DSTs, so-called discrete trigonometric transforms, has been described by Wang and Hunt [4]. The family of discrete trigonometric transforms consists of 8 versions of DCT and corresponding 8 versions of DST [13, 14]. Each transform is identified as even or odd and of type I, II, III, and IV. All present digital signal and image processing applications (mainly transform coding and digital filtering of signals) involve only even types of the DCT and DST. Therefore, this chapter considers four even types of DCT and DST.

4.2.1   Definitions of DCTs and DSTs

In subsequent sections, N is assumed to be an integer power of 2, i.e., $N = 2^m$. A subscript of a matrix denotes its order, while a superscript denotes the version number.

Four normalized even types of DCT in the matrix form are defined as [4]

$$DCT - I : \quad \left[C_{N+1}^{I} \right]_{nk} = \sqrt{\frac{2}{N}} \left[\epsilon_n \, \epsilon_k \cos \frac{\pi n k}{N} \right], \tag{4.1a}$$
$$n, k = 0, 1, \ldots, N,$$

$$DCT - II : \quad \left[C_{N}^{II} \right]_{nk} = \sqrt{\frac{2}{N}} \left[\epsilon_k \cos \frac{\pi (2n + 1)k}{2N} \right], \tag{4.1b}$$
$$n, k = 0, 1, \ldots, N - 1,$$

$$DCT - III : \quad \left[C_{N}^{III} \right]_{nk} = \sqrt{\frac{2}{N}} \left[\epsilon_n \cos \frac{\pi (2k + 1)n}{2N} \right], \tag{4.1c}$$
$$n, k = 0, 1, \ldots, N - 1,$$

$$DCT - IV : \quad \left[C_{N}^{IV} \right]_{nk} = \sqrt{\frac{2}{N}} \left[\cos \frac{\pi (2n + 1)(2k + 1)}{4N} \right], \tag{4.1d}$$
$$n, k = 0, 1, \ldots, N - 1,$$

where

$$\epsilon_p = \begin{cases} \frac{1}{\sqrt{2}} & p = 0 \text{ or } p = N \\ 1 & \text{otherwise} \end{cases}$$

and the corresponding four normalized even types of the DST are defined as [4]

$$DST-I: \quad \left[S_{N-1}^{I} \right]_{nk} = \sqrt{\frac{2}{N}} \left[\sin \frac{\pi(n+1)(k+1)}{N} \right], \qquad (4.2a)$$

$$n, k = 0, 1, \ldots, N-2,$$

$$DST-II: \quad \left[S_{N}^{II} \right]_{nk} = \sqrt{\frac{2}{N}} \left[\epsilon_k \sin \frac{\pi(2n+1)(k+1)}{2N} \right], \qquad (4.2b)$$

$$n, k = 0, 1, \ldots, N-1,$$

$$DST-III: \quad \left[S_{N}^{III} \right]_{nk} = \sqrt{\frac{2}{N}} \left[\epsilon_n \sin \frac{\pi(2k+1)(n+1)}{2N} \right], \qquad (4.2c)$$

$$n, k = 0, 1, \ldots, N-1,$$

$$DST-IV: \quad \left[S_{N}^{IV} \right]_{nk} = \sqrt{\frac{2}{N}} \left[\sin \frac{\pi(2n+1)(2k+1)}{4N} \right], \qquad (4.2d)$$

$$n, k = 0, 1, \ldots, N-1,$$

where

$$\epsilon_q = \begin{cases} \frac{1}{\sqrt{2}} & q = N-1 \\ 1 & \text{otherwise.} \end{cases}$$

The DCT-I introduced by Wang and Hunt [5] is defined for the order $N+1$. It can be considered a special case of symmetric cosine transform introduced by Kitajima [6]. The DST-I introduced by Jain [7] is defined for the order $N-1$ and constitutes the basis of a technique called recursive block coding [35]. The DCT-II and its inverse, DCT-III, first reported by Ahmed, Natarajan, and Rao [8], has an excellent energy compaction property, and among the currently known unitary transforms it is the best approximation for the optimal KLT. The DST-II and its inverse, DST-III, have been introduced by Kekre and Solanki [9]. DST-II is a complementary or alternative transform to DCT II used in transform coding. DCT-IV and DST-IV introduced by Jain [1] have found applications in the fast implementation of lapped orthogonal transform for the efficient transform/subband coding [12].

4.2.2 Mathematical Properties

The basic mathematical properties of discrete transforms are fundamental for their use in practical applications. Thus, properties such as scaling, shifting, and convolution are readily applied in the discrete transform domain. In the following, we briefly summarize the most relevant mathematical properties of the family of DCTs and DSTs.

DCT and DST matrices are real and orthogonal. All DCTs and DSTs are separable transforms; the multidimensional transform can be decomposed into successive application of one-dimensional (1-D) transforms in the appropriate directions.

The Unitarity Property

The following relations hold for inverse DCT matrices

$$\left[C_{N+1}^{I} \right]^{-1} = \left[C_{N+1}^{I} \right]^{T} = \left[C_{N+1}^{I} \right] \tag{4.3a}$$

$$\left[C_{N}^{II} \right]^{-1} = \left[C_{N}^{II} \right]^{T} = \left[C_{N}^{III} \right] \tag{4.3b}$$

$$\left[C_{N}^{III} \right]^{-1} = \left[C_{N}^{III} \right]^{T} = \left[C_{N}^{II} \right] \tag{4.3c}$$

$$\left[C_{N}^{IV} \right]^{-1} = \left[C_{N}^{IV} \right]^{T} = \left[C_{N}^{IV} \right] \tag{4.3d}$$

and for inverse DST matrices

$$\left[S_{N-1}^{I} \right]^{-1} = \left[S_{N-1}^{I} \right]^{T} = \left[S_{N-1}^{I} \right] \tag{4.4a}$$

$$\left[S_{N}^{II} \right]^{-1} = \left[S_{N}^{II} \right]^{T} = \left[S_{N}^{III} \right] \tag{4.4b}$$

$$\left[S_{N}^{III} \right]^{-1} = \left[S_{N}^{III} \right]^{T} = \left[S_{N}^{II} \right] \tag{4.4c}$$

$$\left[S_{N}^{IV} \right]^{-1} = \left[S_{N}^{IV} \right]^{T} = \left[S_{N}^{IV} \right] \tag{4.4d}$$

If the nonsingular matrix is real and orthogonal, its inverse is obtained as its transpose. In the definitions of DCT and DST, matrices given by Eqs. (4.1a)–(4.1d) and Eqs. (4.2a)–(4.2d), respectively, the normalization factors $\sqrt{(2/N)}$ can be merged as $2/N$, and it can be moved to either the forward or inverse transform. By merging these normalization factors, the family of DCT and DST matrices are no longer orthonormal. They are, however, still orthogonal. The DCT-I, DCT-IV, DST-I, and DST-IV matrices are involutory, i.e., they are orthogonal and symmetric. The symmetry of an orthogonal matrix indicates that algorithms for the forward and inverse transform computation will be the same except for the normalization. On the other hand, DCT-II and DCT-III are inverses of each other. The same property holds for DST-II and DST-III.

The Linearity Property

Since matrix multiplication is a linear operation, i.e.,

$$M \left(\alpha \, \mathbf{g} + \beta \, \mathbf{f} \right) = \alpha \, M \, \mathbf{g} + \beta \, M \, \mathbf{f} \tag{4.5}$$

for a matrix M, constants α and β, and vectors $\mathbf{g}$ and $\mathbf{f}$, all DCTs and DSTs are linear transforms.

The Convolution-Multiplication Property

All DCTs and DSTs possess convolution — multiplication property which is a powerful tool for performing digital filtering in the transform domain. The convolution operation in the transform domain realized by taking an inverse transform of the product of forward transforms of two data sequences is equivalent to symmetric convolution of those symmetrically extended sequences in the spatial domain [13, 14]. Let $\{x_n\}$ and $\{y_n\}$ be two input data sequences to be convolved. Generally, the relation between the symmetric convolution and transform domain convolution-multiplication property can be expressed by the following equation

$$\{x_n\} < \mathbf{sc} > \{y_n\} = \mathcal{T}_c^{-1} \left[\mathcal{T}_a \{x_n\} \times \mathcal{T}_b \{y_n\} \right] , \qquad (4.6)$$

where $< \mathbf{sc} >$ is the operator of symmetric convolution, $\times$ denotes element-by-element multiplication of its operands, and $\mathcal{T}_a\{x_n\}$ denotes a specified transform $\mathcal{T}_a$ of the sequence $\{x_n\}$. As an example, the convolution-multiplication property for the DCT-II is obtained by substituting $\mathcal{T}_a = \mathcal{T}_b = [C_N^{II}]$ and $\mathcal{T}_c = [C_{N+1}^{I}]^{-1}$ into Eq. (4.6). Definition of the symmetric convolution and convolution-multiplication properties for the entire family of discrete trigonometric transforms are given in references [13, 14], and [15].

The Shift Property, Scaling, and Difference Property

For the family of DCTs and DSTs, the reader can find the complete derivations of the shift property in references [10, 11], and [30] and scaling in time and the difference property in [30].

4.2.3 Relations to the KLT

The performance of DCTs and DSTs, particularly important in transform coding, is associated with the KLT. KLT is an optimal transform for data compression in a statistical sense because it decorrelates a signal in the transform domain, packs the most information in a few coefficients, and minimizes mean-square error between the reconstructed and original signal compared to any other transform. However, KLT is constructed from the eigenvalues and the corresponding eigenvectors of a covariance matrix of the data to be transformed; it is signal-dependent, and there is no general algorithm for its fast computation. There is asymptotic equivalence of the family of DCTs and DSTs with respect to KLT for a first-order stationary Markov process in terms of the transform size and the adjacent (interelement) correlation coefficient ρ. For finite length data, DCTs and DSTs provide different approximations to KLT, and the best approximating transform varies with the value of correlation coefficient ρ. For example, when $\rho = 1$ the KLT is reduced to DCT-II (DCT-III) [16, 17, 30], for $\rho = 0$ the KLT is reduced to DST-I [7, 17, 18], and for $\rho = -1$ it is reduced to DST-II (DST-III) [19]. On the other hand, if the transform size N increases (i.e., N tends to infinity), it can be shown that KLT is reduced to DCT-I or DCT-IV [30]. This asymptotic behavior implies that DCTs and DSTs can be used as substitutes for KLT of certain random processes.

In general, there are several characteristics that are desirable in a transform when it is used for the purpose of data compression [36]:

- **Data decorrelation:** The ideal transform completely decorrelates the data in a sequence/block; i.e., it packs the most amount of energy in the fewest number of coefficients. In this way, many coefficients can be discarded after quantization and prior to encoding. It is important to note that the transform operation itself does not achieve any compression. It aims at decorrelating the original data and compacting a large fraction of the signal energy into relatively few transform coefficients.

- **Data-independent basis functions:** Owing to the large statistical variations among data, the optimum transform usually depends on the data, and finding the basis functions of such transform is a computationally intensive task. This is particularly a problem if the data blocks are highly nonstationary, which necessitates the use of more than one set of basis functions to achieve high decorrelation. Therefore, it is desirable to trade optimum performance for a transform whose basis functions are data-independent.

- **Fast implementation:** The number of operations required for an n-point transform is generally of the order $\mathcal{O}(n^2)$. Some transforms have fast implementations, which reduce the number of operations to $\mathcal{O}(n \log n)$. For a separable $n \times n$ 2-D transform, performing the row and column 1-D transforms successively reduces the number of operations from $\mathcal{O}(n^4)$ to $\mathcal{O}(2n^2 \log n)$.

Among the family of DCTs and DSTs, the performance of DCT-II is closest to the statistically optimal KLT based on a number of performance criteria (variance distribution, energy packing efficiency, residual correlation, rate distortion, and maximum reducible bits and generalized Wiener filtering) [30]. The importance of DCT-II is further accentuated by its superiority in bandwidth compression (redundancy reduction) of a wide range of signals and by existence of fast algorithms for its implementation. Owing to powerful performance in the bit-rate reduction, DCT-II and its inversion, DCT-III, have been employed in the international image/video coding standards: JPEG for compression of still images, MPEG for compression of motion video including HDTV (High Definition Television), H.261 for compression of video telephony and teleconferencing, and H.263 for visual communication over ordinary telephone lines [31].

4.3   A Unified Fast Computation of DCTs and DSTs

The DCT and DST matrices defined in Section 4.2 are orthonormal. The normalization factor $\sqrt{(2/N)}$ in the forward and the inverse transforms can be merged as $2/N$ and moved to the forward transform. By merging these normalization factors the family of DCT and DST matrices are orthogonal. Without loss of generality, in this section orthogonal DCT and DST matrices will be considered.

A unified fast computation of even types of DCT (DCT-I, -II, -III, -IV) and DST (DST-I, -II, -III, -IV) is based on a universal computational structure both for DCT-

II/DST-II and DCT-III/DST-III computation [26]. This DCT-II/DST-II (DCT-III/DST-III) universal computational structure is used as the basic computational unit (a potential DCT/DST processor) in fast algorithms defined by sparse matrix factorizations. The fast algorithms are simple, numerically stable and efficient. For each type of the DCT and DST computation, the corresponding regular generalized signal flow graph is shown. Generalized signal flow graphs are enabled to realize computation of given DCT and DST for any $N = 2^m$, $m > 0$ (N being the length of the data sequence). The unified fast computation of DCTs and DSTs provides simple and compact transform building blocks. Finally, computer programs for each even type of the DCT and DST computation are presented.

4.3.1 Definitions of Even-Odd Matrices

Even-Odd Transform Matrix

$$A_J = \begin{bmatrix} I_{\frac{J-1}{2}} & & \bar{I}_{\frac{J-1}{2}} \\ & 1 & \\ \bar{I}_{\frac{J-1}{2}} & & -I_{\frac{J-1}{2}} \end{bmatrix} \quad \text{for } J \text{ odd} , \tag{4.7}$$

where I_N is the identity matrix. Blanks in the even-odd transform matrix Eq. (4.7) represent null submatrices and

$$\bar{I}_N = \begin{bmatrix} 0 & \cdots & 0 & 0 & 0 & 1 \\ 0 & \cdots & 0 & 0 & 1 & 0 \\ 0 & \cdots & 0 & 1 & 0 & 0 \\ 0 & \cdots & 1 & 0 & 0 & 0 \\ \vdots & & \vdots & \vdots & \vdots & \vdots \\ 1 & \cdots & 0 & 0 & 0 & 0 \end{bmatrix} \tag{4.8}$$

is the reflection matrix. The orthogonal even-odd transform matrix Eq. (4.7) converts data sequences into their symmetric (even) and anti-symmetric (odd) parts.

Even-Odd Permutation Matrices

$$P_J = \begin{bmatrix} 1 & 0 & 0 & 0 & 0 & \cdots & 0 & 0 & 0 \\ 0 & 0 & 1 & 0 & 0 & \cdots & 0 & 0 & 0 \\ 0 & 0 & 0 & 0 & 1 & \cdots & 0 & 0 & 0 \\ & & \vdots & & & & & \vdots & \\ 0 & 0 & 0 & 0 & 0 & \cdots & 0 & 1 & 0 \\ 0 & 0 & 0 & 0 & 0 & \cdots & 0 & 0 & 1 \\ 0 & 0 & 0 & 0 & 0 & \cdots & 1 & 0 & 0 \\ & & \vdots & & & & & \vdots & \\ 0 & 0 & 0 & 1 & 0 & \cdots & 0 & 0 & 0 \\ 0 & 1 & 0 & 0 & 0 & \cdots & 0 & 0 & 0 \end{bmatrix} \quad \text{for } J \text{ even} , \tag{4.9a}$$

$$P_J = \begin{bmatrix} 1 & 0 & 0 & 0 & 0 & \cdots & 0 & 0 & 0 \\ 0 & 0 & 1 & 0 & 0 & \cdots & 0 & 0 & 0 \\ 0 & 0 & 0 & 0 & 1 & \cdots & 0 & 0 & 0 \\ & & \vdots & & & & & \vdots & \\ 0 & 0 & 0 & 0 & 0 & \cdots & 1 & 0 & 0 \\ 0 & 0 & 0 & 0 & 0 & \cdots & 0 & 0 & 1 \\ 0 & 0 & 0 & 0 & 0 & \cdots & 0 & 1 & 0 \\ & & \vdots & & & & & \vdots & \\ 0 & 0 & 0 & 1 & 0 & \cdots & 0 & 0 & 0 \\ 0 & 1 & 0 & 0 & 0 & \cdots & 0 & 0 & 0 \end{bmatrix} \quad \text{for } J \text{ odd.} \quad (4.9b)$$

The permutation matrix P_J reorders the data sequence such that the first half of even-numbered data is arranged in the natural order, while the last half of odd-numbered data is arranged in the reversed order.

4.3.2   DCT-II/DST-II and DCT-III/DST-III Computation

The DCT-II for a given data sequence $\{x_n\}$, $n = 0, 1, \ldots, N-1$ is defined as [8]

$$z_k^{II} = \frac{2\epsilon_k}{N} \sum_{n=0}^{N-1} x_n \cos\left[\frac{\pi(2n+1)k}{2N}\right], \qquad k = 0, 1, \ldots, N-1 \qquad (4.10)$$

and the inverse DCT-II (DCT-III) is defined by

$$x_n = \sum_{k=0}^{N-1} \epsilon_k z_k^{II} \cos\left[\frac{\pi(2n+1)k}{2N}\right], \qquad n = 0, 1, \ldots, N-1, \qquad (4.11)$$

where

$$\epsilon_k = \begin{cases} \frac{1}{\sqrt{2}} & k = 0 \\ 1 & \text{otherwise.} \end{cases}$$

DCT-II and its inverse, DCT-III, given by Eqs. (4.10) and (4.11), respectively, can be rewritten as [23]

$$z_k^{II} = \frac{2\epsilon_k}{N} \sum_{n=0}^{N-1} \tilde{x}_n \cos\left[\frac{\pi(4n+1)k}{2N}\right], \qquad k = 0, 1, \ldots, N-1, \qquad (4.12)$$

$$\tilde{x}_n = \sum_{k=0}^{N-1} \epsilon_k z_k^{II} \cos\left[\frac{\pi(4n+1)k}{2N}\right], \qquad n = 0, 1, \ldots, N-1, \qquad (4.13)$$

where

$$\tilde{x}_n = x_{2n}$$

$$\tilde{x}_{N-n-1} = x_{2n+1}, \qquad n = 0, 1, \ldots, \frac{N}{2} - 1. \qquad (4.14)$$

The reordering in Eq. (4.14) corresponds to the permutation matrix P_N given by Eq. (4.9a) applied to the input data vector.

Let C_N^{II} be the $N \times N$ orthogonal DCT-II matrix. Then a reordered DCT-II matrix $\hat{C}_N^{II}$ with permuted rows and columns is given by

$$\hat{C}_N^{II} = R_N \, C_N^{II} \, [P_N]^T , \tag{4.15}$$

where R_N is the bit reversal permutation matrix and $[P_N]^T$ is the transpose of the permutation matrix P_N. A fast, recursive algorithm for DCT-II (DCT-III) computation with a regular structure is based on a block matrix factorization of the reordered DCT-II matrix $\hat{C}_N^{II}$. The reordered DCT-II matrix $\hat{C}_N^{II}$ has a recursive structure; higher order matrices can be generated from lower order ones, and its block matrix factorization has the following form [28, 30]

$$\hat{C}_N^{II} = \begin{bmatrix} I_{\frac{N}{2}} & 0 \\ 0 & K_{\frac{N}{2}} \end{bmatrix} \begin{bmatrix} \hat{C}_{\frac{N}{2}}^{II} & 0 \\ 0 & \hat{C}_{\frac{N}{2}}^{II} \end{bmatrix} \begin{bmatrix} I_{\frac{N}{2}} & 0 \\ 0 & Q_{\frac{N}{2}} \end{bmatrix} \begin{bmatrix} I_{\frac{N}{2}} & I_{\frac{N}{2}} \\ I_{\frac{N}{2}} & -I_{\frac{N}{2}} \end{bmatrix} , \tag{4.16}$$

where $K_{\frac{N}{2}}$ is an $\frac{N}{2} \times \frac{N}{2}$ matrix given by

$$K_{\frac{N}{2}} = R_{\frac{N}{2}} L_{\frac{N}{2}} R_{\frac{N}{2}} , \tag{4.17}$$

where $R_{\frac{N}{2}}$ is the bit reversal permutation matrix, $L_{\frac{N}{2}}$ is the lower triangular matrix

$$L_{\frac{N}{2}} = \begin{bmatrix} 1 & 0 & 0 & 0 & \cdots & 0 \\ -1 & 2 & 0 & 0 & \cdots & 0 \\ 1 & -2 & 2 & 0 & \cdots & 0 \\ -1 & 2 & -2 & 2 & \cdots & 0 \\ \vdots & \vdots & \vdots & \vdots & & \vdots \\ -1 & 2 & -2 & 2 & \cdots & 2 \end{bmatrix} ,$$

$Q_{\frac{N}{2}}$ is the $\frac{N}{2} \times \frac{N}{2}$ diagonal matrix

$$Q_{\frac{N}{2}} = diag \, [\cos \phi_m] ,$$
$$\phi_m = \left(m + \frac{1}{4} \right) \left(\frac{2\pi}{N} \right) , \quad m = 0, 1, \ldots, \frac{N}{2} - 1 . \tag{4.18}$$

The block matrix factorization Eq. (4.16) defines Hou's fast, recursive, and numerically stable algorithm for DCT-II (DCT-III) computation which can be represented in the matrix form as [23]

$$\begin{bmatrix} \hat{z}_e^{II} \\ \hat{z}_o^{II} \end{bmatrix} = \frac{2}{N} \, \hat{C}_N^{II} \begin{bmatrix} \tilde{x}_p \\ \tilde{x}_r \end{bmatrix} , \tag{4.19}$$

where

$$\tilde{\mathbf{x}}_p = \left[x_0, x_2, x_4, \ldots, x_{N-4}, x_{N-2} \right]^T ,$$

$$\tilde{\mathbf{x}}_r = \left[x_{N-1}, x_{N-3}, x_{N-5}, \ldots, x_3, x_1 \right]^T ,$$

$$\hat{\mathbf{z}}_e^{II} = R_{\frac{N}{2}} \, \mathbf{z}_e^{II} ,$$

$$\hat{\mathbf{z}}_o^{II} = R_{\frac{N}{2}} \, \mathbf{z}_o^{II} ,$$

$$\mathbf{z}_e^{II} = \left[z_0, z_2, z_4, \ldots, z_{N-4}, z_{N-2} \right]^T ,$$

$$\mathbf{z}_o^{II} = \left[z_1, z_3, z_5, \ldots, z_{N-3}, z_{N-1} \right]^T ,$$

where $\mathbf{z}_e^{II}$ is the even half and $\mathbf{z}_o^{II}$ is the odd half of the DCT-II transformed sequence both arranged in the natural order. T denotes transposition.

A regular generalized signal flow graph based on this algorithm for DCT-II and its inverse, DCT-III, for any $N = 2^m$, $m > 0$ has been described by Britanak [24]. It is shown for $N = 16$ in Fig. 4.1. Full lines represent transfer factors $+1$, while broken lines represent transfer factors -1. $\bigcirc$ represents addition, $\downarrow$ represents multiplication by cosine coefficients $C_n^k = \cos(k\phi_n)$, $\phi_n = \frac{\pi(4n+1)}{2N}$, and $\rightarrow$ represents multiplication by 2. The normalization factor is not included in the signal flow graph. The generalized signal flow graph consists of two regular parts. The first part is related to the butterfly structure, and the second one, after bit-reversal permutation, is mapped into a pipeline structure. This pipeline structure is related to a simple recurrent relation for any $N = 2^m$, $m > 0$ [24].

The DST-II for a given data sequence $\{x_n\}$, $n = 0, 1, \ldots, N-1$ is defined as [9]

$$s_k^{II} = \frac{2\epsilon_k}{N} \sum_{n=0}^{N-1} x_n \sin\left[\frac{\pi(2n+1)(k+1)}{2N} \right], \qquad k = 0, 1, \ldots, N-1 \qquad (4.20)$$

and the inverse DST-II (DST-III) is defined as

$$x_n = \sum_{k=0}^{N-1} \epsilon_k s_k^{II} \sin\left[\frac{\pi(2n+1)(k+1)}{2N} \right], \qquad n = 0, 1, \ldots, N-1 , \qquad (4.21)$$

where

$$\epsilon_k = \begin{cases} \frac{1}{\sqrt{2}} & k = N-1 \\ 1 & \text{otherwise.} \end{cases}$$

Let C_N^{II} and S_N^{II} be orthogonal $N \times N$ DCT-II and DST-II matrices, respectively. According to Wang [20] S_N^{II} is related to C_N^{II} by

$$S_N^{II} = \bar{I}_N \, C_N^{II} \, D_N , \qquad (4.22)$$

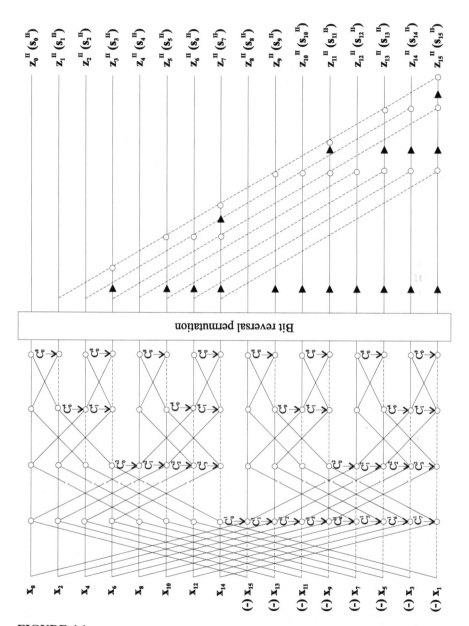

FIGURE 4.1
DCT-II/DST-II (DCT-III/DST-III) universal computational structure for $N =$
16. ©**Slovak Academic Press Ltd.**

where D_N is the diagonal odd sign-changing matrix

$$
D_N = \begin{bmatrix}
1 & 0 & 0 & 0 & \cdots & 0 \\
0 & -1 & 0 & 0 & \cdots & 0 \\
0 & 0 & 1 & 0 & \cdots & 0 \\
0 & 0 & 0 & -1 & \cdots & 0 \\
\vdots & \vdots & \vdots & \vdots & & \vdots \\
0 & 0 & 0 & 0 & \cdots & -1
\end{bmatrix} . \tag{4.23}
$$

The D_N matrix applied to the input data sequence given by Eq. (4.14) corresponds to the reordering and sign changes:

$$
\tilde{x}_n = x_{2n}
$$

$$
\tilde{x}_{N-n-1} = -x_{2n+1}, \quad n = 0, 1, \ldots, \frac{N}{2} - 1 . \tag{4.24}
$$

From Eq. (4.22) it follows that the generalized signal flow graph for the DCT-II computation can also be used for the DST-II computation for any $N = 2^m$, $m > 0$. The output DST-II transformed sequence, after the DCT-II computation for the input data sequence given by Eq. (4.24), is in reversed order; i.e., the final DST-II transformed sequence is obtained as

$$
s_k^{II} = z_{N-1-k}^{II}, \quad k = 0, 1, \ldots, N - 1 . \tag{4.25}
$$

Hence, by the same computational structure, both the DCT-II and DST-II computation can be effectively realized for any $N = 2^m$, $m > 0$ simply by changing the input and output data sequences. Because both DCT-II and DST-II are orthogonal transforms, the algorithm for DST-III computation is obtained by transposing of Eq. (4.22). The generalized signal flow graph for DCT-II/DST-II and their inverse computations, so called DCT-II/DST-II (DCT-III/DST-III) universal computational structure, is shown for $N = 16$ in Fig. 4.1. The symbols in brackets correspond to DST-II (DST-III) computation. DCT-II/DST-II (DCT-III/DST-III) universal computational structure [25] represents the unified DCT-II/DST-II and their inverse computations, DCT-III/DST-III for any $N = 2^m$, $m > 0$. The universality of DCT-II/DST-II computational structure is related to the fact that it can be used as the basic computational unit for the fast implementation of the entire class of discrete sinusoidal transforms, i.e., generalized discrete Fourier transform, generalized discrete Hartley transforms, and the other types of the DCT and DST, respectively [26, 27]. We note that for fast computation of other discrete sinusoidal transforms, the bidirectional DCT-II/DST-II (DCT-III/DST-III) universal computational structure is used without the proper normalization. If DCT-II (DCT-III) or DST-II (DST-III) computation is required, the proper normalization should be applied to the input and output data sequences.

4.3.3 DCT-I and DST-I Computation

DCT-I for a given data sequence $\{x_n\}$, $n = 0, 1, \ldots, N$ is defined as [5]

$$z_k^I = \frac{2\epsilon_k}{N} \sum_{n=0}^{N} \epsilon_n x_n \cos\left[\frac{\pi n k}{N}\right], \qquad k = 0, 1, \ldots, N \qquad (4.26)$$

and the inverse DCT-I (IDCT-I) is defined by

$$x_n = \epsilon_n \sum_{k=0}^{N} \epsilon_k z_k^I \cos\left[\frac{\pi n k}{N}\right], \qquad n = 0, 1, \ldots, N, \qquad (4.27)$$

where

$$\epsilon_p = \begin{cases} \frac{1}{\sqrt{2}} & p = 0 \text{ or } p = N \\ 1 & \text{otherwise.} \end{cases}$$

DCT-I and IDCT-I are defined for data sequences of length $N + 1$. Let C_{N+1}^I be the orthogonal DCT-I matrix of order $N + 1$. Then for $N = 2^m$, $m \geq 1$, C_{N+1}^I can be decomposed into the following recursive matrix form [21]

$$C_{N+1}^I = P_{N+1} \begin{bmatrix} C_{\frac{N}{2}+1}^I & 0 \\ 0 & \bar{I}_{\frac{N}{2}} C_{\frac{N}{2}}^{III} \bar{I}_{\frac{N}{2}} \end{bmatrix} A_{N+1}, \qquad (4.28)$$

where A_{N+1} and P_{N+1} are matrices given by Eq. (4.7) and Eq. (4.9b), respectively. $C_{\frac{N}{2}+1}^I$ is the DCT-I matrix of order $\frac{N}{2} + 1$. The matrix product $\bar{I}_{\frac{N}{2}} C_{\frac{N}{2}}^{III} \bar{I}_{\frac{N}{2}}$ denotes $\frac{N}{2} \times \frac{N}{2}$ DCT-III matrix with reversed order for both its rows and columns. The permutation matrix P_{N+1} applied to the data vector corresponds to the reordering:

$$\tilde{x}_0 = x_0$$
$$\tilde{x}_{n+1} = x_{2n+2}$$
$$\tilde{x}_{N-n} = x_{2n+1}, \qquad n = 0, 1, \ldots, \frac{N}{2} - 1. \qquad (4.29)$$

Because C_{N+1}^I is a symmetric matrix, the algorithms for the DCT-I and IDCT-I computation are the same except for the normalization. The generalized signal flow graph for the DCT-I and IDCT-I computation for $N + 1 = 17$ is shown in Fig. 4.2. Here $\alpha = \frac{1}{\sqrt{2}}$, and the normalization factor is again not included in the signal flow graph.

The DST-I for a given data sequence $\{x_n\}$, $n = 0, 1, \ldots, N - 2$ is defined as [7]

$$s_k^I = \frac{2}{N} \sum_{n=0}^{N-2} x_n \sin\left[\frac{\pi (n + 1)(k + 1)}{N}\right], \qquad k = 0, 1, \ldots, N - 2, \qquad (4.30)$$

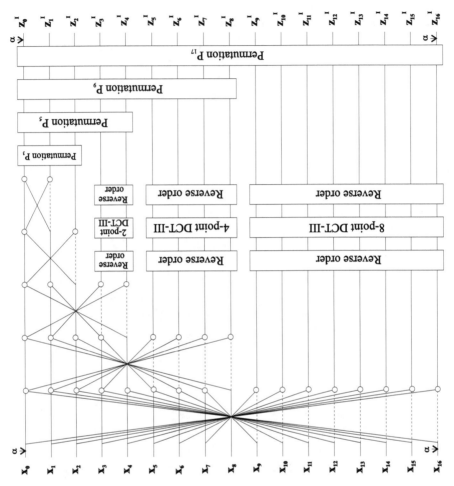

FIGURE 4.2
DCT-I and IDCT-I generalized signal flow graph for $N + 1 = 17$. **©Slovak Academic Press Ltd.**

and the inverse DST-I (IDST-I) is defined by

$$x_n = \sum_{k=0}^{N-2} s_k^I \sin\left[\frac{\pi(n+1)(k+1)}{N}\right], \qquad n = 0, 1, \ldots, N-2. \qquad (4.31)$$

DST-I and IDST-I are defined for data sequences of length $N - 1$. Let S_{N-1}^I be the orthogonal DST-I matrix of order $N - 1$. Then for $N = 2^m$, $m > 1$, S_{N-1}^I can be decomposed into the following recursive matrix form [21]

$$
S_{N-1}^{I} = P_{N-1}
\begin{bmatrix}
S_{\frac{N}{2}}^{III} & 0 \\
0 & \bar{I}_{\frac{N}{2}-1} S_{\frac{N}{2}-1}^{I} \bar{I}_{\frac{N}{2}-1}
\end{bmatrix}
A_{N-1} ,
\tag{4.32}
$$

where $S_{\frac{N}{2}}^{III}$ is the $\frac{N}{2} \times \frac{N}{2}$ DST-III matrix. The matrix product $\bar{I}_{\frac{N}{2}-1} S_{\frac{N}{2}-1}^{I} \bar{I}_{\frac{N}{2}-1}$ denotes the DST-I matrix of order $\frac{N}{2} - 1$ with reversed order for both its rows and columns. The permutation matrix P_{N-1} applied to the data vector corresponds to the reordering

$$
\begin{aligned}
\tilde{x}_0 &= x_0 \\
\tilde{x}_{n+1} &= x_{2n+2} \\
\tilde{x}_{N-2-n} &= x_{2n+1}, \quad n = 0, 1, \ldots, \frac{N}{2} - 2 .
\end{aligned}
\tag{4.33}
$$

Because S_{N-1}^{I} is a symmetric matrix, the algorithms for the DST-I and IDST-I are the same except for the normalization. The generalized signal flow graph for the DST-I and IDST-I computation for $N - 1 = 15$ is shown in Fig. 4.3. The normalization factor is not included in the signal flow graph.

4.3.4 DCT-IV/DST-IV Computation

The DCT-IV for a given data sequence $\{x_n\}, n = 0, 1, \ldots, N - 1$ is defined as [1]

$$
z_k^{IV} = \frac{2}{N} \sum_{n=0}^{N-1} x_n \cos\left[\frac{\pi(2n+1)(2k+1)}{4N} \right], \quad k = 0, 1, \ldots, N - 1
\tag{4.34}
$$

and the inverse DCT-IV (IDCT-IV) is defined by

$$
x_n = \sum_{k=0}^{N-1} z_k^{IV} \cos\left[\frac{\pi(2n+1)(2k+1)}{4N} \right], \quad n = 0, 1, \ldots, N - 1 .
\tag{4.35}
$$

Let C_N^{IV} be the orthogonal $N \times N$ DCT-IV matrix. Then for $N = 2^m$, $m \geq 1$, C_N^{IV} can be decomposed into the following sparse matrix product [22]

$$
C_N^{IV} = T_N
\begin{bmatrix}
C_{\frac{N}{2}}^{III} & 0 \\
0 & \bar{I}_{\frac{N}{2}} S_{\frac{N}{2}}^{III} \bar{I}_{\frac{N}{2}}
\end{bmatrix}
P_N B_N ,
\tag{4.36}
$$

where $C_{\frac{N}{2}}^{III}$ is the $\frac{N}{2} \times \frac{N}{2}$ DCT-III matrix. The matrix product $\bar{I}_{\frac{N}{2}} S_{\frac{N}{2}}^{III} \bar{I}_{\frac{N}{2}}$ denotes $\frac{N}{2} \times \frac{N}{2}$ DST-III matrix with reversed order for both its rows and columns. T_N is the

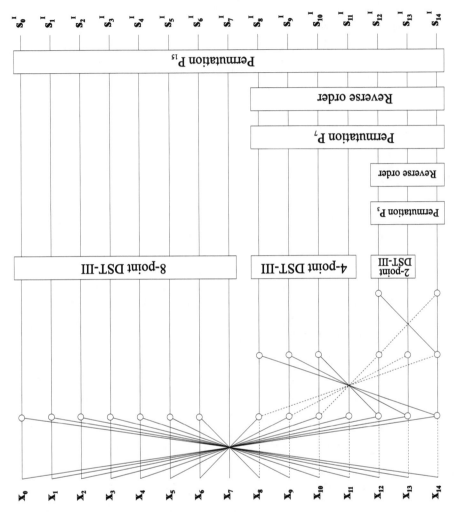

FIGURE 4.3
DST-I and IDST-I generalized signal flow graph for $N - 1 = 15$. ©Slovak
Academic Press Ltd.

rotation matrix given by

$$
T_N = \begin{bmatrix}
\cos\frac{\pi}{4N} & & & & & \sin\frac{\pi}{4N} \\
& \cos\frac{(N-1)\pi}{4N} & \sin\frac{(N-1)\pi}{4N} & & \\
& \sin\frac{(N-1)\pi}{4N} & -\cos\frac{(N-1)\pi}{4N} & & \\
\sin\frac{\pi}{4N} & & & & & -\cos\frac{\pi}{4N}
\end{bmatrix} \tag{4.37}
$$

and B_N is the tridiagonal matrix given by

$$
B_N = \begin{bmatrix}
1 & 0 & 0 & 0 & \cdots & & & 0 \\
0 & 1 & -1 & 0 & \cdots & & & 0 \\
0 & 1 & 1 & 0 & \cdots & & & 0 \\
\vdots & & & & \ddots & & & \vdots \\
0 & \cdots & & 0 & 1 & -1 & & 0 \\
0 & \cdots & & 0 & 1 & 1 & & 0 \\
0 & \cdots & & 0 & 0 & 0 & & 1
\end{bmatrix} . \tag{4.38}
$$

As can be seen, the decomposition of the matrix C_N^{IV} depends on the DCT-III and DST-III matrices of half size. Because C_N^{IV} is a symmetric matrix, the algorithms for the DCT-IV and IDCT-IV computation are the same except for the normalization. The generalized signal flow graph for the DCT-IV and IDCT-IV computation for $N = 16$ is shown in Fig. 4.4. The normalization factor is not included in the signal flow graph. The matrix product $P_N B_N$ can be realized by one butterfly stage in the generalized signal flow graph.

The DST-IV for a given data sequence $\{x_n\}$, $n = 0, 1, \ldots, N - 1$ is defined as [1]

$$
s_k^{IV} = \frac{2}{N} \sum_{n=0}^{N-1} x_n \sin\left[\frac{\pi(2n+1)(2k+1)}{4N}\right], \qquad k = 0, 1, \ldots, N - 1, \tag{4.39}
$$

and the inverse DST-IV (IDST-IV) is defined by

$$
x_n = \sum_{k=0}^{N-1} s_k^{IV} \sin\left[\frac{\pi(2n+1)(2k+1)}{4N}\right], \qquad n = 0, 1, \ldots, N - 1. \tag{4.40}
$$

Let C_N^{IV} and S_N^{IV} be the $N \times N$ DCT-IV and DST-IV matrices, respectively. The matrix S_N^{IV} is related to C_N^{IV} by [21]

$$
S_N^{IV} = \bar{I}_N \, C_N^{IV} \, D_N . \tag{4.41}
$$

Because S_N^{IV} is also a symmetric matrix, the algorithms for the DST-IV and IDST-IV computation are the same except for the normalization. From relation Eq. (4.41) it follows that the generalized signal flow graph for the DCT-IV computation can be also used for the DST-IV computation by changing only the input and output data sequences. The output DST-IV transformed sequence, after the DCT-IV computation for the input data sequence given by Eq. (4.24), is order reversed; the final DST-IV transformed data sequence is obtained as

$$
s_k^{IV} = z_{N-1-k}^{IV}, \qquad k = 0, 1, \ldots, N - 1. \tag{4.42}
$$

The generalized signal flow graph for the DCT-IV/DST-IV and IDCT-IV/IDST-IV computation for $N = 16$ is shown in Fig. 4.4, where the symbols in brackets correspond to DST-IV/IDST-IV computation. This generalized signal flow graph represents the unified DCT-IV/DST-IV and their inverse computations for any $N = 2^m$, $m > 0$.

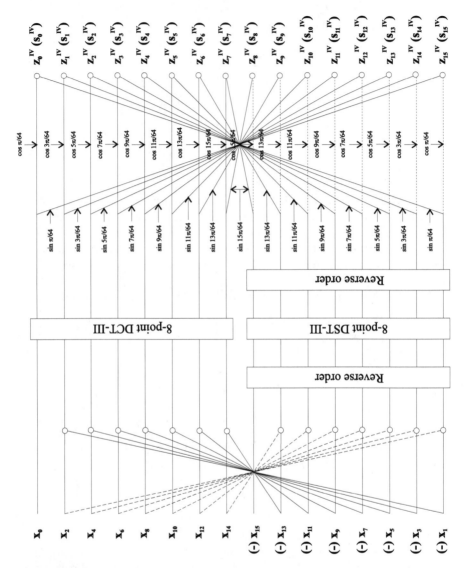

FIGURE 4.4
DCT-IV/DST-IV and IDCT-IV/IDST-IV generalized signal flow graph for $N =$ 16. ©Slovak Academic Press Ltd.

4.3.5 Implementation of the Unified Fast Computation of DCTs and DSTs

All developed algorithms have been implemented in the C language, and they can be used in practical applications. Implemented algorithms are able to compute the DCT/DST orthogonal transform of a given type for real data sequence up to size 1024. By minor modification (macro SIZE and LOG2SIZE) in program modules, any DCT/DST can be computed for the required size. In the implementation of DCT-II/DST-II (DCT-III/DST-III) universal computational structure, the normalization is optional. All computations are performed in double precision.

The orthonormal versions of the DCT and DST have the normalization factor $\sqrt{2/N}$ in both the forward and inverse transforms. Therefore, for the computation of orthonormal DCTs and DSTs, the implemented algorithms can be easily modified.

Computer Program for the Fast DCT-II/DST-II and DCT-III/DST-III Computation

```
/*-------------------------------------------------------------*
 *Module:     The 1-D Fast Discrete Cosine II and III  *
 *            Transform (DCT) and Discrete Sine II     *
 *            and III Transform (DST)                  *
 *                                                     *
 *Algorithm: DCT/DST universal computational           *
 *            structure for the 1-D DCT-II/DST-II and  *
 *            DCT-III/DST-III Transform Computation     *
 *                                                     *
 *Note that the DCT/DST universal computational        *
 *structure in algorithms for discrete sinusoidal      *
 *transforms computation is used without the           *
 *normalization. This module simulates a potential     *
 *DCT/DST processor.                                   *
 *-------------------------------------------------------------*/
/*--- Prototypes to be included in calling program---*/
int dct_processor (
double *pdct, /* input/output vector of length 2**m */
  int m,        /* m = log_2 (N)
                   E.g. for N = 256 --> m = 8        */
  int norm,     /* norm = 0 normalization is disabled
                   norm != 0 normalization is enabled */
  int flag);    /* Transform computation:
                   flag =  1   1-D DCT-II
                   flag = -1   1-D DCT-III
                   flag =  2   1-D DST-II
                   flag = -2   1-D DST-III            */
/* NOTE: Function returns into calling program
   following value:
        0 - successful processing
       -1 - invalid length of input vector or
            invalid type transform computation        */
/*------------------- Includes ---------------------*/
#include        <math.h>
/*------------------- Defines ----------------------*/
#define   SIZE      1024           /* max length 1024 */
#define   LOG2SIZE  10             /* log_2 (SIZE)    */
#define   PI        3.141592653589793  /* pi          */
#define   DCT_II    1
#define   DCT_III   -1
/*--------------- Local Variables ------------------*/
static double ac [SIZE];    /* working vector         */
static double cs [SIZE-1];  /*table of cos coefficients*/
static int length;
/* --- Beginning of the  DCT/DST processor module ---*/
```

```
int dct_processor(double *pdct, int m, int norm,
                  int flag)
{
  int    i,j,k,n,n1,n2,r,s,f0,f1,f2,f3,ip,ic,half,base;
  double arg,fi,scale,tmp,*pc1,*pc2,*pac = &ac [0];
   /* Verification of the input vector length (SIZE) */
   if ( m < 1  ||  m > LOG2SIZE )
       return (-1);
   /* Verification of the transform type computation */
   if ( flag < -2 || flag == 0 || flag > 2 )
       return (-1);
   /* Initialize input vector length and variables  */
   n  = 1 << m;
   n1 = n - 1;
   n2 = n >> 1;
   /* Generate the table of cosine coefficients Table
      is updated for new value of N */
   if ( length != n )
   {
       scale = 1.0 / (double) (n << 1);
       for ( s = base = 0; s < m; s++, base += ip )
       {
           half = n >> s;
           ip   = half >> 1;
           ic   = n / half;
           arg = (double) ic * PI * scale;
           for ( i = 0; i < ip; i++ )
           {
               fi = (double) (4 * i + 1) * arg;
               cs [base+i] = cos (fi);
           }
       }
       length = n;
   }
   /* Test type of computation - Forward or Inverse
      transform */
   if ( flag < 0 )
       goto inv;
/*
   ==========================================================
           THE 1-D FAST DCT-II OR DST-II TRANSFORM
   ==========================================================
*/
   /* Reordering of the original input data sequence */
   for ( i = 0; i < n2; i++ )
   {
       *(pac + i) = *(pdct + 2 * i);
       if ( flag == DCT_II )
           *(pac + n - 1 - i) =    *(pdct + 2 * i + 1);
       else
           *(pac + n - 1 - i) = - *(pdct + 2 * i + 1);
   }
   /* Implementation of the butterfly structure */
   for ( s = base = 0; s < m; s++, base += ip )
   {
       half = n >> s;
       ip   = half >> 1;
       for ( j = 0; j < ip; j++ )
           for ( i = j; i < n; i += half )
```

```
                {
                        pc1 = pac + i;
                        pc2 = pc1 + ip;
                        tmp = *pc1 + *pc2;
                        *pc2 = (*pc1 - *pc2) * cs [base+j];
                        *pc1 = tmp;
                }
        }
        /* Bit reversal permutation */
        for ( i = 1; i < n1; i++ )
        {
                for ( k = j = 0, r = i; k < m; k++ )
                {
                        s = r >> 1;
                        j = j + j + r - s - s;
                        r = s;
                }
                if ( i < j )
                {
                        tmp       = *(pac + i);
                        *(pac + i) = *(pac + j);
                        *(pac + j) = tmp;
                }
        }
        /* Implementation of the pipeline structure */
        if ( m > 1 )
        {
                for ( i = 0; i < m - 1; i++ )
                {
                        f0 = n / (1 << i);
                        f1 = f0 >> 1;
                        f2 = f1 >> 1;
                        f3 = ((1 << i) - 1) << 1;
                        for ( j = 1; j <= f2; j++ )
                        {
                                ip = f0 - j;
                                ic = f1 - j;
                                pc1 = pac + ip;
                                pc2 = pac + ic;
                                *pc1 += *pc1 - *pc2;
                                k = 1;
                                while ( k <= f3 )
                                {
                                        ip += f1;
                                        ic += f1;
                                        pc1 = pac + ip;
                                        pc2 = pac + ic;
                                        *pc1 += *pc1 - *pc2;
                                        k++;
                                }
                        }
                }
        }
/*-----------------------------------------------------------
The normalization of the transformed data sequence.
If DCT-II/DST-II transform is required, then parameter
norm != 0. The block is not used for other discrete
sinusoidal transforms computation. Then norm = 0.
-----------------------------------------------------------*/
```

```
    if ( norm )
    {
        scale = 2.0 / (double) n;
        *pac *= 1.0 / sqrt (2.0);
        for ( i = 0; i < n; i++ )
            *(pac + i) *= scale;
    }
    /* Reverse order of the data sequence for DST-II */
    if ( flag == DCT_II )
        for ( i = 0; i < n; i++ )
            *(pdct + i) = *(pac + i);
    else
        for ( i = 0; i < n; i++ )
            *(pdct + i) = *(pac + n - 1 - i);
    return (0);
/*
  ========================================================
        THE 1-D FAST DCT-III OR DST-III TRANSFORM
  ========================================================
*/
inv:
    /* Reverse order of the data sequence for DST-III  */
    if ( flag == DCT_III )
        for ( i = 0; i < n; i++ )
            *(pac + i) = *(pdct + i);
    else
        for ( i = 0; i < n; i++ )
            *(pac + n - 1 - i) = *(pdct + i);
/*--------------------------------------------------------
  The normalization of the DC term. If DCT-III/DST-III
  transform is required, then parameter norm != 0. The
  block is not used for other discrete sinusoidal
  transforms computation. Then norm = 0.
  --------------------------------------------------------*/
    if ( norm )
        *pac *= 1.0 / sqrt (2.0);
    /* Implementation of the pipeline structure */
    if ( m > 1 )
    {
        for ( i = m - 2; i >= 0; i-- )
        {
            f0 = n / (1 << i);
            f1 = f0 >> 1;
            f2 = f1 >> 1;
            f3 = ((1 << i) - 1) << 1;
            for ( j = f2; j > 0; j-- )
            {
                k   = f3;
                ip = f0 - j + k * f1;
                ic = f1 - j + k * f1;
                pc1 = pac + ic;
                pc2 = pac + ip;
                *pc1 -= *pc2;
                *pc2 += *pc2;
                while ( k > 0 )
                {
                    k--;
                    ip -= f1;
                    ic -= f1;
                    pc1 = pac + ic;
                    pc2 = pac + ip;
```

```
                                *pc1 -= *pc2;
                                *pc2 += *pc2;
                            }
                    }
            }
    }
    /* Bit reversal permutation */
    for ( i = 1; i < n1; i++ )
    {
        for ( k = j = 0, r = i; k < m; k++ )
        {
            s = r >> 1;
            j = j + j + r - s - s;
            r = s;
        }
        if ( i < j )
        {
            tmp         = *(pac + i);
            *(pac + i)  = *(pac + j);
            *(pac + j)  = tmp;
        }
    }
    /* Implementation of the butterfly structure */
    for ( s = 0, base = n - 2; s < m; s++, base -= half )
    {
        half = 1 << (s + 1);
        ip   = half >> 1;
        for ( j = 0; j < ip; j++ )
            for ( i = j; i < n; i += half )
            {
                pc1 = pac + i;
                pc2 = pc1 + ip;
                tmp = *pc2 * cs [base+j];
                *pc2 = *pc1 - tmp;
                *pc1 = *pc1 + tmp;
            }
    }
    /* Reordering of output samples for DCT-III/DST-III */
    for ( i = 0; i < n2; i++ )
    {
        *(pdct + 2 * i) = *(pac + i);
        if ( flag == DCT_III )
            *(pdct + 2 * i + 1) =   *(pac + n - 1 - i);
        else
            *(pdct + 2 * i + 1) = - *(pac + n - 1 - i);
    }
    return (0);
}
/*------- End of the DCT/DST processor module -------*/
```

Computer Program for the Fast DCT-I Computation

```
/*------------------------------------------------------------*
 *Module:     The 1-D Fast Discrete Cosine I Transform *
 *                                                      *
 *Algorithm: The Forward and Inverse 1-D DCT-I          *
 *           Transform Computation                      *
 *                                                      *
 *Note that the DCT-I matrix of order N + 1 and it is   *
 *symmetric. Thus, the forward and inverse transforms   *
```

```
*are the same except for the normalization.            *
*------------------------------------------------------*/
/*--- Prototypes to be included in calling program ---*/
int fdcti1d (
    double *x, /* input/output vector of length 2**m+1 */
    int m,       /* m = log_2 (N)
                    E.g. for N = 256 --> m = 8          */
    int flag); /* Forward or Inverse DCT-I computation:
                    flag = 0  Forward 1-D DCT-I
                    flag = 1  Inverse 1-D DCT-I          */
/*   NOTE: Function returns into calling program
       following
            value:
             0 - successful processing
            -1 - invalid length of input vector
/*------------------- Includes -----------------------*/
#include          <math.h>
/*------------------- Defines ------------------------*/
#define   SIZE        1024        /* max length SIZE+1*/
#define   LOG2SIZE    10          /* log_2 (SIZE)     */
/*   NOTE: Actual transform size is SIZE + 1           */
int dct_processor (double *, int, int, int);
/*------------------ Local Variables -----------------*/
static double y [SIZE+1];
                        /*working vector of length N+1*/
/*------ Beginning of the 1-D Fast DCT-I module-------*/
int fdcti1d (double *x, int m, int flag)
{
    int    i,j,n,n1,n2,n3,nc;
    double scale,tmp;
    /* Verification of the input vector length (SIZE+1)*/
    if ( m < 1  ||  m > LOG2SIZE )
        return (-1);
    /* Initialize the input vector length */
    n = 1 << m;
    /* Multiply x[0] and x [n] by 1 / sqrt(2) */
    scale  = 1.0 / sqrt (2.0);
    x [0] *= scale;
    x [n] *= scale;
    /* Implementation of generalized signal flow graph */
    n1 = n >> 1;
    n2 = n;
    n3 = n << 1;
    nc = m - 1;
    do
    {
    /* Butterflies for even-odd transform matrix A(N) */
        for ( i = 0; i < n1; i++ )
        {
            tmp          = x [i];
            x [i     ]   = tmp + x [n2 - i];
            x [n2 - i]   = tmp - x [n2 - i];
        }
    /* Reverse order of the input data sequence */
        for ( i = n1 + 1, j = i + n1 - 1; i < j; i++, j-- )
        {
            tmp    = x [i];
            x [i]  = x [j];
```

```
                    x [j] = tmp;
        }
/* Compute the DCT-III transform */
        dct_processor (&x [n1+1],nc,0,-1);
/* Reverse order of the transformed data sequence */
   for ( i = n1 + 1, j = i + n1 - 1; i < j; i++, j-- )
        {
                tmp   = x [i];
                x [i] = x [j];
                x [j] = tmp;
        }
        n1 >>= 1;
        n2 >>= 1;
        nc--;
/* The last butterfly - 2x2 transform matrix */
        if ( n2 == 1 )
        {
                tmp   = x [0];
                x [0] = tmp + x [1];
                x [1] = tmp - x [1];
        }
    }
    while ( n2 > 1 );
/* Reorder data sequence by permutation matrix P(N) */
    n2 = 2;
    n1 = n2 >> 1;
    do
    {
        for ( i = 0; i < n2 + 1; i++ )
            y [i] = x [i];
        for ( i = 0; i < n1; i++ )
        {
            x [2 * i + 2] = y [i + 1 ];
            x [2 * i + 1] = y [n2 - i];
        }
        n2 <<= 1;
        n1 <<= 1;
    }
    while ( n2 < n3 );
/* Multiply x[0] and x [n] by 1 / sqrt(2) */
    x [0] *= scale;
    x [n] *= scale;
/* Normalization of the transformed data sequence */
    if ( !flag )
    {
        scale = 2.0 / (double) n;
        for ( i = 0; i < n + 1; i++ )
            x [i] *= scale;
    }
    return (0);
}
/*--------- End of the 1-D Fast DCT-I module--------*/
```

Computer Program for the Fast DST-I Computation

```
/*------------------------------------------------------*
 *Module:    The 1-D Fast Discrete Sine I Transform    *
 *                                                      *
 *Algorithm: The Forward and Inverse 1-D DST-I          *
 *           Transform Computation                      *
```

```
*                                                          *
*Note that the DST-I matrix of order N - 1 and it is  *
*symmetric. Thus, the forward and inverse transforms  *
*are the same except for the normalization.           *
*----------------------------------------------------*/
/*--- Prototypes to be included in calling program ---*/
int fdsti1d (
   double *x,  /* input/output vector of length 2**m-1 */
   int m,      /* m = log_2 (N)
                  E.g. for N = 256 --> m = 8          */
   int flag); /* Forward or Inverse DST-I computation:
                  flag = 0   Forward 1-D DST-I
                  flag = 1   Inverse 1-D DST-I        */
/*NOTE: Function returns into calling program following
          value:
          0 - successful processing
         -1 - invalid length of input vector
/*----------------- Defines ------------------------*/
#define   SIZE       1024        /* max length SIZE-1*/
#define   LOG2SIZE   10          /* log_2 (SIZE)     */
/*  NOTE: Actual transform size is SIZE - 1          */
int dct_processor (double *, int, int, int);
/*--------------- Local Variables -----------------*/
static double y [SIZE-1];
                        /* working vector of length N-1*/
/*------ Beginning of the 1-D Fast DST-I module-------*/
int fdsti1d (double *x, int m, int flag)
{
    int     i,j,n,n1,n2,nb,nc;
    double scale,tmp;
    /* Verification of the input vector length (SIZE-1)*/
    if ( m < 2  ||  m > LOG2SIZE )
        return (-1);
    /* Trivial case m = 1 */
    if ( m == 1 )
        return (0);
    /* Initialize the input vector length */
    n  = 1 << m;
    /* Implementation of generalized signal flow graph */
    n1 = n >> 1;
    n2 = n;
    nc = m - 1;
    nb = 0;
    while ( n2 > 2 )
    {
    /* Butterflies for even-odd transform matrix A(N)   */
        if ( n == n2 )
            for ( i = 0; i < n1 - 1; i++ )
            {
                tmp              = x [i];
                x [i          ] = tmp + x [n - 2 - i];
                x [n - 2 - i] = tmp - x [n - 2 - i];
            }
    /* Butterflies for even-odd transform matrix A(N)
        with reversed order of its columns */
        else
            for ( i = 0; i < n1 - 1; i++ )
            {
```

```
                        tmp              = x [nb + i];
                        x [nb + i    ]  = x [n - 2 - i] + tmp;
                        x [n - 2 - i]   = x [n - 2 - i] - tmp;
                }
        /* Compute the DST-III transform */
                dct_processor (&x [nb],nc,0,-2);
                n2 >>= 1;
                n1 >>= 1;
                nb += (1 << nc);
                nc--;
        }
/* Reorder of data sequence by permutation matrix P(N)*/
        n2 = 2;
        nc = 1;
        nb = n - 2 - (1 << nc);
        while ( n2 < n )
        {
                for ( i = nb; i < n - 1; i++ )
                        y [i] = x [i];
                for ( i = 0; i < n2 - 1; i++ )
                {
                        x [nb + 2 * i + 2] = y [nb + i + 1];
                        x [nb + 2 * i + 1] = y [n - 2 - i];
                }
        /* Reverse order of the permuted data sequence */
                if ( nb != 0 )
                        for ( i = nb, j = n - 2; i < j; i++, j-- )
                        {
                                tmp    = x [i];
                                x [i] = x [j];
                                x [j] = tmp;
                        }
                n2 <<= 1;
                nc++;
                nb -= (1 << nc);
        }
        /* Normalization of the transformed data sequence */
        if ( !flag )
        {
                scale = 2.0 / (double) n;
                for ( i = 0; i < n - 1; i++ )
                        x [i] *= scale;
        }
        return (0);
}
/*--------- End of the 1-D Fast DST-I module---------*/
```

Computer Program for the Fast DCT-IV/DST-IV Computation

```
/*-----------------------------------------------------------*
 *Module:      The 1-D Fast Discrete Cosine IV and          *
 *             Discrete Sine IV Transform                    *
 *                                                           *
 *Algorithm: The Forward and Inverse 1-D DCT-IV/DST-IV*
 *             Transform Computation                         *
 *                                                           *
 *Note that the DCT-IV and DST-IV matrices are              *
 *symmetric. Thus, the forward and inverse transforms       *
 *are the same except for the normalization.                *
 *-----------------------------------------------------------*/
```

```
/* --- Prototypes to be included in calling program --*/
int fdcstiv1d (
    double *x,    /* input/output vector of length 2**m */
    int m,        /* log_2 vector length
                    E.g. N = 256 --> m = 8            */
    int flag);    /* Forward or Inverse DCT-IV/DST-IV
                    computation:
                    flag =  1   Forward 1-D DCT-IV
                    flag = -1   Inverse 1-D DCT-IV
                    flag =  2   Forward 1-D DST-IV
                    flag = -2   Inverse 1-D DST-IV     */
/*   NOTE: Function returns into calling program
     following
           value:
            0 - successful processing
           -1 - invalid length of input vector
                invalid type transform computation    */
/* ------------------- Includes --------------------*/
#include         <math.h>
/* ------------------- Defines --------------------*/
#define   SIZE       1024       /* max length 1024 */
#define   LOG2SIZE   10         /* log_2 (SIZE)    */
#define   PI         3.141592653589793 /* pi       */
int dct_processor (double *, int, int, int);
/* --------------- Local Variables ---------------- */
static double y [SIZE]; /* working vector of length N */
static double as [SIZE/2];/* table of sine values      */
static double cc [SIZE/2];
                       /* table of cosine+sine values*/
static double ss [SIZE/2];
                       /* table of sine-cosine values*/
static int length;
/*-- Beginning of the 1-D Fast DCT-IV/DST-IV module --*/
int fdcstiv1d (double *x, int m, int flag)
{
    int     i,j,n,n2,n4;
    double arg,dev,argc,args,scale,tmp;
    /* Verification of the input vector length (SIZE) */
    if ( m < 1  ||  m > LOG2SIZE )
         return (-1);
  /* Verification of the type transform computation */
     if ( flag < -2 || flag == 0 || flag > 2 )
          return (-1);
  /* Initialize the input vector length and variables */
     n  = 1 << m;
     n2 = n >> 1;
     n4 = n >> 2;
  /* Generate tables of sines and cosines for rotation
     matrix R(N). Table is updated for new value of N */
     if ( length != n )
     {
         arg = PI / (double) (n << 2);
         dev = PI / (double) (n << 1);
         for ( i = 0; i < n2; i++, arg += dev )
         {
             argc    = cos (arg);
             args    = sin (arg);
             as [i]  = args;
```

```
            cc [i] = argc + args;
            ss [i] = args - argc;
        }
        length = n;
    }
/* Reordering of data sequence by permutation matrix
   P(N). For DST-IV computation odd-numbered samples
   are sign-changed */
    for ( i = 0; i < n2; i++ )
    {
        y [i] = x [2 * i];
        if ( flag == 1 || flag == -1 )
            y [n - 1 - i] =   x [2 * i + 1];
        else
            y [n - 1 - i] = - x [2 * i + 1];
    }
/* Butterflies corresponding to the matrix product
   P(N) B(N) */
    for ( i = 1; i < n2; i++ )
    {
        tmp        = y [n - i] - y [i];
        y [i     ] = y [n - i] + y [i];
        y [n - i]  = tmp;
    }
/* Get DCT-III transform of the first n/2 samples */
    dct_processor (&y [0],m-1,0,-1);
/* Reverse order of the last n/2 samples */
    for ( i = n2, j = n - 1; i < j; i++, j-- )
    {
        tmp   = y [i];
        y [i] = y [j];
        y [j] = tmp;
    }
/* Get the DST-III of the last n/2 samples */
    dct_processor (&y [n2],m-1,0,-2);
/* Reverse order of the last n/2 samples */
    for ( i = n2, j = n - 1; i < j; i++, j-- )
    {
        tmp   = y [i];
        y [i] = y [j];
        y [j] = tmp;
    }
/* Butterflies for the rotation matrix T(N) */
    for ( i = 0; i < n2; i++ )
    {
        tmp          = (y [i] - y [n - 1 - i]) * as [i];
        x [i]        = y [i] * cc [i] - tmp;
        x [n - 1 - i] = y [n - 1 - i] * ss [i] + tmp;
    }
/* DST-IV computation -
        reverse order of data sequence */
    if ( flag == 2 || flag == -2 )
        for ( i = 0, j = n - 1; i < j; i++, j-- )
        {
            tmp   = x [i];
            x [i] = x [j];
            x [j] = tmp;
        }
```

```
/* Normalization of the transformed data sequence */
  if ( flag > 0 )
  {
      scale = 2.0 / (double) n;
      for ( i = 0; i < n; i++ )
          x [i] *= scale;
  }
  return (0);
}
/*---- End of the 1-D Fast DCT-IV/DST-IV module ----*/
```

4.4   The 2-D DCT/DST Universal Computational Structure

Section 4.3 presented fast algorithms for 1-D computation of a given type of DCT/DST (I, II, III, IV) together with their implementations. For digital image processing applications, the fast 2-D algorithms are more significant than 1-D ones. For simplicity, in this section DCT and DST refer to types II and III only. The 2-D DCT and its inverse are used as the basic processing elements in international image/video coding standards [31].

Generally, there are two approaches to computation of the 2-D DCT: indirect and direct. In the indirect approach, the 2-D DCT computation can be realized via other 2-D discrete orthogonal transforms, such as the discrete Fourier transform or the Walsh–Hadamard transform [30]. There are two methods of direct approach which is based on direct 2-D DCT computation. The first, a so called row-column method, is based on the separability property of the 2-D DCT kernel, which sequentially uses any fast 1-D DCT algorithm on rows and columns of the input data matrix. The second is a vector radix method which uses a 2-D decomposition process. An algorithm obtained by this method outperforms the conventional row-column method in computational efficiency and works directly on 2-D data sets.

In this section, a generalized signal flow graph, the 2-D DCT/DST universal computational structure, is described. It represents a unified approach to the direct 2-D DCT and 2-D DST computation and their inverses for any square block of size $2^m \times 2^m$. The computer program implementing the direct 2-D DCT/DST is also presented.

4.4.1   The Fast Direct 2-D DCT/DST Computation

The 2-D DCT for an $N \times N$ input data matrix $\{x_{m,n}\}$, $m, n = 0, 1, \ldots, N - 1$ is defined by the following relation [30]

$$z_{k,l} = \frac{4\epsilon_k \epsilon_l}{N^2} \sum_{m=0}^{N-1} \sum_{n=0}^{N-1} x_{m,n} \cos\left[\frac{\pi(2m+1)k}{2N}\right] \cos\left[\frac{\pi(2n+1)l}{2N}\right], \quad (4.43)$$

$$k, l = 0, 1, \ldots, N - 1,$$

and the inverse 2-D DCT (2-D IDCT)

$$x_{m,n} = \sum_{k=0}^{N-1}\sum_{l=0}^{N-1} \epsilon_k \epsilon_l z_{k,l} \cos\left[\frac{\pi(2m+1)k}{2N}\right] \cos\left[\frac{\pi(2n+1)l}{2N}\right], \quad (4.44)$$

$$m, n = 0, 1, \ldots, N-1,$$

where

$$\epsilon_p = \begin{cases} \frac{1}{\sqrt{2}} & p = 0 \\ 1 & \text{otherwise} \end{cases}$$

and N is assumed to be an integer power of 2. The corresponding 2-D DST is defined by

$$s_{k,l} = \frac{4\epsilon_k \epsilon_l}{N^2} \sum_{m=0}^{N-1}\sum_{n=0}^{N-1} x_{m,n} \sin\left[\frac{\pi(2m+1)(k+1)}{2N}\right] \sin\left[\frac{\pi(2n+1)(l+1)}{2N}\right],$$

$$k, l = 0, 1, \ldots, N-1, \quad (4.45)$$

and the inverse 2-D DST (2-D IDST)

$$x_{m,n} = \sum_{k=0}^{N-1}\sum_{l=0}^{N-1} \epsilon_k \epsilon_l s_{k,l} \sin\left[\frac{\pi(2m+1)(k+1)}{2N}\right] \sin\left[\frac{\pi(2n+1)(l+1)}{2N}\right],$$

$$m, n = 0, 1, \ldots, N-1, \quad (4.46)$$

where

$$\epsilon_p = \begin{cases} \frac{1}{\sqrt{2}} & p = N-1, \\ 1 & \text{otherwise.} \end{cases}$$

The recursive 1-D DCT/DST algorithm and its corresponding generalized signal flow graph with regular structure for any value of $N = 2^m$ (1-D DCT/DST universal computational structure) enable the formulation by the vector radix method of direct 2-D DCT/DST fast, recursive algorithm that possesses a regular structure for any $N \times N$ block size. By extension of reordering Eq. (4.14) to a 2-D case, the 2-D DCT and 2-D IDCT defined by Eqs. (4.43) and (4.44), respectively, can be rewritten in the following form [30]

$$z_{k,l} = \frac{4\epsilon_k \epsilon_l}{N^2} \sum_{m=0}^{N-1}\sum_{n=0}^{N-1} \tilde{x}_{m,n} \cos\left[\frac{\pi(4m+1)k}{2N}\right] \cos\left[\frac{\pi(4n+1)l}{2N}\right], \quad (4.47)$$

$$k, l = 0, 1, \ldots, N-1,$$

$$\tilde{x}_{m,n} \;=\; \sum_{k=0}^{N-1}\sum_{l=0}^{N-1} \epsilon_k \epsilon_l z_{k,l} \cos\left[\frac{\pi(4m+1)k}{2N}\right] \cos\left[\frac{\pi(4n+1)l}{2N}\right], \quad (4.48)$$

$$m, n = 0, 1, \ldots, N-1 ,$$

where

$$\tilde{x}_{m,n} = x_{2m,2n}$$
$$\tilde{x}_{m,N-n-1} = x_{2m,2n+1}$$
$$\tilde{x}_{N-m-1,n} = x_{2m+1,2n} \tag{4.49}$$
$$\tilde{x}_{N-m-1,N-n-1} = x_{2m+1,2n+1}, \quad m, n = 0, 1, \ldots, \frac{N}{2} - 1 .$$

By reordering Eq. (4.49) an $N \times N$ input data matrix $\mathbf{X}$ is decomposed into four $\frac{N}{2} \times \frac{N}{2}$ submatrices, as even-even, even-odd, odd-even, and odd-odd indexed elements. After reordering the input data and output transform matrix, a fast recursive algorithm for direct $N \times N$ 2-D DCT/DST computation is given in matrix form as [28]

$$\begin{bmatrix} \hat{\mathbf{z}}_{ee} \\ \hat{\mathbf{z}}_{eo} \\ \hat{\mathbf{z}}_{oe} \\ \hat{\mathbf{z}}_{oo} \end{bmatrix} = (\hat{C}_N \otimes \hat{C}_N) \begin{bmatrix} \tilde{\mathbf{x}}_{pp} \\ \tilde{\mathbf{x}}_{pr} \\ \tilde{\mathbf{x}}_{rp} \\ \tilde{\mathbf{x}}_{rr} \end{bmatrix}, \tag{4.50}$$

where

$$\hat{\mathbf{z}}_e = (R \otimes R)\,\mathbf{z}_e, \quad \hat{\mathbf{z}}_e = \begin{bmatrix} \hat{\mathbf{z}}_{ee} \\ \hat{\mathbf{z}}_{eo} \end{bmatrix}, \quad \mathbf{z}_e = \begin{bmatrix} \mathbf{z}_{ee} \\ \mathbf{z}_{eo} \end{bmatrix},$$

$$\hat{\mathbf{z}}_o = (R \otimes R)\,\mathbf{z}_o, \quad \hat{\mathbf{z}}_o = \begin{bmatrix} \hat{\mathbf{z}}_{oe} \\ \hat{\mathbf{z}}_{oo} \end{bmatrix}, \quad \mathbf{z}_o = \begin{bmatrix} \mathbf{z}_{oe} \\ \mathbf{z}_{oo} \end{bmatrix},$$

$$\tilde{\mathbf{x}} = \begin{bmatrix} \tilde{\mathbf{x}}_{pp} \\ \tilde{\mathbf{x}}_{pr} \\ \tilde{\mathbf{x}}_{rp} \\ \tilde{\mathbf{x}}_{rr} \end{bmatrix} = (P_N \otimes P_N)\,\mathbf{x} .$$

$\otimes$ denotes the Kronecker matrix product. $\mathbf{z}_e$ and $\mathbf{z}_o$ are vectors consisting of transposed even and odd row vectors of the output transform matrix, both of which are arranged in the natural order, respectively. $\mathbf{x}$ denotes the vector consisting of transposed row vectors of the input data matrix. The direct product $R \otimes R$ performs 2-D bit reversal permutation, and $P_N \otimes P_N$ performs 2-D rearrangement defined by

Eq. (4.49). For clarity of Eq. (4.50), an example for $N = 4$ is shown

$$
\begin{bmatrix}
z_{00} \\
z_{02} \\
z_{01} \\
z_{03} \\
-- \\
z_{20} \\
z_{22} \\
z_{21} \\
z_{23} \\
-- \\
z_{10} \\
z_{12} \\
z_{11} \\
z_{13} \\
-- \\
z_{30} \\
z_{32} \\
z_{31} \\
z_{33}
\end{bmatrix}
= \left(\hat{C}_4 \otimes \hat{C}_4 \right)
\begin{bmatrix}
x_{00} \\
x_{02} \\
x_{03} \\
x_{01} \\
-- \\
x_{20} \\
x_{22} \\
x_{23} \\
x_{21} \\
-- \\
x_{30} \\
x_{32} \\
x_{33} \\
x_{31} \\
-- \\
x_{10} \\
x_{12} \\
x_{13} \\
x_{11}
\end{bmatrix} .
$$

Substituting the block matrix factorization of the DCT matrix $\hat{C}_N$ Eq. (4.16) into Eq. (4.50) and using properties of the Kronecker matrix product the direct, fast and recursive 2-D DCT/DST algorithm is developed [28]

$$
\hat{C}_N \otimes \hat{C}_N =
$$

$$
\left\{ \begin{bmatrix} I_{\frac{N}{2}} & 0 \\ 0 & K_{\frac{N}{2}} \end{bmatrix} \otimes \begin{bmatrix} I_{\frac{N}{2}} & 0 \\ 0 & K_{\frac{N}{2}} \end{bmatrix} \right\} \left\{ \begin{bmatrix} \hat{C}_{\frac{N}{2}} & 0 \\ 0 & \hat{C}_{\frac{N}{2}} \end{bmatrix} \otimes \begin{bmatrix} \hat{C}_{\frac{N}{2}} & 0 \\ 0 & \hat{C}_{\frac{N}{2}} \end{bmatrix} \right\}
$$

$$
\left\{ \begin{bmatrix} I_{\frac{N}{2}} & 0 \\ 0 & Q_{\frac{N}{2}} \end{bmatrix} \otimes \begin{bmatrix} I_{\frac{N}{2}} & 0 \\ 0 & Q_{\frac{N}{2}} \end{bmatrix} \right\} \left\{ \begin{bmatrix} I_{\frac{N}{2}} & I_{\frac{N}{2}} \\ I_{\frac{N}{2}} & -I_{\frac{N}{2}} \end{bmatrix} \otimes \begin{bmatrix} I_{\frac{N}{2}} & I_{\frac{N}{2}} \\ I_{\frac{N}{2}} & -I_{\frac{N}{2}} \end{bmatrix} \right\},
$$

$$
\tag{4.51}
$$

where $K_{\frac{N}{2}}$ and $Q_{\frac{N}{2}}$ are $\frac{N}{2} \times \frac{N}{2}$ matrices given by Eqs. (4.17) and (4.18), respectively. From Eq. (4.22) it follows that by this algorithm the direct 2-D DST computation can be realized merely by sign changes on the input data matrix (direct product $D_N \otimes D_N$) and after the 2-D DCT computation, reversing order along both rows and columns of the output transformed DCT data matrix (direct product $\bar{I}_N \otimes \bar{I}_N$).

The detailed analysis of the intrinsic structure of the algorithm given by Eqs. (4.50) and (4.51) results in a highly regular 2-D DCT/DST generalized signal flow graph, the 2-D DCT/DST universal computational structure, representing the unified direct 2-D DCT and 2-D DST computation and their inverses for any $N \times N$ block size [29]. It is shown for a 16×16 block in Fig. 4.6. The 2-D DCT/DST universal computational structure consists of two regular parts. The first part is related to

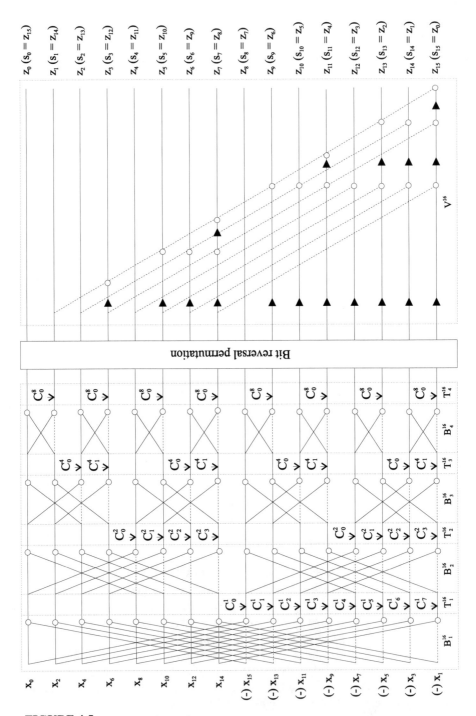

FIGURE 4.5
1-D DCT/DST universal computational structure for $N = 16$. ©Springer–Verlag
London Ltd.

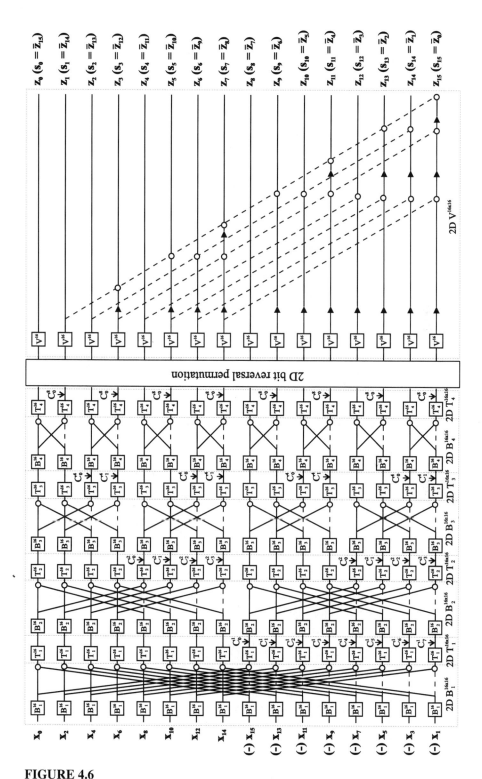

FIGURE 4.6
2-D DCT/DST universal computational structure for 16×16 **block size.**

the 2-D butterfly structure, and the second one, after the 2-D bit reversal permutation, is mapped into a 2-D pipeline structure. This 2-D pipeline structure can be represented by a regular computational scheme of the same type for any block size $2^m \times 2^m$ [29]. In order to show a one-to-one relationship between the 2-D DCT/DST universal computational structure and its 1-D counterpart, for a given $N \times N$ block size it is partitioned into blocks 2-D $B_i^{N \times N}$, 2-D $T_i^{N \times N}$, $i = 1, 2, \dots, \log_2 N$ related to the 2-D butterfly structure and the block 2-D $V^{N \times N}$ related to the 2-D pipeline structure. All blocks indicated by B_i^N, T_i^N, $i = 1, 2, \dots, \log_2 N$ and the block V^N are defined in the 1-D DCT/DST universal computational structure (Fig. 4.5). Heavy lines in Fig. 4.6 denote vector operations on rows of the input data matrix, $\mathbf{x}_i = [x_{i,0}, x_{i,2}, \dots, x_{i,N-2}, x_{i,N-1}, \dots, x_{i,3}, x_{i,1}]^T$ and $\mathbf{z}_i = [z_{i,0}, z_{i,1}, \dots, z_{i,N-2}, z_{i,N-1}]^T$ for $i = 0, 1, \dots, N - 1$. The symbols in brackets correspond to the 2-D DST computation and $\bar{\mathbf{z}} = \bar{I} \mathbf{z}$.

Recall that in the international image/video coding standards [31] the 2-D DCT and its inverse are defined for fixed 8×8 blocks as [43]

$$z_{k,l} = \frac{\epsilon_k \epsilon_l}{4} \sum_{m=0}^{7} \sum_{n=0}^{7} x_{m,n} \cos\left[\frac{\pi(2m+1)k}{16}\right] \cos\left[\frac{\pi(2n+1)l}{16}\right], \quad (4.52)$$

$$k, l = 0, 1, \dots, 7$$

$$x_{m,n} = \frac{1}{4} \sum_{k=0}^{7} \sum_{l=0}^{7} \epsilon_k \epsilon_l z_{k,l} \cos\left[\frac{\pi(2m+1)k}{16}\right] \cos\left[\frac{\pi(2n+1)l}{16}\right], \quad (4.53)$$

$$m, n = 0, 1, \dots, 7$$

The 2-D DCT given by Eq. (4.52) is identical to Eq. (4.43) for $N = 8$ except for a scaling factor of 4.

4.4.2   Implementation of the Direct 2-D DCT/DST Computation

The 2-D DCT/DST universal computational structure has been implemented in C. It can compute 2-D DCT or 2-D DST and their inverses for any square $2^m \times 2^m$, $m > 0$ block size. The cosine coefficients for a given $N = 2^m$ are precomputed and stored in tables. The tables are updated if the program calls for a new value of N. If a larger block size is required for 2-D DCT/DST computation, then macros SIZE and LOG2SIZE should be redefined in the program. In the implementation of the 2-D DCT/DST universal computational structure, the normalization is optional. All computations are performed in double precision.

The transposition of the input data matrix required in Eq. (4.50) and its reordering given by Eq. (4.49) can be realized simultaneously as follows:

$$\tilde{x}_{n,m} = x_{2m,2n}$$

$$\tilde{x}_{n,N-m-1} = x_{2m,2n+1}$$

$$\tilde{x}_{N-n-1,m} = x_{2m+1,2n} \quad\quad (4.54)$$

$$\tilde{x}_{N-n-1,N-m-1} = x_{2m+1,2n+1}, \quad m, n = 0, 1, \dots, \frac{N}{2} - 1.$$

```
/*-----------------------------------------------------------*
 * Module:     The 2-D Fast Discrete Cosine/Sine             *
 *             Transform (2-D DCT/DST Universal              *
 *             Computational Structure)                      *
 *                                                           *
 * Algorithm:  The Forward and Inverse 2-D DCT/DST           *
 *             computation by vector-radix structured        *
 *             approach for block sizes N x N, i.e.,         *
 *             square blocks. N is assumed to be an          *
 *             integer powers of 2.                          *
 *-----------------------------------------------------------*/
/* --- Prototypes to be included in calling program --*/
int fdcst2d (
   double **x,   /* input/output matrix of dimension NxN */
   int m,        /* m = log_2 (N) for N x N block size
                    e.g., length = 8 -> m = 3            */
   int norm,     /* norm  = 0 normalization is disabled
                    norm != 0 normalization is enabled   */
   int flag);    /* Forward or Inverse DCT/DST computation:
                    flag =  1  2-D DCT-II
                    flag = -1  2-D DCT-III
                    flag =  2  2-D DST-II
                    flag = -2  2-D DST-III
          ----------------------------------------------
   DECLARATION OF THE INPUT MATRIX: Let N = 8 --> then
   m = 3. Input matrix 8x8 must be declared in calling
   program as follows:
          double block [8*8]; /declarations
          double *x    [8];
          for ( i = 0; i < 8; i++ )
             x [i] = block + i * 8; /pointers to rows
                                     of the block
          fdcst2d (&x,3,1, 1); / DCT-II computation
          fdcst2d (&x,3,1,-1); /IDCT-II computation
          fdcst2d (&x,3,1, 2); / DST-II computation
          fdcst2d (&x,3,1,-2); /IDST-II computation
          ----------------------------------------------
   NOTE: Function returns into calling program following
         value:
         0 - successful processing
        -1 - invalid dimension of input matrix
        -2 - invalid transform type                      */
/* ------------------- Includes -------------------*/
#include        <math.h>
/* --------------------- Defines -------------------*/
#define    SIZE       32          /* max dimension 32x32 */
#define    LOG2SIZE   5           /* log_2 max dimension */
#define    PI         3.141592653589793 /* pi            */
#define    SQRT2      0.707106781186547 /* sqr (1/2)     */
#define    DCT        1
#define    IDCT      -1
/* ---------------- Local Variables ----------------*/
static double ac [SIZE*SIZE];     /* working array       */
static double *z [SIZE];          /* array of pointers   */
static int ntab_cs = 0;
static double tc1 [SIZE-1];
                          /* tables of cos coefficients*/
static double tc2 [SIZE*SIZE/3];
static int tab1_len = 0;
static int tab2_len = 0;
```

```
/*---- Beginning of the Fast 2-D DCT/DST module ----- */
int fdcst2d (double **x, int m, int norm, int flag)
{
    int     i,j,k,n,n1,n2,r,s,t,u,f0,f1,f2,f3,ip,ic,half;
    int     b1,b2;
    double  arg,fi1,fi2,scale,scl,tmp,*ptr,*z1,*z2;
    /* Verification of the input matrix dimension
       (SIZE x SIZE) */
    if ( m < 0  ||  m > LOG2SIZE )
        return (-1);
    /* Verification of the transform type computation */
    if ( flag < -2 || flag == 0 || flag > 2 )
        return (-2);
    /* Trivial transform if m = 0 */
    if ( m == 0 )
        return (0);
    /* Initialize input matrix dimension and variables */
    n  = 1 << m;
    n1 = n - 1;
    n2 = n >> 1;
    /* Initialize pointers on rows of the input matrix */
    for ( i = 0; i < n; i++ )
        z [i] = ac + i * n;
    /* Compute tables of cosine coefficients for new
       value of N */
    if ( ntab_cs != n )
    {
        b1      = b2 = tab1_len = tab2_len = 0;
        scale = 1.0 / (double) (n << 1);
        for ( s = ip = 1; s <= m; s++, ip <<= 1 )
        {
            ic  = n >> s;
            arg = (double) ip * PI * scale;
            for ( i = 0; i < ic; i++ )
            {
                fi1 = (double) (4 * i + 1) * arg;
                tc1 [b1+i] = cos (fi1);
                tab1_len++;
            }
            for ( i = u = 0; i < ic; i++, u = i * ic )
                for ( j = 0; j < ic; j++ )
                {
                    fi2 = (double) (4 * j + 1) * arg;
                    tc2 [b2+u+j] = tc1 [b1+i] * cos (fi2);
                    tab2_len++;
                }
            b1 += ic;
            b2 += ic * ic;
        }
        ntab_cs = n;
    }
    /* Test type of 2-D DCT/DST computation */
    if ( flag < 0 )
        goto inv;
/*
    ========================================================
    THE 2-D FAST FORWARD DISCRETE COSINE/SINE TRANSFORM
    ========================================================
*/
```

```
/* Reordering and transposition of input data matrix
---------------------------------------------------- */
for ( i = 0; i < n2; i++ )
     for ( j = 0; j < n2; j++ )
     {
          z [j     ] [i     ] = x [2*i  ] [2*j  ];
          z [n-j-1] [n-i-1] = x [2*i+1] [2*j+1];
          if ( flag == DCT )
          {
               z [n-j-1] [i     ] = x [2*i  ] [2*j+1];
               z [j     ] [n-i-1] = x [2*i+1] [2*j  ];
          }
          else
          {
               z [n-j-1] [i     ] = -x [2*i  ] [2*j+1];
               z [j     ] [n-i-1] = -x [2*i+1] [2*j  ];
          }
     }
/* Implementation of the 2-D butterfly structure
---------------------------------------------- */
for ( s = b1 = b2 = 0; s < m; s++ )
{
     half = n >> s;
     ip   = half >> 1;
/* Butterflies along rows of the data matrix */
   for ( i = 0, z1 = z [0]; i < n; i++, z1 = z [i] )
        for ( j = 0; j < ip; j++ )
             for ( k = j; k < n; k += half )
             {
                  tmp        = z1 [k] + z1 [k+ip];
                  z1 [k+ip] = z1 [k] - z1 [k+ip];
                  z1 [k]     = tmp;
             }
/* Butterflies between rows of the data matrix */
     for ( j = u = 0; j < ip; j++, u = j*ip )
          for ( k = j; k < n; k += half )
          {
               z1 = z [k];
               z2 = z [k+ip];
               for ( i = 0; i < n; i++ )
               {
                    tmp      = *z1 + *z2;
                    *z2++ = *z1 - *z2;
                    *z1++ = tmp;
               }
/* Multiplications by cosine coefficients */
               z1 = z [k];
               z2 = z [k+ip];
               for ( r = 0; r < ip; r++ )
                    for ( t = r; t < n; t += half )
                    {
                         z1 [t+ip] *= tc1 [b1+r];
                         z2 [t     ] *= tc1 [b1+j];
                         z2 [t+ip] *= tc2 [b2+u+r];
                    }
          }
     b1 += ip;
     b2 += ip * ip;
}
```

```
/* The 2-D bit reversal permutation
   ------------------------------ */
for ( t = 0, z1 = z [0]; t < n; t++, z1 = z [t] )
    for ( i = 1; i < n1; i++ )
    {
        for ( k = j = 0, r = i; k < m; k++ )
        {
            s = r >> 1;
            j = j + j + r - s - s;
            r = s;
        }
        if ( i < j )
        {
            tmp      = z1 [i];
            z1 [i] = z1 [j];
            z1 [j] = tmp;
        }
    }
for ( i = 1; i < n1; i++ )
{
    for ( k = j = 0, r = i; k < m; k++ )
    {
        s = r >> 1;
        j = j + j + r - s - s;
        r = s;
    }
    if ( i < j )
    {
        ptr    = z [i];
        z [i] = z [j];
        z [j] = ptr;
    }
}
/* Implementation of the 2-D pipeline structure
   --------------------------------------------- */
if ( m > 1 )
{
    /* Pipelines along rows of the data matrix */
    for ( i = 0; i < m - 1; i++ )
    {
        f0 = n / (1 << i);
        f1 = f0 >> 1;
        f2 = f1 >> 1;
        f3 = ((1 << i) - 1) << 1;
        z1 = z [0];
        for ( t = 0; t < n; t++, z1 = z [t] )
            for ( j = 1; j <= f2; j++ )
            {
                ip = f0 - j;
                ic = f1 - j;
                z1 [ip] += z1 [ip] - z1 [ic];
                k = 1;
                while ( k <= f3 )
                {
                    ip += f1;
                    ic += f1;
                    z1 [ip] += z1 [ip] - z1 [ic];
                    k++;
                }
            }
    }
```

```
/* Pipelines between rows of the data matrix */
        for ( j = 1; j <= f2; j++ )
            {
                ip = f0 - j;
                ic = f1 - j;
                z1 = z [ip];
                z2 = z [ic];
                for ( t = 0; t < n; t++, z1++ )
                    *z1 += *z1 - *z2++;
                k = 1;
                while ( k <= f3 )
                    {
                        ip += f1;
                        ic += f1;
                        z1 = z [ip];
                        z2 = z [ic];
                        for ( t = 0; t < n; t++, z1++ )
                            *z1 += *z1 - *z2++;
                        k++;
                    }
            }
    }
/*-----------------------------------------------------------
    The normalization of the transformed data sequence.
    If DCT-II/DST-II is required, then parameter
    norm != 0. The block is not used for other discrete
    sinusoidal transforms computation. Then norm = 0.
    -----------------------------------------------------------*/
    if ( norm )
        {
            scale = 4.0 / ((double) n * (double) n);
            for ( i = 0, z [0] [0] *= SQRT2; i < n; i++ )
                for ( j = 0; j < n; j++ )
                    {
                        z [i] [j] *= scale;
                        if ( i == 0 || j == 0 )
                            z [i] [j] *= SQRT2;
                    }
        }
/* Reverse rows and columns of the transformed data
    matrix for the DST
    ------------------------------------------------- */
    for ( i = 0; i < n2; i++ )
        for ( j = 0; j < n; j++ )
            if ( flag == DCT )
                {
                    x [j     ] [i     ] = z [i     ] [j     ];
                    x [n-1-j] [n-1-i] = z [n-1-i] [n-1-j];
                }
            else
                {
                    x [j     ] [i     ] = z [n-1-i] [n-1-j];
                    x [n-1-j] [n-1-i] = z [i     ] [j     ];
                }
    return (0);
/*
    ===========================================================
    THE 2-D FAST INVERSE DISCRETE COSINE/SINE TRANSFORM
    ===========================================================
```

```
*/
inv:
    /* Reverse rows and columns of the transformed data
       matrix for the IDST
       --------------------------------------------------   */
    for ( i = 0; i < n2; i++ )
        for ( j = 0; j < n; j++ )
            if ( flag == IDCT )
            {
                z [j     ] [i     ] = x [i     ] [j     ];
                z [n-1-j] [n-1-i] = x [n-1-i] [n-1-j];
            }
            else
            {
                z [j     ] [i     ] = x [n-1-i] [n-1-j];
                z [n-1-j] [n-1-i] = x [i     ] [j     ];
            }
/* -----------------------------------------------------------
   The normalization of the DC term. If DCT-III/DST-III
   is required, then parameter norm != 0. The block is
   not used for other discrete sinusoidal transforms
   computation. Then norm = 0.
   -------------------------------------------------------*/
    if ( norm )
        for ( i = 0, z [0] [0] *= SQRT2; i < n; i++ )
            for ( j = 0; j < n; j++ )
                if ( i == 0 || j == 0 )
                    z [i] [j] *= SQRT2;
    /* Implementation of the 2-D pipeline structure
       ------------------------------------------------   */
    if ( m > 1 )
    {
    /* Pipelines between rows of the data matrix */
        for ( i = m - 2; i >= 0; i-- )
        {
            f0 = n / (1 << i);
            f1 = f0 >> 1;
            f2 = f1 >> 1;
            f3 = ((1 << i) - 1) << 1;
            for ( j = f2; j > 0; j-- )
            {
                k = f3;
                u = k * f1;
                ip = f0 - j + u;
                ic = f1 - j + u;
                z1 = z [ip];
                z2 = z [ic];
                for ( t = 0; t < n; t++, z2++ )
                {
                    *z2 -= *z1;
                    *z1 += *z1++;
                }
                while ( k > 0 )
                {
                    k--;
                    ip -= f1;
                    ic -= f1;
                    z1 = z [ip];
                    z2 = z [ic];
                    for ( t = 0; t < n; t++, z2++ )
```

```
                    {
                         *z2  -=  *z1;
                         *z1  +=  *z1++;
                    }
             }
       }
/* Pipelines along rows of the data matrix */
       z1 = z [0];
       for ( t = 0; t < n; t++, z1 = z [t] )
             for ( j = f2; j > 0; j-- )
                   {
                        k = f3;
                        u = k * f1;
                        ip = f0 - j + u;
                        ic = f1 - j + u;
                        z1 [ic] -= z1 [ip];
                        z1 [ip] += z1 [ip];
                        while ( k > 0 )
                              {
                                   k--;
                                   ip -= f1;
                                   ic -= f1;
                                   z1 [ic] -= z1 [ip];
                                   z1 [ip] += z1 [ip];
                              }
                   }
       }
}
/* The 2-D bit reversal permutation
   -------------------------------- */
for ( t = 0, z1 = z [0]; t < n; t++, z1 = z [t] )
     for ( i = 1; i < n1; i++ )
          {
               for ( k = j = 0, r = i; k < m; k++ )
                    {
                         s = r >> 1;
                         j = j + j + r - s - s;
                         r = s;
                    }
               if ( i < j )
                    {
                         tmp     = z1 [i];
                         z1 [i] = z1 [j];
                         z1 [j] = tmp;
                    }
          }
for ( i = 1; i < n1; i++ )
{
     for ( k = j = 0, r = i; k < m; k++ )
          {
               s = r >> 1;
               j = j + j + r - s - s;
               r = s;
          }
     if ( i < j )
          {
               ptr     = z [i];
               z [i] = z [j];
               z [j] = ptr;
          }
}
```

```
        }
/* Implementation of the 2-D Butterfly structure
   --------------------------------------------- */
    b1 = tab1_len;
    b2 = tab2_len;
    for ( s = 0; s < m; s++ )
    {
        half = 1 << (s + 1);
        ip   = half >> 1;
        b1   -= ip;
        b2   -= ip * ip;
/* Multiplications by cosine coefficients */
        for ( j = u = 0; j < ip; j++, u = j*ip )
            for ( k = j; k < n; k += half )
            {
                z1 = z [k];
                z2 = z [k+ip];
                for ( r = 0; r < ip; r++ )
                    for ( t = r; t < n; t += half )
                    {
                        z1 [t+ip] *= tc1 [b1+r];
                        z2 [t   ] *= tc1 [b1+j];
                        z2 [t+ip] *= tc2 [b2+u+r];
                    }
            }
/* Butterflies between rows of the data matrix */
                z1 = z [k];
                z2 = z [k+ip];
                for ( i = 0; i < n; i++ )
                {
                    tmp     = *z2;
                    *z2++ = *z1 - tmp;
                    *z1++ = *z1 + tmp;
                }
            }
/* Butterflies along rows of the data matrix */
            z1 = z [0];
            for ( i = 0; i < n; i++, z1 = z [i] )
                for ( j = 0; j < ip; j++ )
                    for ( k = j; k < n; k += half )
                    {
                        tmp       = z1 [k+ip];
                        z1 [k+ip] = z1 [k] - tmp;
                        z1 [k   ] = z1 [k] + tmp;
                    }
    }
/* Reordering and transposition of DCT/DST output
   data matrix
   --------------------------------------------- */
    for ( i = 0; i < n2; i++ )
        for ( j = 0; j < n2; j++ )
        {
            x [2*i  ] [2*j  ] = z [j    ] [i    ];
            x [2*i+1] [2*j+1] = z [n-j-1] [n-i-1];
            if ( flag == IDCT )
            {
                x [2*i  ] [2*j+1] = z [n-j-1] [i    ];
                x [2*i+1] [2*j  ] = z [j    ] [n-i-1];
            }
            else
```

```
      {
      x [2*i   ] [2*j+1] = -z [n-j-1] [i     ];
      x [2*i+1] [2*j   ] = -z [j      ] [n-i-1];
      }
   }
   return (0);
}
/*-------- End of Fast 2-D DCT/DST module ---------- */
```

4.5   DCT and Data Compression

The amount of information in its many forms (images, text, speech, video, audio, etc.) that is handled is increasing at a phenomenal rate. As a result, the ability to access, store, and transmit information in an efficient manner has become crucial, particularly in the case of digital images. Although representing images in digital form allows visual information to be easily manipulated in useful and novel ways, there is one potential problem with digital images — the large number of bits required to represent even a single digital image directly. In order to utilize digital images effectively, specific techniques are needed to reduce the number of bits required for their representation. Fortunately, digital images in their canonical representation generally contain a significant amount of redundancy (spatial, spectral, or temporal redundancy). Image data compression (the art/science of efficient coding of the picture data) aims at taking advantage of this redundancy to reduce the number of bits required to represent an image. This can result in significantly reducing the memory needed for image storage and channel capacity for image transmission [36].

The need for image compression becomes apparent when we compute the number of bits per image resulting from typical sampling and quantization schemes. We consider the amount of storage for the "Lena" digital image shown in Fig. 4.7. The monochrome (grayscale) version of this image with a resolution $512 \times 512 \times 8$ bits/pixel requires a total of $2,097,152$ bits, or equivalently $262,144$ bytes. The color version of the same image in RGB format (red, green, and blue color bands) with a resolution of 8 bits/color requires a total of $6,291,456$ bits, or $786,432$ bytes. Such an image should be compressed for efficient storage or transmission.

Image compression methods can be classified into two fundamental groups: lossless and lossy [34, 36, 37]. In lossless compression, the reconstructed image after compression is identical to the original image. However, only a modest amount of compression is possible; typically 1:2 or 1:3 compression ratios are achieved. In lossy compression, the reconstructed image contains degradations relative to the original. Generally, more compression is obtained at the expense of more distortion. As a result, much higher compression can be achieved by lossy techniques than by lossless techniques. The most used lossy compression technique is transform coding [32]. A general transform coding scheme involves subdividing an $N \times N$ image into smaller nonoverlapping $n \times n$ sub-image blocks and performing a unitary transform on each

FIGURE 4.7
Monochrome $512 \times 512 \times 8$ **bits/pixel "Lena" digital image. Reproduced by Special Permission of** *Playboy* **magazine. Copyright ©1972, 2000 by Playboy.**

block. The transform operation itself does not achieve any compression. It aims at decorrelating the original data and compacting a large fraction of the signal energy into a relatively small set of transform coefficients (energy packing property). In this way, many coefficients can be discarded after quantization and prior to encoding.

Most practical transform coding systems are based on DCT of types II and III, which provides good compromise between energy packing ability and computational complexity. The energy packing property of DCT is superior to that of any other unitary transform. Transforms that redistribute or pack the most information into the fewest coefficients provide the best sub-image approximations and, consequently, the smallest reconstruction errors. DCT basis images are fixed (image independent) as opposed to the optimal KLT which is data dependent. Moreover, when compared to the other image independent transforms, DCT has the advantages of having been implemented in a single integrated circuit [30] and minimizing the blocklike appearance (blocking artifact) that results when the boundaries between sub-image blocks become visible. This last property is particularly important in comparison with the other sinusoidal transforms [34]. Important properties of DCT have proved to be of practical value, and, therefore, it has become the basic processing unit for data compression in the international image/video coding standards [30, 31, 39, 40, 41, 42].

4.5.1   DCT-Based Image Compression/Decompression

For the purposes of using DCT in real data compression applications, we have selected the JPEG DCT-based image compression and decompression technique. There are several reasons for this selection. JPEG is the first established/emerging international digital compression standard for continuous-tone (multilevel) still images, both monochrome and color [31, 43, 44]. It has been recently recognized as the most popular, simple, and efficient transform coding technique that yields a satisfactory solution to most of the practical image coding problems. Furthermore, the JPEG standard played a considerable role in the development of other international video coding standards. From the methodological viewpoint, the JPEG standard enables one to simply illustrate the compression capability of DCT. Finally, the JPEG DCT-based coding approach is the basis of hybrid intraframe/interframe MC (motion compensated)/DPCM (differential pulse code modulation)/DCT coding scheme used in the international video coding standards: H.261 video coder, MPEG-1 audiovisual coder for digital storage media, MPEG-2/H.262 digital video coder, MPEG-4 and H.263 coders for very low-bit rate video coding, digital HDTV standards, and the CMTT.723 digital broadcasting standard for transmission of television signals [31].

The JPEG standard specifies the basic encoding and decoding operations by means of specific functions and defines the syntax and semantics of encoded bit stream [31, 43, 44]. Detailed requirements such as file format, spatial resolution, and color space are not defined by the standard. It is only necessary that the encoding processes comply with the functions defined by the standard and they produce the valid bit stream. Thus, there is freedom and flexibility in the actual design and development of the JPEG compression and decompression system.

The JPEG standard has four main processing modes: sequential, progressive, lossless, and hierarchical. The sequential mode provides the variability of coding operations from a baseline system to an extended one. For simplicity, we consider the JPEG sequential baseline system. The extended system allows the baseline system to satisfy a broader range of applications. Input and output data precision in the baseline system is limited to 8 bits. RGB color images prior to compression are converted into a monochrome compatible luminance component and two chrominance components. The luminance component contains the shades of gray and is a monochrome image. Two chrominance components together contain the color information. Encoding/decoding operations in the JPEG baseline system are performed for luminance and chrominance components.

All compression systems consist of two distinct structural blocks: an encoder and a decoder. An input image is fed into the encoder, which creates encoded compressed representation of the input data. After transmission over the channel, the encoded representation is fed into the decoder, where the reconstructed output image is generated.

The block diagram of the encoder and decoder for JPEG DCT-based image compression and decompression is shown in Fig. 4.8. For processing the luminance component of an image the algorithm generally consists of the following steps [31, 34, 36, 43, 47]:

- The source image is partitioned into nonoverlapping $n \times n$ pixel blocks which are processed sequentially in a raster scan fashion, left to right and top to bottom. The JPEG standard uses the fixed block size 8×8. Each block is first level shifted and transformed using DCT. In principle, DCT introduces no loss to the source samples, it merely transforms them to a domain in which they can be more efficiently encoded.

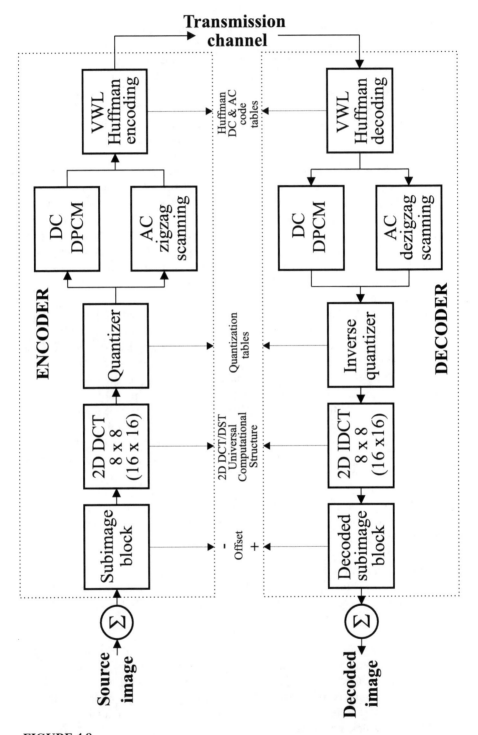

FIGURE 4.8
Block diagram of encoder and decoder for JPEG DCT-based image compression and decompression.

- The 2-D DCT array of coefficients is uniformly quantized. The top left coefficient in the 2-D DCT array with zero frequency in both dimensions is referred to as the *DC coefficient*, and it is proportional to the average brightness of the spatial block. The remaining coefficients are called the *AC coefficients*. Prior to quantization, transform coefficients can be weighted according to their visual importance using HVS (Human Visual System) sensitivity models [47, 48].

- The quantization of the AC coefficients produces many zeros, especially at the higher frequencies. To take advantage of these zeros, the 2-D DCT array of quantized coefficients is reordered using a zigzag pattern [see Fig. 4.9(a)] to form a 1-D sequence. This rearranges the coefficients in approximately decreasing order of their average energy (as well as in order of increasing spatial frequency) with the aim of creating large runs of zero values. The quantization is a key operation because the combination of the quantization and runlength coding contributes to most of the compression.

- The final processing step at the encoder is entropy coding. This step achieves additional compression losslessly by encoding the quantized coefficients more compactly based on their statistical characteristics. The quantized DCT coefficients are variable-length coded using two global different predetermined Huffman coding tables, one for DC and one for AC coefficients.

At the decoder, after the encoded bit stream is Huffman decoded and the 2-D array of quantized DCT coefficients is recovered and dezigzag reordered, each coefficient is inverse quantized. The resulting array is transformed by inverse 2-D DCT and inverse level shifted to yield an approximation of the original sub-image block. The same quantization table and Huffman coding tables are used in both the encoder and decoder.

Each chrominance component of a color image is processed and encoded independently in the same way as the luminance component, except that it is downsampled by a factor of two or four in both horizontal and vertical directions prior to DCT operation. At the decoder, the reconstructed chrominance component is bilinearly interpolated to the original size.

The following sections describe the JPEG DCT-based image compression and decompression system. The description is restricted to one sub-image block only because the same encoding and decoding operations are performed on each block. Although required algorithms in the JPEG standard are based on fixed block size (8×8), the system described in this chapter can use larger blocks. In fact, the 2-D DCT/DST universal computational structure offers the flexibility of computing the 2-D DCT and its inverse for any $2^m \times 2^m$ block size. The encoding and decoding operations are described in detail followed by an implementation in C. Where necessary, the input and output data samples are provided; they can be useful for verification of the correctness of a given program module. Low-level routines — setting quantization table, computation of Huffman coding/decoding tables, Huffman encoding and Huffman decoding — are based on shareware generated by Independent JPEG group (Thomas G. Lane) [49]. Program modules together provide the simple, efficient, and

low-cost image compression and decompression system which the reader can use in
his or her own data compression applications.

4.5.2 Data Structures for Compression/Decompression

One of the most important aspects of image/video coding standards is to define
data structures so that a decoder can decode the received bit stream efficiently and
without any ambiguity. This section shows header files that contain definitions and
declarations of data structures for an image compression and decompression system.

The header file JPEGDEF.H contains macro definitions and the definition of data
structure for the Huffman coding/decoding table.

```
/*
   JPEGDEF.H
*/
#define  SIZE        16  /* max dimension of the block */
#define  I_LEVEL     256 /* the number of gray levels  */
#define  DCT         1   /* 2-D DCT computation         */
#define  DISABLE_NORM 0 /* disable DCT normalization   */
#define  SQRT2       0.707106781186547 /* sqrt (2)     */
#define  LOOKAHEAD   8   /* # of bits of lookahead      */
#define  MIN_GET_BITS 15 /* minimum allowable value     */
/* ---------------------------------------------------- */
/*          Huffman coding and decoding table           */
/* ----------------------------------------------------*/
struct huff_table {
/* bits [k] = # of symbols with codes of length k bits,
    bits [0] is unused                                  */
  unsigned char bits [17];
  /* Symbols in order of incremental code length        */
  unsigned char hufval [256];
  /*              ENCODING TABLES                        */
  unsigned int hufcode [256]; /* code for each symbol   */
  char          hufsize [256]; /* and its length        */
  /*              DECODING TABLES                        */
  /* Basic tables: element [0] of each array is unused  */
  long int mincode [17]; /* smallest code of length k   */
  long int maxcode [18];
                      /* and largest code (-1 if none)  */
    /* Index of 1st symbol of length k                  */
    int valptr [17];
/* Lookahead tables: indexed by the next
  LOOKAHEAD bits of the input data stream. If the next
  Huffman code is no more than LOOKAHEAD bits long, it
  can be obtained its length and the corresponding
  symbol directly from these tables                     */
  int look_nbits [1<<LOOKAHEAD];
   /* # bits,or 0 if too long */
  unsigned int look_sym [1<<LOOKAHEAD];
   /* symbol,or unused */
};
```

The header file JPEGDATA.H contains declarations of variables and arrays for the
image compression and decompression system. Declarations for JPEG luminance
sample quantization table, zigzag, and dezigzag scanning patterns are shown for 8×8

block size only. For larger block sizes, the user must specify the corresponding arrays for a given block size. The JPEG DCT-based image compression and decompression system has two optional parameters: the block size and a quality factor for scaling the quantization table.

```
/*
    JPEGDATA.H
*/
unsigned char out_buffer [256];
            /* output bit stream buffer                    */
int bytes_in_buf;
            /* and # of bytes in it                        */
int encode_bits;
            /* # of bits for compressed block              */
int exp_val;            /* log2 value of block size        */
int blk_size;           /* block size                      */
int center_samp;        /* center sample value             */
int tdc_last;           /* the last DC value for encoder   */
int rdc_last;           /* the last DC value for decoder   */
int q_factor;           /* quality factor                  */
long int total_bits;
            /* total # of bits for original data           */
long int total_bytes;
            /* total # of bytes for original data          */
long int cmprs_bits;
            /* total # of bits for compressed data         */
long int cmprs_bytes;
            /* total # of bytes for compressed data        */
double dct_block [SIZE*SIZE];
            /* 2-D DCT block of coefficients               */
double *dctptr  [SIZE];        /* pointers to its rows     */
double scaling; /* scale factor for DCT normalization      */
double bit_rate; /* the # of bits per pixel (bpp)          */
double cmprs_ratio;     /* compression ratio               */
/* # of symbols with codes of length k bits
    (lumbits [k]) and symbols in order of incremental
    code length (lumval [k]) for DC luminance
    values - valid for 8-bit data precision                */
unsigned char dc_lumbits [17] =
    {0,0,1,5,1,1,1,1,1,1,0,0,0,0,0,0,0};
unsigned int  dc_lumval [12]
    = {0,1,2,3,4,5,6,7,8,9,10,11};
/* # of symbols with codes of length k bits
    (lumbits [k]) and symbols in order of incremental
    code length (lumval [k]) for AC luminance
    values - valid for 8-bit data precision                */
unsigned char ac_lumbits [17] =
    {0,0,2,1,3,3,2,4,3,5,5,4,4,0,0,1,0x7d};
unsigned char ac_lumval [162] =
    { 0x01, 0x02, 0x03, 0x00, 0x04, 0x11, 0x05, 0x12,
      0x21, 0x31, 0x41, 0x06, 0x13, 0x51, 0x61, 0x07,
      0x22, 0x71, 0x14, 0x32, 0x81, 0x91, 0xa1, 0x08,
      0x23, 0x42, 0xb1, 0xc1, 0x15, 0x52, 0xd1, 0xf0,
      0x24, 0x33, 0x62, 0x72, 0x82, 0x09, 0x0a, 0x16,
      0x17, 0x18, 0x19, 0x1a, 0x25, 0x26, 0x27, 0x28,
      0x29, 0x2a, 0x34, 0x35, 0x36, 0x37, 0x38, 0x39,
      0x3a, 0x43, 0x44, 0x45, 0x46, 0x47, 0x48, 0x49,
      0x4a, 0x53, 0x54, 0x55, 0x56, 0x57, 0x58, 0x59,
```

```
      0x5a, 0x63, 0x64, 0x65, 0x66, 0x67, 0x68, 0x69,
      0x6a, 0x73, 0x74, 0x75, 0x76, 0x77, 0x78, 0x79,
      0x7a, 0x83, 0x84, 0x85, 0x86, 0x87, 0x88, 0x89,
      0x8a, 0x92, 0x93, 0x94, 0x95, 0x96, 0x97, 0x98,
      0x99, 0x9a, 0xa2, 0xa3, 0xa4, 0xa5, 0xa6, 0xa7,
      0xa8, 0xa9, 0xaa, 0xb2, 0xb3, 0xb4, 0xb5, 0xb6,
      0xb7, 0xb8, 0xb9, 0xba, 0xc2, 0xc3, 0xc4, 0xc5,
      0xc6, 0xc7, 0xc8, 0xc9, 0xca, 0xd2, 0xd3, 0xd4,
      0xd5, 0xd6, 0xd7, 0xd8, 0xd9, 0xda, 0xe1, 0xe2,
      0xe3, 0xe4, 0xe5, 0xe6, 0xe7, 0xe8, 0xe9, 0xea,
      0xf1, 0xf2, 0xf3, 0xf4, 0xf5, 0xf6, 0xf7, 0xf8,
      0xf9, 0xfa };
struct huff_table dc_table;   /* Huffman DC code table */
struct huff_table ac_table;   /* Huffman AC code table */
/* luminance sample quantization table for 8 x 8 DCT   */
int qbase8_tbl [8*8] =
    {  16,  11,  10,  16,  24,  40,  51,  61,
       12,  12,  14,  19,  26,  58,  60,  55,
       14,  13,  16,  24,  40,  57,  69,  56,
       14,  17,  22,  29,  51,  87,  80,  62,
       18,  22,  37,  56,  68, 109, 103,  77,
       24,  35,  59,  64,  81, 104, 113,  92,
       49,  64,  78,  87, 103, 121, 120, 101,
       72,  92,  95,  98, 112, 100, 103,  99 };
/* zigzag scanning pattern for an 8 x 8 DCT transform */
int zag8 [8*8] =
    {   0,   1,   5,   6,  14,  15,  27,  28,
        2,   4,   7,  13,  16,  26,  29,  42,
        3,   8,  12,  17,  25,  30,  41,  43,
        9,  11,  18,  24,  31,  40,  44,  53,
       10,  19,  23,  32,  39,  45,  52,  54,
       20,  22,  33,  38,  46,  51,  55,  60,
       21,  34,  37,  47,  50,  56,  59,  61,
       35,  36,  48,  49,  57,  58,  62,  63 };
/* dezigzag scanning pattern for an
   8 x 8 DCT transform                                 */
int dezag8 [8*8] =
    {   0,   1,   8,  16,   9,   2,   3,  10,
       17,  24,  32,  25,  18,  11,   4,   5,
       12,  19,  26,  33,  40,  48,  41,  34,
       27,  20,  13,   6,   7,  14,  21,  28,
       35,  42,  49,  56,  57,  50,  43,  36,
       29,  22,  15,  23,  30,  37,  44,  51,
       58,  59,  52,  45,  38,  31,  39,  46,
       53,  60,  61,  54,  47,  55,  62,  63 };
```

4.5.3   Setting the Quantization Table

JPEG gives simple and easy quantization methods and suggests informative tables for DC and AC coefficients [31]. One such informative quantization table for the luminance component is shown in the header file JPEGDATA.H. Although default quantization tables are provided by the JPEG standard for both luminance and chrominance processing, the user is free to design custom tables which can be adapted to the characteristics of the image to be compressed.

The quantization of the DCT coefficients is based on properties of the HVS which tolerates more quantization errors at higher frequencies than at lower frequencies. It means that the transform coefficients have different visual sensitivities; visual perception is less sensitive to the high frequency coefficients and more sensitive to low frequency coefficients. Thus, the weighting factors are selected to produce coarser quantization of high frequency coefficients and finer quantization of the low frequency coefficients.

The quantization table can be scaled to provide a variety of compression levels. JPEG specifies the following possible bit rates and quality rates [31]:

$0.25 \sim 0.50$ bpp: moderate to good quality
$0.50 \sim 0.75$ bpp: good to very good quality
$0.75 \sim 1.50$ bpp: excellent images
$1.50 \sim 2.00$ bpp: indistinguishable images (visually lossless)

The quantization table in the JPEG DCT-based image compression and decompression system is scaled according to a specified quality factor. The quality factor takes values in the range 0–100 (given as percentage) with the scaling value of 50 corresponding to the basic quantization table. The value of 100 will cause elements of the quantization table to be equal to 1 for an 8×8 block size and to equal to 2 for a 16×16 block size. The elements of the quantization table are in the range from 1 to 255.

The following program sets the user quantization table according to the specified quality factor.

```
/*
-----------------------------------------------------------
SET USER QUANTIZATION TABLE ACCORDING TO DEFINED
'QUALITY'
Set a quantization table equal to the basic table times
a scale factor (given as a percentage). The basic table
is used as-is (scaling 100) for a quality of 50. Values
of the basic table produce "good" quality, and when
divided by 2, "very good" quality. These two settings
are selected by quality = 50 and quality = 75,
respectively. Qualities 50 ... 100 are converted to
scaling percentage 200 - 2*Q. Note that at Q = 100 the
scaling is 0, which will cause qnt_tbl to make all the
table entries 1 (no quantization loss).
-----------------------------------------------------------
*/
#include    "jpegdef.h"
void set_qtable (
     int   *qnt_tbl,    /* user quantization table */
     int   blksize,     /* block size */
     int   *qbase_tbl,  /* basic quantization table */
     int   quality )    /* quality factor */
{
     int       i;
     long int temp;
/*   Safety limit on quality factor (convert 0 to 1 to
     avoid zero divide) */
     if ( quality <= 0 )
         quality = 1;
     else
         if ( quality > 100 )
```

```
                    quality = 100;
/*   Convert a user-specified quality rating 0-100 to a
     percentage scaling factor. Qualities 1 ... 50 are
     converted to scaling percentage 5000/Q */
     if ( quality < 50 )
        quality = 5000 / quality;
     else
        quality = 200 - quality * 2;
/*   Set quantization table equal to the qbasic_tbl
     times a scale factor. Limit the values to the
     valid range   */
     for ( i = 0; i < blksize * blksize; i++ )
     {
     temp = ((long int) qbase_tbl [i]
        * quality + 50L) / 100L;
        if ( temp <= 0L )
        {
            temp = 1L;
            if ( blksize == SIZE )
                temp = 2L;
        }
        if ( temp > 255L )
            temp = 255L;
        qnt_tbl [i] = (int) temp;
     }
}
```

4.5.4   Standard Huffman Coding/Decoding Tables

The JPEG baseline system uses only the Huffman coding method for encoding the quantized DCT coefficients, and it suggests standard Huffman coding tables for the luminance and chrominance DCT coefficients, two DC and two AC Huffman coding tables [31].

Based on data structures defined in the header file JPEGDATA.H for DC and AC luminance values (structures specifying the number of symbols with codes of length k bits and code symbols), the following program generates standard Huffman coding/decoding tables. The program must be called separately for the DC and AC coding tables (see Section 4.5.7). These DC and AC Huffman coding/decoding tables are valid for 8-bit data precision and can be found in Rao and Hwang [31].

```
/*-------------------------------------------------------
            COMPUTE HUFFMAN CODING AND DECODING TABLES
*/
#include     <string.h>
#include     "jpegdef.h"
void fix_huftbl (
    struct huff_table *htbl ) /* Huffman code table */
{
    int          p,i,j,k,lastp,size,lookbits;
    char         huffsize [257];
    unsigned int huffcode [257],code;
/*   Make table of Huffman code length for each symbol
     in code-length order */
     for ( k = 1, p = 0; k <= 16; k++ )
```

```
            for ( i = 1; i <= (int) htbl->bits [k]; i++ )
                huffsize [p++] = (char) k;
        huffsize [p] = 0;
        lastp        = p;
/* Generate the codes themselves in code-length order */
        code = p = 0;
        size = huffsize [0];
        while ( huffsize [p] )
        {
            while ( ((int) huffsize [p]) == size )
            {
                huffcode [p++] = code;
                code++;
            }
            code <<= 1;
            size++;
        }
/*  Generate encoding tables. These are code and size
    indexed by symbol value. Set any codeless symbols
    to have code length 0. This allows emit_bits () to
    detect any attempt to emit such symbols */
        memset (htbl->hufsize,0,sizeof (htbl->hufsize));
        for ( p = 0; p < lastp; p++ )
        {
            htbl->hufcode [htbl->hufval [p]] = huffcode [p];
            htbl->hufsize [htbl->hufval [p]] = huffsize [p];
        }
/* Generate decoding tables for bit-sequential
   decoding */
        for ( k = 1, p = 0; k <= 16; k++ )
            if ( htbl->bits [k] )
            {
            htbl->valptr   [k] = p;
            htbl->mincode  [k] = huffcode [p];   /* min code */
            p                 += htbl->bits [k];
            htbl->maxcode  [k] = huffcode [p-1];/* max code */
            }
            else
                htbl->maxcode [k] = -1;          /* -1 if no codes */
/* Ensures that huff_decode () terminates */
        htbl->maxcode [17] = 0xFFFFFL;
/* Compute lookahead tables to speed up decoding.
   First set all the table entries to 0, indicating
   "too long"; then iterate through the Huffman codes
   that are short enough and fill in all the entries
   that correspond to bit sequences starting with that
   code; k = current code's length, p = its index in
   hufcode [] & hufval []. Generate left-justified code
   followed by all possible bit sequences */
        memset (htbl->look_nbits,0,sizeof (htbl->look_nbits));
        for ( k = 1, p = 0; k <= LOOKAHEAD; k++ )
            for ( i = 1; i <= (int) htbl->bits [k]; i++, p++ )
            {
                lookbits = huffcode [p] << (LOOKAHEAD - k);
                for ( j = 1 << (LOOKAHEAD - k); j > 0; j-- )
                {
                htbl->look_nbits [lookbits] = k;
                htbl->look_sym   [lookbits] = htbl->hufval [p];
```

```
      lookbits++;
    }
  }
}
```

4.5.5 Compression of One Sub-Image Block

Having defined and prepared all required data structures, we can concentrate on the image compression process. For simplicity, we consider the compression of one sub-image block because the same operations are performed for each extracted block from the source image. For processing the luminance component of the image, the following steps are performed at the encoder for each block.

1. The data in the block is first level shifted by subtracting the quantity 2^{p-1}, where 2^p is the maximum number of gray levels and p is the precision parameter of the image intensity in bits. In the JPEG baseline system, $p = 8$ and the level shift is 128.

2. The level-shifted block is transformed by the forward 2-D DCT.

3. The 2-D DCT array of coefficients is uniformly quantized by rounding to the nearest integer. Specifically, the quantized DCT coefficients, $\bar{C}_{uv}$, are defined by the following equation:

$$\bar{C}_{uv} = \text{nearest integer} \left(\frac{C_{uv}}{Q_{uv}} \right) , \qquad (4.55)$$

where C_{uv} is the DCT coefficient and Q_{uv} is the corresponding element in the quantization table.

4. The 2-D array of quantized DCT coefficients is scanned and formatted into a 1-D sequence using the zigzag pattern shown in Fig. 4.9(a). The DC coefficient is sensitive to spatial frequency response of the HVS and is treated separately from the remaining AC coefficients. Prior to encoding, the DC coefficient is differenced by the following first-order prediction:

$$DIFF = DC_i - DC_{i-1} , \qquad (4.56)$$

where DC_i and DC_{i-1} are DC coefficients in the current and previous blocks, respectively. The initial starting DC value at the beginning of the image is set to zero.

We note that in international image/video coding standards two scan methods of quantized DCT coefficients are used: the zigzag scan [Fig. 4.9(a)] which is typical for progressive (noninterlaced) mode processing (in JPEG, MPEG-1, and H.261 standards) and alternate scan [Fig. 4.9(b)] which is more efficient for interlaced video format (adopted in MPEG-2 and HDTV standards). The structure of an alternate scan seems like a vertical scan since the correlation along the horizontal direction is higher than along the vertical direction [31].

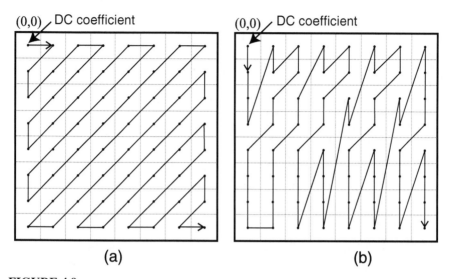

FIGURE 4.9
Scanning patterns of quantized DCT coefficients: (a) zigzag scan; (b) alternate scan.

The following program compresses one sub-image block according to steps described previously.

```
/* --------------------------------------------------------------
              COMPRESSION OF ONE SUB-IMAGE BLOCK
Level shifting, forward 2-D DCT, quantization, zigzag
reordering and Huffman encoding the quantized
coefficients. ----------------------------------------------------
*/
#include    "jpegdef.h"
extern int     exp_val;     /* log2 value of block size*/
extern int     center_samp; /* center sample value     */
extern double  scaling;/* scaling for DCT normalization*/
extern int     tdc_last;/*the last DC value for encoder*/
extern int     encode_bits;
                    /* # of bits for compressed block */
void cmprs_blk (
     int              *qnt_blk,
                /* input/quantized data block */
     int              blksize,
                /* block size */
     int              *qnt_tbl,
                /* user quantization table */
     int              *zigzag,        /* zigzag pattern */
     double           **dctb,         /* 2-D DCT block */
     struct huff_table *dctbl,/* DC Huffman code table */
     struct huff_table *actbl )/* AC Huffman code table*/
{
     int     i,j,k,temp,*q_ptr;
```

```
          double coef,*dctptr;
/*   Level shift of samples in the sub-image block */
     for ( i = 0, q_ptr = qnt_blk; i < blksize; i++ )
        for ( j = 0, dctptr = dctb [i]; j < blksize; j++ )
            *dctptr++ = (double) (*q_ptr++ - center_samp);
/*   Perform forward 2-D DCT computation and
     normalization of transform coefficients */
     fdcst2d (dctb,exp_val,DISABLE_NORM,DCT);
     for ( i = 0, dctb [0] [0] *= SQRT2; i < blksize; i++ )
        for ( j = 0; j < blksize; j++ )
        {
            dctb [i] [j] *= scaling;
            if ( (i == 0) || (j == 0) )
                dctb [i] [j] *= SQRT2;
        }
/*   Quantization of the transform DCT coefficients
     and zigzag reordering */
     for ( i = k = 0; i < blksize; i++ )
        for ( j = 0, dctptr = dctb [i]; j < blksize; j++ )
            if ( (coef = *dctptr++ / *qnt_tbl++) > 0.0 )
                qnt_blk [zigzag [k++]] = (int) (coef + 0.5);
            else
                qnt_blk [zigzag [k++]] = (int) (coef - 0.5);
/*   Huffman encoding the quantized coefficients. The DC
     coefficient is converted to a difference value */
     temp        = qnt_blk [0];
     qnt_blk [0] -= tdc_last;
     tdc_last    = temp;
     encode_bits = encode_blk(qnt_blk, blksize,
                              dctbl,actbl);
}
```

As an example, the following 8×8 data block is selected from the "Lena" digital image [31]:

```
79 75 79 82 82 86 94 94
76 78 76 82 83 86 85 94
72 75 67 78 80 78 74 82
74 76 75 75 86 80 81 79
73 70 75 67 78 78 79 85
69 63 68 69 75 78 82 80
76 76 71 71 67 79 80 83
72 77 78 69 75 75 78 78
```

After level shifting, this block transformed by the forward 2-D 8×8 DCT is given by

```
-404.375 -29.971   8.623   1.909   1.625 -3.936   0.893   1.516
  23.226  -7.184  -4.327  -0.438   7.346  0.010  -2.266  -3.186
  11.798  -0.278   5.197  -4.772  -3.572  4.160  -0.261  -3.507
   2.299 -10.742   5.495   0.791  -1.029  7.603   3.791   2.820
   6.375   2.511  -1.549  -1.074  -3.625 -0.797   0.506   8.723
   0.739   2.612   0.717   2.530  -0.926  3.206  -2.945  -2.792
```

```
-9.081  -1.660 -4.511   1.743   2.156   1.549 -1.697   2.055
-3.626   2.241  5.355  -1.960   0.899  -1.370  1.828  -3.314
```

By applying the basic luminance quantization table (quality factor is equal to 50), the 2-D array of quantized coefficients is

```
-25  -3   1   0   0   0   0   0
  2  -1   0   0   0   0   0   0
  1   0   0   0   0   0   0   0
  0  -1   0   0   0   0   0   0
  0   0   0   0   0   0   0   0
  0   0   0   0   0   0   0   0
  0   0   0   0   0   0   0   0
  0   0   0   0   0   0   0   0
```

Assuming that the quantized DC coefficient of the previous block is 34, the DC differencing and the reordering 2-D array of quantized coefficients into a 1-D sequence based on zigzag scan result in

```
-59 -3   2   1  -1   1   0   0   0   0   0  -1  EOB
```

The 1-D sequence of quantized DCT coefficients is prepared for Huffman encoding. The encoder employs one DC and one AC Huffman table lookups for luminance DCT coefficients. All codes consist of a set of Huffman codes with a maximum length of 16 bits followed by appended additional bits for representing the exact value of the coefficient.

Coding the DC and AC Coefficients

The DIFF values as defined by Eq. (4.56) are classified into 12 categories, each category written by two's complement expression. A Huffman DC coding/decoding table is generated for each category. The difference values in category k are in the range $< -2^k + 1, 2^k - 1 >$, where $0 \le k \le 11$. Thus, k denotes the number of bits needed for the magnitude of the coefficient. In the case of $k = 0$ (DIFF = 0), the current DC coefficient is the same as the previous DC coefficient, and additional bits are not required. For the other categories, extra bits are needed to express the exact value in the category, consisting of the sign and magnitude of the DIFF value. When DIFF is positive, the sign bit is 1 and k low-order bits of DIFF are appended to the Huffman code. When DIFF is negative, the sign bit is 0 and k low-order bits of (DIFF-1) are appended to the Huffman code. A (DIFF-1) operation implies one's complement representation to avoid all 1 bits of two's complement operation. This procedure for appending the additional bits is also applied to encoding the AC coefficients.

To encode the AC coefficients, each nonzero coefficient is first described by a composite 8-bit value of the form "RRRRSSSS" in binary notation. The Huffman AC coding/decoding table is generated for each composite value. The four least significant bits, "SSSS," define a category for the coefficient magnitude. The values in category k are in the range $< -2^k + 1, 2^k - 1 >$, where $1 \leq k \leq 10$ resulting in 10 categories. The four most significant bits in the composite value, "RRRR," give the position of the current coefficient relative to the previous nonzero coefficient, i.e., the runlength of zero coefficients between successive nonzero coefficients. The runlenghts specified by "RRRR" can range from 0 to 15, and a separate symbol "11110000" (11-bits ZRL code = 11111111001) is defined to represent a runlength of 16 zero coefficients. If the runlength is greater than 16, it is coded by using multiple symbols. In addition, if all remaining coefficients in the block are zero, a special symbol "00000000" is used to code the end of block (4-bits EOB code = 1010).

By the following program, the 1-D sequence of quantized coefficients is Huffman encoded. The result is stored in the output bit stream buffer.

```
/*---------------------------------------------------------------
-------------HUFFMAN_ENTROPY_ENCODING_ROUTINES----------
*/
#include    "jpegdef.h"
extern unsigned char out_buffer [];
                              /* bit stream buffer    */
extern int          bytes_in_buf;
                              /* # of bytes in it     */
static long int hufput_buf   = 0L;
                              /* bit accumulator buffer */
static int      hufput_bits = 0;
                              /* # of bits in buffer   */
static void emit_bits (unsigned int, int);
/* --------------------------------------------------------
   ENCODE A SINGLE BLOCK OF COEFFICIENTS
   It is assumed that DC coefficient in a block was
   converted to a difference value. Function returns the
   total number of bits for encoded block of
   coefficients.-----------------------------------------
*/
int encode_blk (
  int                *block,   /* quantized data block  */
  int                blksize,  /* block size */
  struct huff_table *dctbl,    /* DC Huffman code table */
  struct huff_table *actbl )   /* AC Huffman code table */
{
    int i,k,nbits,run,temp,temp2,num_bits = 0;
/*
   =======================================================
                ENCODE THE DC COEFFICIENT
   =======================================================
*/
    if ( (temp = temp2 = block [0]) < 0 )
    {
      temp = -temp;                 /* abs value of input   */
      temp2--;
            /* negative value is bitwise complement  */
    }
/* Find the number of bits for magnitude of the
   coefficient */
    nbits = 0;
    while ( temp )
```

```
        {
            nbits++;
            temp >>= 1;
        }
/*  Emit the Huffman coded symbol for the number
    of bits */
        emit_bits (dctbl->hufcode [nbits],
            dctbl->hufsize [nbits]);
        num_bits += dctbl->hufsize [nbits];
/*  Emit the number of bits of the coefficient value
    (positive value) or complement of its magnitude
    (negative value). Reject if nbits = 0 */
        if ( nbits )
        {
            emit_bits ((unsigned int) temp2,nbits);
            num_bits += nbits;
        }
/*
  ==========================================================
                  ENCODE THE AC COEFFICIENTS
  ==========================================================
*/
        for ( k = 1, run = 0; k < blksize * blksize; k++ )
        {
            if ( (temp = block [k]) == 0 )
                run++;
            else
            {
/*  If run length > 15 then emit special run-length
    codes (0xF0) */
                while ( run > 15 )
                {
                    emit_bits (actbl->hufcode [0xF0],
                        actbl->hufsize [0xF0]);
                    num_bits += actbl->hufsize [0xF0];
                    run -= 16;
                }
                if ( (temp2 = temp) < 0 )
                {
                    temp = -temp;
                    temp2--;
                }
/* Find the number of bits needed for the magnitude of
   the coefficient. The number of bits must be at least
   1 bit */
                nbits = 1;
                while ( temp >>= 1 )
                    nbits++;
/* Emit the Huffman symbol for
   (run length / number of bits) */
                i = (run << 4) + nbits;
                emit_bits (actbl->hufcode [i],
                    actbl->hufsize [i]);
                num_bits += actbl->hufsize [i];
/*  Emit the number of bits of the coefficient value
    (positive value) or complement of its magnitude
    (negative value) */
                emit_bits ((unsigned int) temp2,nbits);
                num_bits += nbits;
                run = 0;
            }
```

```
        }
/*  If the last coefficients were zero, emit EOB code */
    if ( run > 0 )
        {
            emit_bits (actbl->hufcode [0],
                actbl->hufsize [0]);
            num_bits += actbl->hufsize [0];
        }
/*  Fill any partial byte with ones and reset
    bit-buffer */
    emit_bits (0x7F,7);
    hufput_buf  = 0L;
    hufput_bits = 0;
    return (num_bits);
}
/* ------------------------------------------------------------
    OUTPUT HUFFMAN COMPRESSED COEFFICIENTS
    Only the right 24 bits of hufput_buf are used.
    The valid bits are left justified. At most 16 bits
    can be passed to emit_bits () in one call and is
    never retained more than 7 bits in accumulator buffer
    between calls. ----------------------------------------------
*/
static void emit_bits (
    unsigned int code,
    int          size )
{
    long int put_buffer = code;
    int      put_bits   = hufput_bits,byte;
/*  Mask off excess bits in put_buffer */
    put_buffer &= (((long int) 1) << size) - 1;
    put_bits   += size;       /* new # of bits in buffer */
    put_buffer <<= 24 - put_bits;/* align incoming bits */
    put_buffer |= hufput_buf; /* merge with old buffer */
/*  Load byte into output bit stream buffer and count
    the number of bytes. Update bit accumulator buffer */
    while ( put_bits >= 8 )
        {
        byte = (unsigned int) ((put_buffer >> 16) & 0xFF);
        out_buffer [bytes_in_buf++] = (unsigned char) byte;
          put_buffer <<= 8;
          put_bits    -= 8;
        }
    hufput_buf  = put_buffer;
    hufput_bits = put_bits;
}
```

For our example of 1-D sequence of the quantized DCT coefficients, the program generates the following output-encoded bit stream (last unused bits are set to 1):

```
    The number of bits        39 (5 bytes)
    Bit stream buffer (hex) E1 11 88 3E 95
    11100001 00010001 10001000 00111110 1001010/1
```

4.5.6 Decompression of One Sub-Image Block

At the decoder (see Fig. 4.8) for each sub-image block, the inverse operations of the encoder are followed but in reverse order. The quantization table and Huffman coding/decoding tables are the same at both the encoder and decoder.

Each of the Huffman codes is uniquely defined and the quantized DCT coefficients are decoded by the Huffman decoding procedure. The DC coefficient is reconstructed from the differential value. The initial starting DC value at the beginning is set to zero. The reconstructed 1-D sequence of quantized coefficients is dezigzag reordered to form a 2-D array. Each DCT coefficient, $\bar{C}_{uv}$, in the 2-D array is inverse quantized by multiplying it by the corresponding element of the quantization table as follows:

$$\hat{C}_{uv} = \bar{C}_{uv} \cdot Q_{uv} . \tag{4.57}$$

The resulting array is transformed by the inverse 2-D DCT. Inverse level shift restores the samples in the original block to the unsigned 8-bit representation.

With the following program, the sub-image block is reconstructed from the encoded bit stream.

```
/* ------------------------------------------------------------
      DECOMPRESSION OF ONE SUB-IMAGE BLOCK
Huffman decoding, inverse quantization, inverse 2-D DCT,
and reconstruction of the original sub-image block.
 */
#include    "jpegdef.h"
extern int     exp_val;   /* log2 value of block size    */
extern int     center_samp; /* center sample value       */
extern double scaling;/* scaling for DCT normalization*/
extern int     rdc_last;/* last DC value for decoder    */
void decmprs_blk (
        int          *qnt_blk,
                          /* quantized/output data block */
        int          blksize,   /* block size */
        int          *qnt_tbl,
                            /* user quantization table */
        int          *dezigzag,  /* dezigzag pattern */
        double       **dctb,        /* 2-D IDCT block */
        struct huff_table *dctbl,/* DC Huffman code table */
        struct huff_table *actbl )/* AC Huffman code table*/
{
        int     i,j,k,*q_ptr;
        double pixel,*dctptr;
/*      Huffman decoding the quantized coefficients and
        dezigzag ordering. Convert DC difference to actual
        value and update the last DC value */
        decode_blk (qnt_blk,blksize,dezigzag,dctbl,actbl);
        qnt_blk [0]  += rdc_last;
        rdc_last      = qnt_blk [0];
/*      Inverse quantization of the coefficients */
    for ( i = k = 0; i < blksize; i++ )
        for ( j = 0, dctptr = dctb [i]; j < blksize; j++ )
           *dctptr++ = (double) (qnt_blk [k++] * *qnt_tbl++);
/*      Perform denormalization and inverse 2-D DCT
```

```
        computation */
    for ( i = 0, dctb [0] [0] *= SQRT2; i < blksize; i++ )
        for ( j = 0; j < blksize; j++ )
            {
                dctb [i] [j] *= scaling;
                if ( (i == 0) || (j == 0) )
                    dctb [i] [j] *= SQRT2;
            }
        fdcst2d (dctb,exp_val,DISABLE_NORM,-DCT);
/*  Reconstruction of the original sub-image block */
        for ( i = 0, q_ptr = qnt_blk; i < blksize; i++ )
            {
                dctptr = dctb [i];
            for ( j = 0; j < blksize; j++, q_ptr++ )
                if ( (pixel = *dctptr++ + center_samp) > 0.0 )
                    {
                        if ( (*q_ptr = (int) (pixel + 0.5))
                                > I_LEVEL - 1 )
                            *q_ptr = I_LEVEL - 1;
                    }
                else
                        *q_ptr = 0;
            }
    }
}
```

For our example the inverse quantized block is

400	-33	10	0	0	0	0	0
24	-12	0	0	0	0	0	0
14	0	0	0	0	0	0	0
0	-17	0	0	0	0	0	0
0	0	0	0	0	0	0	0
0	0	0	0	0	0	0	0
0	0	0	0	0	0	0	0
0	0	0	0	0	0	0	0

This 2-D array transformed by the inverse 2-D 8×8 DCT is given by

```
-53.992 -53.111 -51.068 -47.587 -42.784 -37.390 -32.641 -29.846
-51.247 -51.084 -50.368 -48.621 -45.696 -42.036 -38.614 -36.537
-50.225 -50.684 -51.118 -50.873 -49.573 -47.415 -45.143 -43.689
-53.805 -54.156 -54.390 -53.884 -52.300 -49.881 -47.408 -45.846
-58.944 -58.765 -58.018 -56.232 -53.263 -49.563 -46.111 -44.018
-59.846 -59.558 -58.611 -56.564 -53.311 -49.350 -45.697 -43.496
-55.036 -55.370 -55.573 -55.027 -53.401 -50.941 -48.438 -46.859
-49.611 -50.664 -52.194 -53.382 -53.633 -52.908 -51.732 -50.872
```

and after inverse level shift the reconstructed sub-image block is (for easy comparison the original sub-image block is also given)

74	75	77	80	85	91	95	98		79	75	79	82	82	86	94	94
77	77	78	79	82	86	89	91		76	78	76	82	83	86	85	94
78	77	77	77	78	81	83	84		72	75	67	78	80	78	74	82

```
74 74 74 74 76 78 81 82      74 76 75 75 86 80 81 79
69 69 70 72 75 78 82 84      73 70 75 67 78 78 79 85
68 68 69 71 75 79 82 85      69 63 68 69 75 78 82 80
73 73 72 73 75 77 80 81      76 76 71 71 67 79 80 83
78 77 76 75 74 75 76 77      72 77 78 69 75 75 78 78
```

The following program module contains routines for Huffman decoding the quantized
DCT coefficients from the encoded bit stream.

```c
/* ------------------------------------------------------------
--------HUFFMAN_ENTROPY_DECODING_ROUTINES------------
*/
#include           "jpegdef.h"
extern unsigned char out_buffer [];
                              /* bit stream buffer    */
static unsigned char *out_buf;  /* and pointer to it    */
static long int get_buffer = 0L;
                              /* bit-extraction buffer */
static int        bits_left  = 0; /* # of unused bits     */
static void fill_buf     (int);
static int  huff_decode (struct huff_table *);
static int  slow_decode (struct huff_table *, int);
/*
    ++++++++++++++++++++++++++++++++++++++++++++++++++++
    DECODE A SINGLE BLOCK OF COEFFICIENTS
    Data block for the coefficients should be zeroed
    before. Output coefficients are in dezigzagged
    (natural) order.
    ++++++++++++++++++++++++++++++++++++++++++++++++++++
*/
void decode_blk (
    int              *block,          /* decoded block */
    int              blksize,         /* block size */
    int              *dezigzag,       /* dezigzag pattern */
    struct huff_table *dctbl,  /* DC Huffman code table */
    struct huff_table *actbl ) /* AC Huffman code table */
{
    int k,s,r;
    out_buf = out_buffer;
/*
    ===================================================
    DECODE THE DC COEFFICIENT
    Extract Huffman symbol from input bit stream and
    get the number of bits of DC coefficient difference.
    Extract bits of the DC coefficient difference and
    extend sign.
    ===================================================
*/
    if ( s = huff_decode (dctbl) )
    {
        if ( bits_left < s )
            fill_buf (s);
        bits_left -= s;
        r = (int) ((get_buffer >> bits_left))
            & ((1 << s) - 1);
        s = ( r < (1 << (s - 1)) ) ? r
            + ((-1 << s) + 1) : r;
    }
    block [0] = s;
```

```
/*
       ===================================================
                    DECODE THE AC COEFFICIENTS
       Extract Huffman symbol from input bit stream and
       get value of (run length / number of bits).
       ===================================================
*/
       for ( k = 1; k < blksize * blksize; k++ )
       {
           s = huff_decode (actbl);
           r = s >> 4;
           if ( s &= 15 )
           {
               k += r;
/*     Extract bits of AC coefficient magnitude and
       extend sign */
               if ( bits_left < s )
                   fill_buf (s);
               bits_left -= s;
               r = (int) ((get_buffer >> bits_left))
                   & ((1 << s) - 1);
               s = ( r < (1 << (s - 1)) ) ? r
                   + ((-1 << s) + 1) : r;
               block [dezigzag [k]] = s;
           }
/*     The code EOB was detected - the last coefficients
       are zeros */
           else
           {
               if ( r != 15 )
                   break;
               k += 15;
           }
       }
/*     Reset bit-extraction buffer to empty */
       get_buffer = 0L;
       bits_left = 0;
}
/*
       ---------------------------------------------------
       LOAD UP THE BIT BUFFER TO A DEPTH OF AT LEAST nbits
       Source bytes are read into get_buffer and bits are
       doled out as needed. If get_buffer already contains
       enough bits, they are fetched in-line. When there
       are not enough bits, fill_buf () is called.
       ---------------------------------------------------
*/
static void fill_buf (
       int nbits )
{
       int c;
/*     Attempt to load at least MIN_GET_BITS into
       get_buffer */
       while ( bits_left < MIN_GET_BITS )
       {
/*     There are enough bits still left in get_buffer */
           if ( nbits > 0 && bits_left >= nbits )
               break;
/*     Load byte from input bit stream buffer into
       get_buffer */
           c          = *out_buf++;
```

```
            get_buffer = (get_buffer << 8) | c;
            bits_left += 8;
        }
    }
/*
------------------------------------------------------------
EXTRACT NEXT HUFFMAN-CODED SYMBOL FROM INPUT BIT STREAM
Lookahead table is used to process codes of up to
LOOKAHEAD bits without looping. Usually, more than 95%
of the Huffman codes will be 8 or fewer bits long. The
few overlength codes are handled with a loop.
------------------------------------------------------------
*/
static int huff_decode (
    struct huff_table *htbl )
{
    int nb,look,result,b = LOOKAHEAD;
/*  1.The first if-test is coded to call fill_buf ()
      only when necessary.
    2.If the lookahead succeeds, is needed only
      decrement bits_left to remove the proper number
      of bits from get_buffer.
    3.If the lookahead table contains no entry, the
      next code must be more than LOOKAHEAD bits long */
    if ( bits_left >= LOOKAHEAD ||
        (fill_buf (0),bits_left >= LOOKAHEAD) )
    {
     nb   = bits_left - b;
     look = (int) ((get_buffer >> nb)) & ((1 << b) - 1);
        if ( (nb = htbl->look_nbits [look]) != 0 )
        {
            bits_left -= nb;
            result    = htbl->look_sym [look];
        }
        else
            result = slow_decode (htbl,LOOKAHEAD+1);
    }
    else
        result = slow_decode (htbl,1);
    return (result);
}
static int slow_decode (
    struct huff_table *htbl,
    int                min_bits )
{
    int    k = min_bits,rs;
    long int code;
/*  huff_decode () has determined that the code is at
    least min_bits long, so fetch that many bits in one
    swoop */
    if ( bits_left < k )
        fill_buf (k);
    bits_left -= k;
    code = (int) ((get_buffer >> bits_left))
           & ((1 << k) - 1);
/*  Collect the rest of the Huffman code one bit at
    a time */
    while ( code > htbl->maxcode [k] )
    {
        code <<= 1;
```

```
        if ( bits_left < 1 )
            fill_buf (1);
        code |= (int) ((get_buffer >> (--bits_left))) & 1;
        k++;
    }
    rs = htbl->valptr [k] + (int)
            (code - htbl->mincode [k]);
    return (htbl->hufval [rs]);
}
```

4.5.7   Image Compression/Decompression

This section shows a sample program for compression and decompression of an image. It performs all described steps of the JPEG DCT-based coding technique for image compression and decompression. One extracted sub-image block is first compressed, immediately decompressed, and displayed on the screen. The displaying routine is not shown. If any dimension of the processed image is not a multiple of the block size, the remaining elements in the block are set to zeros. These additional elements are removed during decompression. No file for the compressed image is created.

```
/*
----------------------------------------------------------
---------IMAGE_COMPRESSION_AND_DECOMPRESSION---------
*/
#include    <string.h>
#include    "jpegdef.h"
extern unsigned char _huge *img_ptr [];
    /* ptrs to image rows */
extern unsigned char dc_lumbits [17];
extern unsigned int  dc_lumval  [12];
extern unsigned char ac_lumbits [17];
extern unsigned int  ac_lumval [162];
extern double        dct_block [SIZE*SIZE];
extern double        *dctptr   [SIZE];
extern unsigned char out_buffer [];
extern int           bytes_in_buf;
extern int           encode_bits;
extern long int      cmprs_bits;
extern int           tdc_last;
extern int           rdc_last;
extern struct huff_table dc_table;
extern struct huff_table ac_table;
static int  q_blk [SIZE*SIZE];
    /* input/quantized/output block */
static int  q_tbl [SIZE*SIZE];
    /* user quantization table        */
void process_img (
    int xsize,          /* image xsize */
    int ysize,          /* image ysize */
    int *qbase_tbl,     /* basic quantization table */
    int *zag,           /* zigzag pattern */
    int *dezag,         /* dezigzag pattern */
    int blksize,        /* block size */
```

```
        int quality )           /* quality factor */
{
        int             i,j,k,l,m,n,*q_ptr;
        int             xp,yp,xpos,ypos,hblk,hrest,vblk;
        struct huff_table *dctbl = &dc_table;
        struct huff_table *actbl = &ac_table;
/*  Set up quantization table according to user
    specified 'quality' factor */
        set_qtable (q_tbl,blksize,qbase_tbl,quality);
/*  Compute standard Huffman DC and AC code tables */
 memcpy (&dctbl->bits  ,dc_lumbits,sizeof (dc_lumbits));
 memcpy (&dctbl->hufval,dc_lumval ,sizeof (dc_lumval));
 memcpy (&actbl->bits  ,ac_lumbits,sizeof (ac_lumbits));
 memcpy (&actbl->hufval,ac_lumval ,sizeof (ac_lumval));
 fix_huftbl (dctbl);
 fix_huftbl (actbl);
/*  Initialize variables for compression/decompression */
        tdc_last     = 0; /* the last DC value for encoder */
        rdc_last     = 0; /* the last DC value for decoder */
        cmprs_bits   = 0L;/* # of bits for compressed data */
        bytes_in_buf = 0; /* # of bytes in output buffer  */
/*  Set the number of subblocks horizontally
    and vertically */
        hblk = xsize / blksize;
        if ( (hrest = xsize % blksize) )
            hblk++;
        vblk = ysize / blksize;
        if ( ysize % blksize )
            vblk++;
        for ( i = 0; i < blksize; i++ )
            dctptr [i] = dct_block + i * blksize;
/*  Extract the 2-D blocks from source image, one block
    at the time and do compression/decompression */
        for ( i = 0; i < vblk; i++ )
        {
            ypos = i * blksize;
            for ( j = 0; j < hblk; j++ )
            {
                xpos = j * blksize;
                memset (q_blk,0,blksize * blksize
                    * sizeof (int));
                q_ptr = &q_blk [0];
                for ( m = 0, yp = ypos; m < blksize;
                    m++, yp++ )
                    if ( yp < ysize )
                        for ( n = 0, xp = xpos; n
                                < blksize; n++, xp++ )
                            if ( xp < xsize )
                                *q_ptr++ = (int)
                                    *(img_ptr [yp] + xp);
                            else
                            {
                                q_ptr += (blksize - hrest);
                                break;
                            }
/*  Compression of a single sub-image block */
                cmprs_blk (q_blk,blksize,q_tbl,zag,
                        &dctptr [0],dctbl,actbl);
```

```
                    cmprs_bits += encode_bits;
          /*  Decompression of the single sub-image block      */
                    memset (q_blk,0,blksize * blksize
                          * sizeof (int));
                    decmprs_blk (q_blk,blksize,q_tbl,dezag,
                          &dctptr
[0],dctbl,actbl);
          /*  Display reconstructed sub-image block              */
              display_block (xsize,ysize,xpos,ypos,hrest,q_blk);
          /*  Clear output bit stream buffer and byte counter    */
                    while ( bytes_in_buf > 0 )
                          out_buffer [bytes_in_buf--] = 0;
              }
          }
}
```

4.5.8 Compression of Color Images

In many imaging applications, it is necessary to deal with color images. Although the RGB representation of images is typical of color displays, it is not the best representation from the viewpoint of compression. RGB images are converted into more suitable YC_BC_R color format using the following equations [31]:

$$
\begin{aligned}
Y &= \quad 0.299\ R + 0.587\ G + 0.114\ B \\
C_B &= \ -0.169\ R - 0.331\ G + 0.500\ B = 0.564\ (B - Y) \qquad (4.58) \\
C_R &= \quad 0.500\ R - 0.419\ G - 0.081\ B = 0.713\ (R - Y),
\end{aligned}
$$

where Y represents a monochrome compatible luminance component, and C_B, C_R represent chrominance components containing color information. Most of image/video coding standards adopt YC_BC_R color format as an input image signal [31]. This color conversion has the desirable property of packing most of the signal energy into Y and significantly less energy into the chrominance components. Furthermore, the HVS is much more sensitive to variations in the luminance component. These properties suggest a compression scheme for color images. The luminance component is encoded with high fidelity while larger errors are allowed in the chrominance components.

We have described the JPEG DCT-based coding technique for the luminance component. The chrominance components are similarly processed except for some minor modifications. Each chrominance component is subsampled by a factor of 2 or 4 in both the horizontal and vertical directions prior to compression. At the decoder, the reconstructed chrominance components are bilinearly interpolated back to their original size. Then, the image in YC_BC_R color format is transformed into RGB format using the following equations:

$$
\begin{aligned}
R &= Y & & + & 1.402\ C_R \\
G &= Y & - \quad 0.344\ C_B & - & 0.714\ C_R \qquad (4.59) \\
B &= Y & + \quad 1.772\ C_B & &
\end{aligned}
$$

For compression and decompression of color images, the user needs the header file JPEGCOLOR.H containing the definitions of data structures for computation of standard Huffman chrominance DC and AC coding and decoding tables and the definition of the chrominance sample quantization table.

```
/*
    JPEGCOLOR.H
*/
/* # of symbols with codes of length k bits (lumbits
   [k]) and symbols in order of incremental code length
   (lumval [k]) for DC chrominance values - valid
   for 8-bit data precision */
unsigned char dc_chrombits [17] =
   {0,0,3,1,1,1,1,1,1,1,1,1,0,0,0,0,0};
unsigned char dc_chromval  [12]
   = {0,1,2,3,4,5,6,7,8,9,10,11};
/* # of symbols with codes of length k bits (lumbits
   [k]) and symbols in order of incremental code length
   (lumval [k]) for AC chrominance values - valid for
   8-bit data precision */
unsigned char ac_chrombits [17] =
   {0,0,2,1,2,4,4,3,4,7,5,4,4,0,1,2,0x77};
unsigned char ac_chromval [162] =
{ 0x00, 0x01, 0x02, 0x03, 0x11, 0x04, 0x05, 0x21,
  0x31, 0x06, 0x12, 0x41, 0x51, 0x07, 0x61, 0x71,
  0x13, 0x22, 0x32, 0x81, 0x08, 0x14, 0x42, 0x91,
  0xa1, 0xb1, 0xc1, 0x09, 0x23, 0x33, 0x52, 0xf0,
  0x15, 0x62, 0x72, 0xd1, 0x0a, 0x16, 0x24, 0x34,
  0xe1, 0x25, 0xf1, 0x17, 0x18, 0x19, 0x1a, 0x26,
  0x27, 0x28, 0x29, 0x2a, 0x35, 0x36, 0x37, 0x38,
  0x39, 0x3a, 0x43, 0x44, 0x45, 0x46, 0x47, 0x48,
  0x49, 0x4a, 0x53, 0x54, 0x55, 0x56, 0x57, 0x58,
  0x59, 0x5a, 0x63, 0x64, 0x65, 0x66, 0x67, 0x68,
  0x69, 0x6a, 0x73, 0x74, 0x75, 0x76, 0x77, 0x78,
  0x79, 0x7a, 0x82, 0x83, 0x84, 0x85, 0x86, 0x87,
  0x88, 0x89, 0x8a, 0x92, 0x93, 0x94, 0x95, 0x96,
  0x97, 0x98, 0x99, 0x9a, 0xa2, 0xa3, 0xa4, 0xa5,
  0xa6, 0xa7, 0xa8, 0xa9, 0xaa, 0xb2, 0xb3, 0xb4,
  0xb5, 0xb6, 0xb7, 0xb8, 0xb9, 0xba, 0xc2, 0xc3,
  0xc4, 0xc5, 0xc6, 0xc7, 0xc8, 0xc9, 0xca, 0xd2,
  0xd3, 0xd4, 0xd5, 0xd6, 0xd7, 0xd8, 0xd9, 0xda,
  0xe2, 0xe3, 0xe4, 0xe5, 0xe6, 0xe7, 0xe8, 0xe9,
  0xea, 0xf2, 0xf3, 0xf4, 0xf5, 0xf6, 0xf7, 0xf8,
  0xf9, 0xfa };
struct huff_table dc_ctable; /* Huffman DC code table */
struct huff_table ac_ctable; /* Huffman AC code table */
/* chrominance sample quantization table for
   an 8 x 8 DCT */
int qcbase8_tbl [8*8] =
{ 17, 18, 24, 47, 99, 99, 99, 99,
  18, 21, 26, 66, 99, 99, 99, 99,
  24, 26, 56, 99, 99, 99, 99, 99,
  47, 99, 99, 99, 99, 99, 99, 99,
  99, 99, 99, 99, 99, 99, 99, 99,
  99, 99, 99, 99, 99, 99, 99, 99,
  99, 99, 99, 99, 99, 99, 99, 99,
  99, 99, 99, 99, 99, 99, 99, 99 };
```

Assuming that each chrominance component was subsampled, the program from Section 4.5.7 can be used for compression and decompression. It is necessary only to substitute proper identifiers for the quantization table and Huffman coding/decoding tables.

4.5.9   Results of Image Compression

The performance of a compression algorithm can be evaluated in a number of different ways:

- Implementation complexity (algorithm complexity, computational speed, and memory requirements).

- The amount of compression expressed by the compression ratio.

- The average number of bits required to represent a single sample; this is generally referred to as *bit rate*.

- How closely the reconstruction resembles the original; this is related to the reconstructed image quality.

In evaluating the reconstructed image quality, a frequently used measure is the root-mean-square-error (RMSE) as an error metric [36]. Denoting the original $N \times N$ image by f_{ij} and the compressed/decompressed image by $\hat{f}_{ij}$, RMSE is given by

$$RMSE = \sqrt{\frac{1}{N^2} \sum_{i=0}^{N-1} \sum_{j=0}^{N-1} \left(f_{ij} - \hat{f}_{ij} \right)^2},$$

and represents the standard deviation of the error image. Error images represent the difference between the original and reconstructed images. The error image, g_{ij}, can be generated using

$$g_{ij} = k \left| f_{ij} - \hat{f}_{ij} \right|,$$

where the scaling factor k is included to make any error more visible.

The results of applying the JPEG DCT-based coding technique are summarized in Tables 4.1 and 4.2 for the monochrome "Lena" image. Recall that the original 512×512 monochrome "Lena" image requires a total of 2,097,152 bits, or 262,144 bytes. In the JPEG DCT-based image compression and decompression system, two block sizes have been used — 8×8 and 16×16. Table 4.1 summarizes results of compression using the 8×8 block size, and Table 4.2 summarizes results of compression using the 16×16 block size for several values of the quality factor. From the tables it is evident that for 16×16 block size the compression ratios are about two-fold better than for 8×8 block size. On the other hand, at very low bit rates the blocking artifact is more visible for the larger block size. Actual reconstructed and corresponding error images using 8×8 and 16×16 blocks for some values of the quality factor (its definition is given in Section 4.5.3) are shown in Figs. 4.10 and 4.11, respectively.

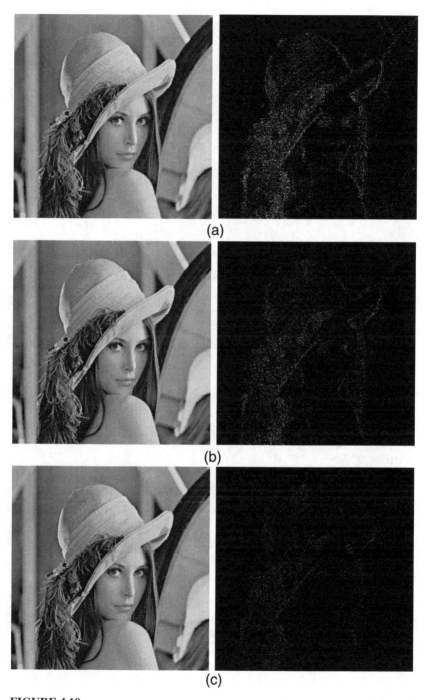

(a)

(b)

(c)

FIGURE 4.10
Reconstructed and corresponding error images using 8×8 DCT block size for the quality factor: (a) 25%, (b) 50%, (c) 75%. Error images are magnified by a factor of 8. Reproduced by Special Permission of *Playboy* magazine. Copyright ©1972, 2000 by Playboy.

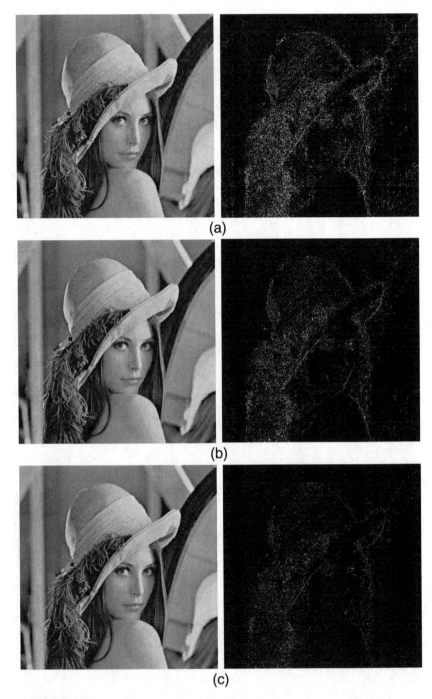

FIGURE 4.11
Reconstructed and corresponding error images using 16×16 DCT block size for the quality factor: (a) 25%, (b) 50%, (c) 75%. Error images are magnified by factor of 8. Reproduced by Special Permission of *Playboy* magazine. Copyrig ©1972, 2000 by Playboy.

Table 4.1   Results of the "Lena" Image Compression for 8 × 8 Block Size

Quality factor	The number of compressed bits (bytes)	Bit rate	Compression ratio	RMSE error
25%	96 351 (12 044)	0.368	21.766	4.774
50%	148 132 (18 517)	0.565	14.157	3.734
60%	170 878 (21 360)	0.652	12.273	3.470
75%	230 924 (28 866)	0.881	9.082	2.965
90%	422 392 (52 799)	1.611	4.965	2.124
100%	1 212 625 (151 579)	4.626	1.729	0.289

Table 4.2   Results of "Lena" Image Compression for 16 × 16 Block Size

Quality factor	The number of compressed bits (bytes)	Bit rate	Compression ratio	RMSE error
25%	47 101 (5 888)	0.180	44.525	6.234
50%	77 868 (9 734)	0.297	26.932	4.778
60%	91 272 (11 409)	0.348	22.977	4.428
75%	126 850 (15 857)	0.484	16.533	3.817
90%	244 865 (30 609)	0.934	8.565	2.808
100%	926 585 (115 824)	3.535	2.263	0.644

4.6   Summary

The definitions and properties of four types of the even DCT and corresponding even DST have been discussed and the unified fast computation of DCTs and DSTs has been presented. For each type of DCT and DST, the fast computational algorithm was described and the corresponding regular generalized signal flow graph was shown followed by its implementation in C. Among the DCTs, DCT of types II and III have been employed as the main compression tool in the international image/video coding standards. To illustrate the compression capability of DCT, a real data compression application is considered. The JPEG DCT-based image compression and decompression system with its implementation is described in detail. This simple, efficient, and low-cost image compression and decompression system can be used in real data compression applications. Finally, the results of image compression were presented. We believe that all implemented algorithms will be useful in any other DCT- and DST-related applications.

References

[1] Jain, A.K., A sinusoidal family of unitary transforms, *IEEE Trans. on Pattern Analysis and Machine Intelligence,* 1, 356, 1979.

[2] Bongiovanni, G., Corsini, P. and Frosini, G., One-dimensional and two-dimensional generalized discrete Fourier transform, *IEEE Trans. on Acoustics, Speech and Signal Processing,* 24, 97, 1976.

[3] Wang, Z., Comments on generalized discrete Hartley transforms, *IEEE Trans. on Signal Processing,* 43, 1711, 1995.

[4] Wang, Z. and Hunt, B.R., The discrete W transform, *Applied Mathematics and Computation,* 16, 19, 1985.

[5] Wang, Z. and Hunt, B.R., The discrete cosine transform — a new version, *Proc. IEEE ICASSP,* Boston, MA, 1256, 1983.

[6] Kitajima, H., A symmetric cosine transform, *IEEE Trans. on Computers,* 29, 317, 1980.

[7] Jain, A.K., A fast Karhunen–Loève transform for a class of random processes, *IEEE Trans. on Communications,* 24, 1023, 1976.

[8] Ahmed, N., Natarajan, T., and Rao, K.R., Discrete cosine transform, *IEEE Trans. on Computers,* 23, 90, 1974.

[9] Kekre, H.B. and Solanki, J.K., Comparative performance of various trigonometric unitary transforms for transform image coding, *Int. J. Electronics,* 44, 305, 1978.

[10] Yip, P. and Rao, K.R., On shift property of DCTs and DSTs, *IEEE Trans. on Acoustics, Speech, and Signal Processing,* 35, 404, 1987.

[11] Wu, L.N., Comments on shift property of DCTs and DSTs, *IEEE Trans. on Acoustics, Speech, and Signal Processing,* 38, 186, 1990.

[12] Malvar, H.S., Lapped transforms for efficient transform/subband coding, *IEEE Trans. on Acoustics, Speech, and Signal Processing,* 38, 969, 1990.

[13] Martucci, S.A., Convolution-multiplication properties for the entire family of discrete sine and cosine transforms, *Proc. Twenty-Sixth Annual Conf. on Information Sciences and Systems,* Princeton, NJ, 399, 1992.

[14] Martucci, S.A., Symmetric convolution and the discrete sine and cosine transforms, *IEEE Trans. on Signal Processing,* 42, 1038, 1994.

[15] Martucci, S.A., Digital filtering of images using the discrete sine or cosine transform, *Optical Engineering,* 35, 119, 1996.

[16] Clarke, R.J., Relation between the Karhunen–Loève and cosine transforms, *IEEE Proc. Part F: Communications, Radar and Signal Processing,* 128, 359, 1981.

[17] Ahmed, N. and Flickner, M., Some considerations of the discrete cosine transform, *16th Asilomar Conf. on Circuits, Systems and Computers,* Pacific Grove, CA, 295, 1982.

[18] Clarke, R.J., Relation between the Karhunen–Loève and sine transforms, *Electronics Letters,* 20, 12, 1984.

[19] Zou, F. and Gallagher, R.R., A new transform with symmetrical coding performance for Markov(1) signals, *IEEE Trans. on Signal Processing,* 43, 2195, 1995.

[20] Wang, Z., A fast algorithm for the discrete sine transform implemented by the fast cosine transform, *IEEE Trans. on Acoustics, Speech, and Signal Processing,* 30, 814, 1982.

[21] Wang, Z., Fast algorithms for the discrete W transform and discrete Fourier transform, *IEEE Trans. on Acoustics, Speech, and Signal Processing,* 32, 803, 1984.

[22] Wang, Z., On computing the discrete Fourier and cosine transform, *IEEE Trans. on Acoustics, Speech, and Signal Processing,* 33, 1341, 1985.

[23] Hou, H.S., A fast recursive algorithm for computing the discrete cosine transform, *IEEE Trans. on Acoustics, Speech, and Signal Processing,* 35, 1455, 1987.

[24] Britanak, V., On the discrete cosine transform computation, *Signal Processing,* 40, 183, 1994.

[25] Britanak, V., A unified discrete cosine and discrete sine transform computation, *Signal Processing,* 43, 333, 1995.

[26] Britanak, V., A unified approach to fast computation of discrete sinusoidal transforms I: DCT and DST transforms, *Computers and Artificial Intelligence,* 17, 583, 1998.

[27] Britanak, V., A unified approach to fast computation of discrete sinusoidal transforms II: DFT and DWT transforms, *Computers and Artificial Intelligence,* 18, 19, 1999.

[28] Wu, H.R. and Paoloni, F.J., A two-dimensional fast cosine transform algorithm based on Hou's approach, *IEEE Trans. on Signal Processing,* 39, 544, 1991.

[29] Britanak, V., A generalized signal flow graph for the 2-D DCT computation, *Applied Signal Processing,* 1, 76, 1994.

[30] Rao, K.R. and Yip, P., *Discrete Cosine Transform: Algorithms, Advantages, Applications,* Academic Press, Boston, 1990.

[31] Rao, K.R. and Hwang, J.J., *Techniques and Standards for Image, Video and Audio Coding,* Prentice-Hall, Upper Saddle River, NJ, 1996.

[32] Clarke, R.J., *Transform Coding of Images,* Academic Press, London, 1990.

[33] Poularikas, A.D., *The Transforms and Application Handbook,* CRC Press and IEEE Press, Boca Raton, FL, 1996.

[34] Gonzalez, R.C. and Woods, R.E., *Digital Image Processing,* Addison–Wesley, Reading, MA, chap. 6, 1992.

[35] Jain, A.K., *Fundamentals of Digital Image Processing,* Prentice-Hall, Englewood Cliffs, NJ, chap. 5 and 11, 1989.

[36] Rabbani, M. and Jones, P.W., *Digital Image Compression Techniques,* Volume TT7 of Tutorial Texts Series, SPIE Optical Engineering Press, Bellingham, WA, 1991.

[37] Sayood, K., *Introduction to Data Compression,* Morgan Kaufmann, San Francisco, CA, 1996.

[38] Madisetty, V.K. and Williams, D.B., *The Digital Signal Processing Handbook,* CRC Press and IEEE Press, Boca Raton, FL, 1998.

[39] Bhaskaran, V. and Konstantinides, K., *Image and Video Compression Standards: Algorithms and Architecture,* Kluwer Academic, Norwell, MA, 1995.

[40] Kou, W., *Digital Image Compression — Algorithms and Standards,* Kluwer Academic, Hingham, MA, 1995.

[41] Gibson, J.D., Berger, T., Lookabaugh, T., Lindbergh, D., and Baker, R.L., *Digital Compression for Multimedia: Principles and Standards,* Morgan Kaufmann, San Francisco, CA, 1998.

[42] Symes, P.D., *Video Compression: Fundamental Compression Techniques and Overview of the JPEG and MPEG Compression Systems,* McGraw-Hill, New York, 1998.

[43] Wallace, G.K., *JPEG Technical Specification, Revision 5,* JPEG Joint Photographic Experts Group ISO/IEC JTC1/SC2/WG8 CCITT SGVIII, JPEG-8-R5, January 2, 1990.

[44] Wallace, G.K., The JPEG still picture compression standard, *Communications of the ACM,* 34, 31, 1991.

[45] Le Gall, D., MPEG: A video compression standard for multimedia applications, *Communications of the ACM,* 34, 47, 1991.

[46] Liou, M., Overview of the $p \times 64 kbit/s$ video coding standard, *Communications of the ACM,* 34, 60, 1991.

[47] Rabbani, M. and Dally, S., An optimized image data compression technique utilized in the Kodak SV9600 still video transceiver, *SPIE Proc. Vol. 1071 Optical Sensors and Electronic Photography,* Bellingham, WA, 246, 1989.

[48] Chitprasert, B. and Rao, K.R., Human visual weighted progressive image transmission, *IEEE Trans. on Communications,* 38, 1040, 1990.

[49] Murray, J.D. and VanRyper, W., *Encyclopedia of Graphics File Formats,*
O'Reilly and Associates, Sebastopol, CA, 1994.

Chapter 5

Lapped Transforms for Image Compression

Ricardo L. de Queiroz
Xerox Corporation

Trac D. Tran
The Johns Hopkins University

5.1   Introduction

This chapter covers the basic aspects of lapped transforms and their applications to image compression. It is a subject that has been extensively studied mainly because lapped transforms are closely related to filter banks, wavelets, and time-frequency transformations. Some of these topics are also covered in other chapters in this handbook. In any case it is certainly impractical to reference all the contributions in the field. Therefore, the presentation will be more focused rather than general. We refer the reader to excellent texts such as Malvar [26], Strang and Nguyen [55], Vaidyanathan [63], and Vetterli and Kovacevic [66] for a more detailed treatment of filter banks.

For the rest of this introductory section we will cover the basic notation, give a brief history of lapped transforms, and introduce block-based transforms. We will describe the principles of a block transform and its corresponding transform matrix along with its factorization. We will also introduce multi-input multi-output systems and relate them to block transforms. In Section 5.2, lapped transforms are introduced. Basic theory and concepts are presented for both orthogonal and nonorthogonal cases. In Section 5.3 lapped transforms are related to multi-input multi-output discrete systems with memory laying the theoretical basis for the understanding of the factorization of a lapped transform. Such a factorization is then presented in Section 5.4. Section 5.5 introduces hierarchical lapped transforms (which are constructed by connecting transforms hierarchically in a tree path), briefly introducing time-frequency diagrams and concepts such as the exchange of resolution between time and frequency. Another concept is also introduced in Section 5.5: variable length lapped transforms, which are also found through hierarchical connection of systems. Practical transforms are then presented. Transforms with symmetric bases including the popular lapped orthogonal transform, its bi-orthogonal, and generalized versions are described in Section 5.6, while fast transforms with variable-length are presented in Section 5.7. The transforms based on cosine modulation are presented in Section 5.8. In order to apply lapped transforms to images, one has to be able to transform signal segments of finite-length. Several methods for doing so are discussed in Section 5.9. Design issues for lapped transforms are discussed in Section 5.10, wherein the emphasis is given to compression applications. In Section 5.11, image compression systems are briefly introduced, including JPEG and other methods based on wavelet transforms. The performance analysis of lapped transforms in image compression is carried in Section 5.12 for different compression systems and several transforms. Finally, the conclusions are presented in Section 5.13.

5.1.1   Notation

Notation conventions used here: $\mathbf{I}_n$ is the $n \times n$ identity matrix; $\mathbf{0}_n$ is the $n \times n$ null matrix, while $\mathbf{0}_{n \times m}$ stands for the $n \times m$ null matrix. $\mathbf{J}_n$ is the $n \times n$ counter-identity, or exchange, or reversing matrix, illustrated by the following example:

$$\mathbf{J}_3 = \begin{bmatrix} 0 & 0 & 1 \\ 0 & 1 & 0 \\ 1 & 0 & 0 \end{bmatrix}.$$

$\mathbf{J}$ reverses the ordering of elements of a vector. $[\ ]^T$ means transposition. $[\ ]^H$ means transposition combined with complex conjugation, where this combination is usually called the Hermitian conjugation of the vector or matrix. Unidimensional concatenation of matrices and vectors is indicated by a comma. In general, capital bold face letters are reserved for matrices, so that $\mathbf{a}$ represents a (column) vector while $\mathbf{A}$ represents a matrix.

5.1.2   Brief History

In the early 1980s, transform coding was maturing, and the discrete cosine transform (DCT) [45] was the preferred transformation method. At that time, DCT-based image compression was state-of-the-art, but researchers were uncomfortable with the blocking artifacts which are common (and annoying) artifacts found in images compressed at low bit rates using block transforms. To resolve the problem, the idea of a lapped transform (LT, for short) was developed in the early 1980s at M.I.T. The idea was to extend the basis function beyond the block boundaries, creating an overlap, in order to eliminate the blocking effect. This idea was not new at that time. However, the new ingredient was to preserve the number of transform coefficients and orthogonality, just as in the nonoverlapped case. Cassereau [5] introduced the lapped orthogonal transform (LOT). It was Malvar [18, 19, 20] who gave the LOT an elegant design strategy and a fast algorithm, thus making the LOT practical and a serious contender to replace the DCT for image compression.

It was also Malvar [22] who pointed out the equivalence between an LT and a multirate filter bank, which is now a very popular signal processing tool [63]. Based on cosine-modulated filter banks [33], modulated lapped transforms were designed [21, 48]. Modulated transforms were later generalized for an arbitrary overlap, creating the class of extended lapped transforms (ELT) [24]–[27]. Recently a new class of LTs with symmetric bases were developed yielding the class of generalized LOTs (GenLOT) [35, 37, 40]. The GenLOTs were made to have basis functions of arbitrary length (not a multiple of the block size) [57], extended to the nonorthogonal case [61], and even made to have filters of different lengths [60]. As mentioned before, filter banks and LTs are the same, although studied independently in the past. Because of this duality, it would be impractical to mention all related work in the field. Nevertheless, Vaidyanathan's book [63] is considered an excellent text on filter banks, while Malvar [26] is a good reference to bridge the gap between lapped transforms and filter banks. We usually refer to LTs as uniform critically sampled FIR filter banks with fast implementation algorithms based on special factorizations of the basis functions, with particular design attention for signal (mainly image) coding.

5.1.3 Block Transforms

We assume a one-dimensional input sequence $\{x(n)\}$ which is transformed into several coefficients sequences $\{y_i(n)\}$, where $y_i(n)$ would belong to the i-th subband. In traditional block-transform processing, the signal is divided into blocks of M samples, and each block is processed independently [6, 12, 26, 32, 43, 45, 46]. Let the samples in the m-th block be denoted as

$$\mathbf{x}_m^T = \left[x_0(m), x_1(m), \ldots, x_{M-1}(m)\right] , \qquad (5.1)$$

with $x_k(m) = x(mM + k)$, and let the corresponding transform vector be

$$\mathbf{y}_m^T = \left[y_0(m), y_1(m), \ldots, y_{M-1}(m)\right] . \qquad (5.2)$$

For a real unitary transform $\mathbf{A}$, $\mathbf{A}^T = \mathbf{A}^{-1}$. The forward and inverse transforms for the m-th block are, respectively,

$$\mathbf{y}_m = \mathbf{A}\mathbf{x}_m , \qquad (5.3)$$

and

$$\mathbf{x}_m = \mathbf{A}^T\mathbf{y}_m . \qquad (5.4)$$

The rows of $\mathbf{A}$, denoted $\mathbf{a}_n^T$ ($0 \leq n \leq M - 1$), are called the basis vectors because they form an orthogonal basis for the M-tuples over the real field [46]. The transform coefficients $[y_0(m), y_1(m), \ldots, y_{M-1}(m)]$ represent the corresponding weights of vector $\mathbf{x}_m$ with respect to the above basis.

If the input signal is represented by vector $\mathbf{x}$ while the subbands are grouped into blocks in vector $\mathbf{y}$, we can represent the transform $\mathbf{H}$ which operates over the entire signal as a block diagonal matrix:

$$\mathbf{H} = diag\{\ldots, \mathbf{A}, \mathbf{A}, \mathbf{A}, \ldots\} , \qquad (5.5)$$

where, of course, $\mathbf{H}$ is an orthogonal matrix if $\mathbf{A}$ is also. In summary, a signal is transformed by block segmentation followed by block transformation, which amounts to transforming the signal with a sparse matrix. Also, it is well known that the signal energy is preserved under a unitary transformation [12, 45], assuming stationary signals, i.e.,

$$M\sigma_x^2 = \sum_{i=0}^{M-1} \sigma_i^2 , \qquad (5.6)$$

where σ_i^2 is the variance of $y_i(m)$ and σ_x^2 is the variance of the input samples.

5.1.4   Factorization of Discrete Transforms

For our purposes, discrete transforms of interest are linear and governed by a square matrix with real entries. Square matrices can be factorized into a product of sparse matrices of the same size. Notably, orthogonal matrices can be factorized into a product of plane (Givens) rotations [10]. Let $\mathbf{A}$ be an $M \times M$ real orthogonal matrix, and let $\mathbf{\Theta}(i, j, \theta_n)$ be a matrix with entries Θ_{kl}, which is like the identity matrix $\mathbf{I}_M$ except for four entries:

$$\Theta_{ii} = \cos(\theta_n) \quad \Theta_{jj} = \cos(\theta_n) \quad \Theta_{ij} = \sin(\theta_n) \quad \Theta_{ji} = -\sin(\theta_n) . \qquad (5.7)$$

$\mathbf{\Theta}(i, j, \theta_n)$ corresponds to a rotation by the angle θ_n about an axis normal to the i-th and the j-th axes. Then, $\mathbf{A}$ can be factorized as

$$\mathbf{A} = \mathbf{S} \prod_{i=0}^{M-2} \prod_{j=i+1}^{M-1} \mathbf{\Theta}(i, j, \theta_n) \qquad (5.8)$$

where n is increased by one for every matrix, and $\mathbf{S}$ is a diagonal matrix with entries ± 1 to correct for any sign error [10]. This correction is not necessary in most cases and is not required if we can apply variations of the rotation matrix defined in Eq. (5.7) as

$$\Theta_{ii} = \cos(\theta_n) \quad \Theta_{jj} = -\cos(\theta_n) \quad \Theta_{ij} = \sin(\theta_n) \quad \Theta_{ji} = \sin(\theta_n) . \qquad (5.9)$$

All combinations of pairs of axes shall be used for a complete factorization. Fig. 5.1(a) shows an example of the factorization of a 4×4 orthogonal matrix into plane rotations [the sequence of rotations is slightly different than the one in Eq. (5.8) but it is equally complete]. If the matrix is not orthogonal, we can always decompose the matrix using singular value decomposition (SVD) [10]. $\mathbf{A}$ is decomposed through SVD as

$$\mathbf{A} = \mathbf{U\Lambda V} \qquad (5.10)$$

where $\mathbf{U}$ and $\mathbf{V}$ are orthogonal matrices and $\mathbf{\Lambda}$ is a diagonal matrix containing the singular values of $\mathbf{A}$. While $\mathbf{\Lambda}$ is already a sparse matrix, we can further decompose the orthogonal matrices using Eq. (5.8):

$$\mathbf{A} = \mathbf{S} \left(\prod_{i=0}^{M-2} \prod_{j=i+1}^{M-1} \Theta\left(i, j, \theta_n^U\right) \right) \mathbf{\Lambda} \left(\prod_{i=0}^{M-2} \prod_{j=i+1}^{M-1} \Theta\left(i, j, \theta_n^V\right) \right) \qquad (5.11)$$

where $\{\theta_n^U \text{ and } \theta_n^V\}$ compose the set of angles for $\mathbf{U}$ and $\mathbf{V}$, respectively. Fig. 5.1(c) illustrates the factorization for a 4×4 nonorthogonal matrix, where α_i are the singular values.

The reader will later see that the factorization above is an invaluable tool for the design of block and lapped transforms. In the orthogonal case, all of the degrees of freedom are contained in the rotation angles. In an $M \times M$ orthogonal matrix, there are $M(M - 1)/2$ angles, and by spanning all the angles' space (0 to 2π for each one) one spans the space of all $M \times M$ orthogonal matrices. The idea is to span the space of all possible orthogonal matrices through varying arbitrarily and freely the rotation angles in an unconstrained optimization. In the general case, there are M^2 degrees of freedom, and we can either utilize the matrix entries directly or employ the SVD decomposition. However, we are mainly concerned with invertible matrices. Hence, using the SVD-based method, one can stay in the invertible matrix space by freely spanning the angles. The only mild constraint here is to assure that all singular values in the diagonal matrix are nonzero. The authors commonly use the unconstrained nonlinear optimization based on the simplex search provided by MATLAB^{TM} to search for the optimal rotation angles and singular values.

5.1.5 Discrete MIMO Linear Systems

Let a multi-input multi-output (MIMO) [63] discrete linear FIR system have M input and M output sequences with respective Z-transforms $X_i(z)$ and $Y_i(z)$, for $0 \leq i \leq M - 1$. Then, $X_i(z)$ and $Y_i(z)$ are related by

$$\begin{bmatrix} Y_0(z) \\ Y_1(z) \\ \vdots \\ Y_{M-1}(z) \end{bmatrix} = \begin{bmatrix} E_{0,0}(z) & E_{0,1}(z) & \cdots & E_{0,M-1}(z) \\ E_{1,0}(z) & E_{1,1}(z) & \cdots & E_{1,M-1}(z) \\ \vdots & \vdots & \ddots & \vdots \\ E_{M-1,0}(z) & E_{M-1,1}(z) & \cdots & E_{M-1,M-1}(z) \end{bmatrix} \begin{bmatrix} X_0(z) \\ X_1(z) \\ \vdots \\ X_{M-1}(z) \end{bmatrix} \qquad (5.12)$$

where $E_{ij}(z)$ are entries of the given MIMO system $\mathbf{E}(z)$. $\mathbf{E}(z)$ is called the transfer matrix of the system, and we have chosen it to be square for simplicity. It is a regular matrix whose entries are polynomials. Of relevance to us is the case wherein the

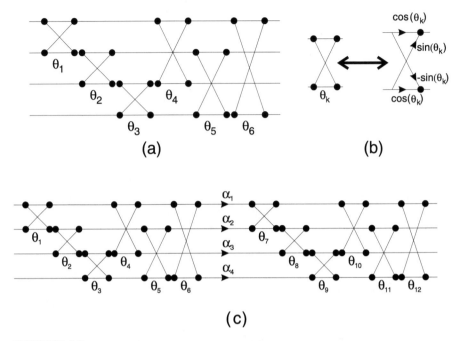

FIGURE 5.1
**Factorization of a 4x4 matrix. (a) Orthogonal factorization into Givens rotations.
(b) Details of the rotation element. (c) Factorization of a nonorthogonal matrix
through SVD with the respective factorization of SVD's orthogonal factors into
rotations.**

entries belong to the field of real-coefficient polynomials of z^{-1}; i.e., the entries
represent real-coefficient FIR filters. The degree of $\mathbf{E}(z)$ (or the McMillan degree,
N_z) is the minimum number of delays necessary to implement the system. The order
of $\mathbf{E}(z)$ is the maximum degree among all $E_{ij}(z)$. In both cases we assume that the
filters are causal and FIR.

A special subset of great interest is comprised of the transfer matrices that are
normalized paraunitary. In the paraunitary case, $\mathbf{E}(z)$ becomes a unitary matrix when
evaluated on the unit circle:

$$\mathbf{E}^H\left(e^{j\omega}\right)\mathbf{E}\left(e^{j\omega}\right) = \mathbf{E}\left(e^{j\omega}\right)\mathbf{E}^H\left(e^{j\omega}\right) = \mathbf{I}_M .\qquad(5.13)$$

Furthermore:

$$\mathbf{E}^{-1}(z) = \mathbf{E}^T\left(z^{-1}\right) .\qquad(5.14)$$

For causal inverses of paraunitary systems,

$$\mathbf{E}'(z) = z^{-n}\mathbf{E}^T\left(z^{-1}\right)\qquad(5.15)$$

is often used, where n is the order of $\mathbf{E}(z)$, since $\mathbf{E}'(z)\mathbf{E}(z) = z^{-n}\mathbf{I}_M$.

For paraunitary systems, the determinant of $\mathbf{E}(z)$ is of the form az^{-N_z} for a real constant a [63], where we recall that N_z is the McMillan degree of the system. For FIR causal entries, they are also said to be lossless systems [63]. In fact, a familiar orthogonal matrix is one where all $E_{ij}(z)$ are constant for all z.

We also have interest in invertible, although nonparaunitary, transfer matrices. In this case, it is required that the matrix be invertible on the unit circle, i.e., for all $z = e^{j\omega}$ and real ω. Nonparaunitary systems are also called bi-orthogonal or perfect reconstruction (PR) [63].

5.1.6   Block Transform as a MIMO System

The sequences $\{x_i(m)\}$ in Eq. (5.1) are called the polyphase components of the input signal $\{x(n)\}$. On the other hand, the sequences $\{y_i(m)\}$ in Eq. (5.2) are the subbands resulting from the transform process. In an alternative view of the transformation process, the signal samples are "blocked" or parallelized into polyphase components through a sequence of delays and decimators as shown in Fig. 5.2. Each block is transformed by system $\mathbf{A}$ into M subband samples (transformed samples). Inverse transform (for orthogonal transforms) is accomplished by system $\mathbf{A}^T$ whose outputs are polyphase components of the reconstructed signal, which are then serialized by a sequence of upsamplers and delays. In this system, blocks are processed independently. Therefore, the transform can be viewed as a MIMO system of order 0, i.e., $\mathbf{E}(z) = \mathbf{A}$, and if $\mathbf{A}$ is unitary, so is $\mathbf{E}(z)$ which is obviously also paraunitary. The system matrix relating the polyphase components to the subbands is referred to as the polyphase transfer matrix (PTM).

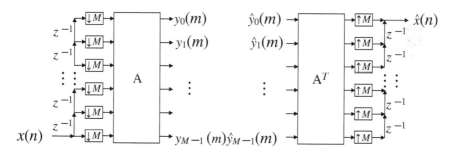

FIGURE 5.2

The signal samples are parallelized into polyphase components through a sequence of delays and decimators ($\downarrow M$ means subsampling by a factor of M). The signal is "blocked," and each block is transformed by system A into M subband samples (transformed samples). Inverse transform (for orthogonal transforms) is accomplished by system A^T whose outputs are polyphase components of the reconstructed signal, which are then serialized by a sequence of upsamplers ($\uparrow M$ means upsampling by a factor of M, padding the signal with $M-1$ zeros) and delays.

5.2   Lapped Transforms

The motivation for a transform with overlap, as mentioned in the introduction, is to try to improve the performance of block (nonoverlapped) transforms for image and signal compression. Compression commonly implies signal losses due to quantization [12]. As the bases of block transforms do not overlap, there may be discontinuities along the boundary regions of the blocks. Different approximations of those boundary regions on each side of the border may cause an artificial "edge" between blocks, the so-called *blocking* effect. Fig. 5.3 shows an example signal that is to be projected into bases, by segmenting the signal into blocks and projecting each segment into the desired bases. Alternatively, one can view the process as projecting the whole signal into several translated bases (one translation per block). Fig. 5.3 shows, on the left, translated versions of the first basis of the DCT in order to account for all the different blocks. The same figure, on the right, shows the same diagram for the first basis of a typical short LT. Note that the bases overlap spatially. The idea is that overlap would help decrease, if not eliminate, the *blocking* effect.

There are M basis functions for either the DCT or the LT, although Fig. 5.3 shows only one of them. An example of the bases for $M = 8$ is shown in Fig. 5.4 which plots the bases for the DCT and for the LOT, a particular LT discussed later. The reader may note that not only are the LOT bases longer but they are also smoother than the DCT counterpart. Fig. 5.5(a) is an example of an image compressed using the standard JPEG baseline coder [32], and the blocking artifacts at the boundaries of 8×8 pixels blocks are readily seen. By replacing the DCT with the LOT and keeping the same compression ratio, we obtain the image shown in Fig. 5.5(b), where blocking is largely reduced. This brief introduction to the motivation behind the development of LTs illustrates only the overall problem. We have not described the details on how to apply LTs. The following section develops the LT framework.

5.2.1   Orthogonal Lapped Transforms

A lapped transform [26] can be generally defined as any transform whose basis vectors have length L, such that $L > M$, extending across traditional block boundaries. Thus, the transform matrix is no longer square, and most of the equations valid for block transforms do not apply to an LT. We will concentrate our efforts on *orthogonal* LTs [26] and consider $L = NM$, where N is the overlap factor. Note that N, M, and hence L are all integers. As in the case of block transforms, we define the transform matrix as containing the orthonormal basis vectors as its rows. A lapped transform matrix $\mathbf{P}$ of dimensions $M \times L$ can be divided into square $M \times M$ submatrices $\mathbf{P}_i$ $(i = 0, 1, \dots, N - 1)$ as follows:

$$\mathbf{P} = \begin{bmatrix} \mathbf{P}_0 \ \mathbf{P}_1 \ \cdots \ \mathbf{P}_{N-1} \end{bmatrix} . \tag{5.16}$$

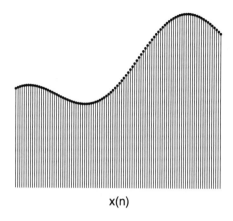

x(n)

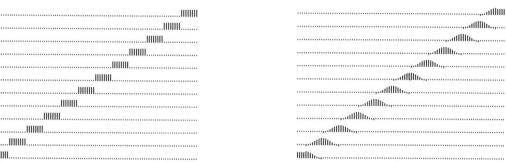

FIGURE 5.3
The example discrete signal $x(n)$ is to be projected onto a number of bases. Left: spatially displaced versions of the first DCT basis. Right: spatially displaced versions of the first basis of a typical short LT.

The orthogonality property does not hold because **P** is no longer a square matrix, and it is replaced by the perfect reconstruction (PR) property [26], defined by

$$\sum_{i=0}^{N-1-l} \mathbf{P}_i \mathbf{P}_{i+l}^T = \sum_{i=0}^{N-1-l} \mathbf{P}_{i+l}^T \mathbf{P}_i = \delta(l)\mathbf{I}_M , \tag{5.17}$$

for $l = 0, 1, \dots , N-1$, where $\delta(l)$ is the Kronecker delta; i.e., $\delta(0) = 1$ and $\delta(l) = 0$ for $l \neq 0$. As will be seen later, Eq. (5.17) states the PR conditions and orthogonality of the transform operating over the entire signal.

If we divide the signal into blocks, each of size M, we would have vectors $\mathbf{x}_m$ and $\mathbf{y}_m$ as in Eqs. (5.1) and (5.2). These blocks are not used by LTs in a straightforward manner. The actual vector that is transformed by the matrix **P** has to have L samples, and, at block number m, it is composed of the samples of $\mathbf{x}_m$ plus $L - M$ samples from the neighboring blocks. These samples are chosen by picking $(L - M)/2$ samples on each side of the block $\mathbf{x}_m$, as shown in Fig. 5.6, for $N = 2$. However, the number

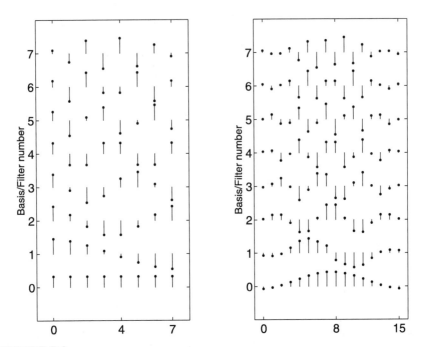

FIGURE 5.4
Bases for the 8-point DCT ($M = 8$) (left) and for the LOT (right) with $M = 8$. The LOT is a particular LT which will be explained later.

of transform coefficients at each step is M, and, in this respect, there is no change in the way we represent the transform-domain blocks $\mathbf{y}_m$.

The input vector of length L is denoted as $\mathbf{v}_m$, which is centered around the block $\mathbf{x}_m$, and is defined as

$$\mathbf{v}_m^T = \left[x\left(mM - (N-1)\frac{M}{2} \right) \cdots x\left(mM + (N+1)\frac{M}{2} - 1 \right) \right]. \qquad (5.18)$$

Then, we have

$$\mathbf{y}_m = \mathbf{P}\mathbf{v}_m. \qquad (5.19)$$

The inverse transform is not direct as in the case of block transforms; with the knowledge of $\mathbf{y}_m$ we know neither the samples in the support region of $\mathbf{v}_m$, nor those in the support region of $\mathbf{x}_m$. We can reconstruct a vector $\hat{\mathbf{v}}_m$ from $\mathbf{y}_m$, as

$$\hat{\mathbf{v}}_m = \mathbf{P}^T \mathbf{y}_m, \qquad (5.20)$$

where $\hat{\mathbf{v}}_m \neq \mathbf{v}_m$. To reconstruct the original sequence, it is necessary to accumulate the results of the vectors $\hat{\mathbf{v}}_m$, in a sense that a particular sample $x(n)$ will be reconstructed from the sum of the contributions it receives from all $\hat{\mathbf{v}}_m$. This additional

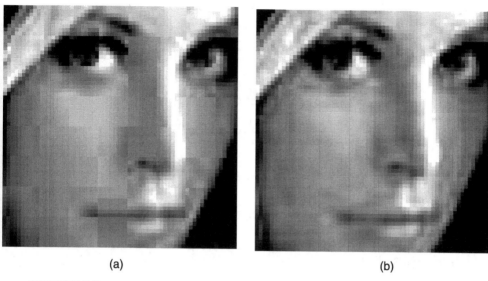

(a)   (b)

FIGURE 5.5
Zoom of image compressed using JPEG at 0.5 bits/per pixel. (a) DCT, (b) LOT.
Reproduced by Special Permission of *Playboy* magazine. Copyright ©1972, 2000
by Playboy.

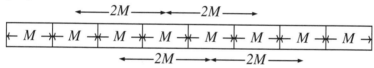

FIGURE 5.6
The signal samples are divided into blocks of M samples. The lapped transform
uses neighboring blocks samples, as in this example for $N = 2$; i.e., $L = 2M$,
yielding an overlap of $(L - M)/2 = M/2$ samples on either side of a block.

complication comes from the fact that **P** is not a square matrix [26]. However, the entire analysis-synthesis system (applied to the entire input vector) is still orthogonal, assuring the PR property using Eq. (5.20).

We can also describe the above process using a sliding rectangular window applied over the samples of $\{x(n)\}$. As an M-sample block, $\mathbf{y}_m$ is computed using $\mathbf{v}_m$, and $\mathbf{y}_{m+1}$ is computed from $\mathbf{v}_{m+1}$ which is obtained by shifting the window to the right by M samples, as shown in Fig. 5.7.

As the reader may have noticed, the region of support of all vectors $\mathbf{v}_m$ is greater than the region of support of the input vector. Hence, a special treatment has to be given to the transform at the borders. We will discuss this operation later and assume infinite-length signals until then. We can also assume that the signal length is very large and the borders of the signal are far enough from the region on which we are focusing our attention.

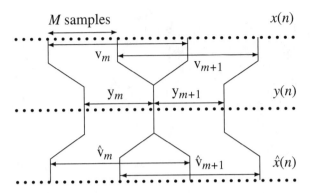

FIGURE 5.7
Illustration of a lapped transform with $N = 2$ applied to signal $x(n)$, yielding transform domain signal $y(n)$. The input L-tuple as vector $\mathbf{v}_m$ is obtained by a sliding window advancing M samples, generating $\mathbf{y}_m$. This sliding is also valid for the synthesis side.

If we denote by $\mathbf{x}$ the input vector and by $\mathbf{y}$ the transform-domain vector, we can be consistent with our notation of transform matrices by defining a matrix $\mathbf{H}$ such that $\mathbf{y} = \mathbf{Hx}$ and $\hat{\mathbf{x}} = \mathbf{H}^T \mathbf{y}$. In this case, we have

$$\mathbf{H} = \begin{bmatrix} \ddots & & & & \mathbf{0} \\ & \mathbf{P} & & & \\ & & \mathbf{P} & & \\ & & & \mathbf{P} & \\ \mathbf{0} & & & & \ddots \end{bmatrix}, \qquad (5.21)$$

where the displacement of the matrices $\mathbf{P}$ obeys the following:

$$\mathbf{H} = \begin{bmatrix} \ddots & \ddots & & \ddots & & \mathbf{0} \\ & \mathbf{P}_0 & \mathbf{P}_1 & \cdots & \mathbf{P}_{N-1} & \\ & & \mathbf{P}_0 & \mathbf{P}_1 & \cdots & \mathbf{P}_{N-1} \\ \mathbf{0} & & & \ddots & \ddots & \ddots \end{bmatrix}. \qquad (5.22)$$

$\mathbf{H}$ has as many block-rows as transform operations over each vector $\mathbf{v}_m$.

Let the rows of $\mathbf{P}$ be denoted by $1 \times L$ vectors $\mathbf{p}_i^T$ ($0 \leq i \leq M - 1$), so that $\mathbf{P}^T = [\mathbf{p}_0, \ldots, \mathbf{p}_{M-1}]$. In an analogy to the block transform case, we have

$$y_i(m) = \mathbf{p}_i^T \mathbf{v}_m . \qquad (5.23)$$

The vectors $\mathbf{p}_i$ are the basis vectors of the lapped transform. They form an orthogonal basis for an M-dimensional subspace (there are only M vectors) of the L-tuples over the real field. As a remark, assuming infinite length signals, from the orthogonality of the basis vectors and from the PR property in Eq. (5.17), the energy is preserved, such that Eq. (5.6) is valid.

In order to compute the variance of the subband signals of a block or lapped transform, assume that $\{x(n)\}$ is a zero-mean stationary process with a given autocorrelation function. Let its $L \times L$ autocorrelation matrix be $\mathbf{R}_{xx}$. Then, from Eq. (5.23)

$$E\left[y_i(m)\right] = \mathbf{p}_i^T E\left[\mathbf{v}_m\right] = \mathbf{p}_i^T \mathbf{0}_{L \times 1} = 0 , \tag{5.24}$$

so that

$$\sigma_i^2 = E\left[y_i^2(m)\right] = \mathbf{p}_i^T E\left[\mathbf{v}_m \mathbf{v}_m^T\right] \mathbf{p}_i = \mathbf{p}_i^T \mathbf{R}_{xx} \mathbf{p}_i ; \tag{5.25}$$

i.e., the output variance is easily computed from the input autocorrelation matrix for a given set of bases $\mathbf{P}$.

Assuming that the entire input and output signals are represented by the vectors $\mathbf{x}$ and $\mathbf{y}$, respectively, and that the signals have infinite length, we have, from Eq. (5.21),

$$\mathbf{y} = \mathbf{H}\mathbf{x} \tag{5.26}$$

and, if $\mathbf{H}$ is orthogonal,

$$\mathbf{x} = \mathbf{H}^T \mathbf{y} . \tag{5.27}$$

Note that $\mathbf{H}$ is orthogonal if and only if Eq. (5.17) is satisfied. Thus, the meaning for Eq. (5.17) becomes clear, as it forces the transform operating over the entire input-output signals to be orthogonal. Hence, the resulting LT is called orthogonal. For block transforms, as there is no overlap, it is sufficient to state the orthogonality of $\mathbf{A}$ because $\mathbf{H}$ will be a block-diagonal matrix.

These formulations for LTs are general, and if the transform satisfies the PR property described in Eq. (5.17), then the LTs are independent of the contents of the matrix $\mathbf{P}$. The definition of $\mathbf{P}$ with a given N can accommodate any lapped transform whose length of the basis vectors lies between M and NM. For the case of block transforms, $N = 1$; i.e., there is no overlap.

Causal notation — If one is not concerned with particular localization of the transform with respect to the origin $x(0)$ of the signal $\{x(n)\}$, it is possible to change the notation to apply a causal representation. In this case, we can represent $\mathbf{v}_m$ as

$$\mathbf{v}_m^T = \left[\mathbf{x}_{m-N+1}^T, \ldots, \mathbf{x}_{m-1}^T, \mathbf{x}_m^T\right] , \tag{5.28}$$

which is identical to the previous representation, except for a shift in the origin to maintain causality. The block $\mathbf{y}_m$ is found in a similar fashion as

$$\mathbf{y}_m = \mathbf{P}\mathbf{v}_m = \sum_{i=0}^{N-1} \mathbf{P}_{N-1-i}\mathbf{x}_{m-i} . \tag{5.29}$$

Similarly, $\hat{\mathbf{v}}_m$ can be reconstructed as in Eq. (5.20) where the support region for the vector is the same, except that the relation between it and the blocks $\hat{\mathbf{x}}_m$ will be changed accordingly.

5.2.2   Nonorthogonal Lapped Transforms

So far, we have discussed orthogonal LTs where a segment of the input signal is projected onto the basis functions of $\mathbf{P}$, yielding the coefficients (subband samples). The signal is reconstructed by the overlapped projection of the same bases weighted by the subband samples. In the nonorthogonal case, we define another LT matrix $\mathbf{Q}$ as

$$\mathbf{Q} = \begin{bmatrix} \mathbf{Q}_0 & \mathbf{Q}_1 & \cdots & \mathbf{Q}_{N-1} \end{bmatrix}, \tag{5.30}$$

in the same way as we did for $\mathbf{P}$ with the same size. The difference is that $\mathbf{Q}$ instead of $\mathbf{P}$ is used in the reconstruction process so that Eq. (5.20) is replaced by

$$\hat{\mathbf{v}}_m = \mathbf{Q}^T \mathbf{y}_m . \tag{5.31}$$

We also define another transform matrix as

$$\mathbf{H}' = \begin{bmatrix} \ddots & \ddots & & \ddots & & \mathbf{0} \\ & \mathbf{Q}_0 & \mathbf{Q}_1 & \cdots & \mathbf{Q}_{N-1} & \\ & & \mathbf{Q}_0 & \mathbf{Q}_1 & \cdots & \mathbf{Q}_{N-1} \\ \mathbf{0} & & & \ddots & \ddots & & \ddots \end{bmatrix} . \tag{5.32}$$

The forward and inverse transformations are now

$$\mathbf{y} = \mathbf{H}_F \mathbf{x} , \quad \mathbf{x} = \mathbf{H}_I \mathbf{y} . \tag{5.33}$$

In the orthonormal case, $\mathbf{H}_F = \mathbf{H}$ and $\mathbf{H}_I = \mathbf{H}^T$. In the general case, it is required that $\mathbf{H}_I = \mathbf{H}_F^{-1}$. With the choice of $\mathbf{Q}$ as the inverse LT, then $\mathbf{H}_I = \mathbf{H}'^T$, while $\mathbf{H}_F = \mathbf{H}$. Therefore the perfect reconstruction condition is

$$\mathbf{H}'^T \mathbf{H} = \mathbf{I}_\infty . \tag{5.34}$$

The reader can check that the above equation can also be expressed in terms of the LTs $\mathbf{P}$ and $\mathbf{Q}$ as

$$\sum_{k=0}^{N-1-m} \mathbf{Q}_k^T \mathbf{P}_{k+m} = \sum_{k=0}^{N-1-m} \mathbf{Q}_{k+m}^T \mathbf{P}_k = \delta(m)\mathbf{I}_M , \tag{5.35}$$

which establish the general necessary and sufficient conditions for the perfect reconstruction of the signal by using $\mathbf{P}$ as the forward LT and $\mathbf{Q}$ as the inverse LT. Unlike the orthogonal case in Eq. (5.17), here both sets are necessary conditions; there is a total of $2N - 1$ matrix equations.

5.3   LTs as MIMO Systems

As previously discussed in Sections 5.1.3 and 5.1.6, the input signal can be decomposed into M polyphase signals $\{x_i(m)\}$, each sequence having one M-th of the original rate. As there are M subbands $\{y_i(m)\}$ under the same circumstances, and only linear operations are used to transform the signal, there is a MIMO system $\mathbf{F}(z)$ that converts the M polyphase signals to the M subband signals. Those transfer matrices are also called PTM (Section 5.1.6). The same is true for the inverse transform (from subbands $\{\hat{y}_i(m)\}$ to polyphase $\{\hat{x}_i(m)\}$). Therefore, we can use the diagram shown in Fig. 5.8 to represent the forward and inverse transforms. Note that Fig. 5.8

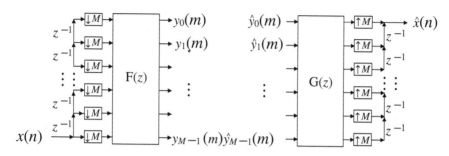

FIGURE 5.8
The filter bank represented as a MIMO system is applied to the polyphase components of the signal. The matrices $\mathbf{F}(z)$ and $\mathbf{G}(z)$ are called polyphase transfer matrices. For a PR system, both must be inverses of each other and for paraunitary filter banks they must be paraunitary matrices; i.e., $\mathbf{G}(z) = \mathbf{F}^{-1}(z) = \mathbf{F}^T(z^{-1})$. For a PR paraunitary causal system of order N, we must choose $\mathbf{G}(z) = z^{-(N-1)}\mathbf{F}^T(z^{-1})$.

is identical to Fig. 5.2 except that the transforms have memory; they depend not only on the present input vector, but on past input vectors also. One can view the system as a clocked one, in which at every clock a block is input, transformed, and output. The parallelization and serialization of blocks are performed by the chain of delays, upsamplers, and downsamplers shown in Fig. 5.8. If we express the forward and inverse PTM as matrix polynomials

$$\mathbf{F}(z) = \sum_{i=0}^{N-1} \mathbf{F}_i z^{-1} , \tag{5.36}$$

$$\mathbf{G}(z) = \sum_{i=0}^{N-1} \mathbf{G}_i z^{-1} , \tag{5.37}$$

then the forward and inverse transforms are given by

$$\mathbf{y}_m = \sum_{i=0}^{N-1} \mathbf{F}_i \mathbf{x}_{m-i} , \tag{5.38}$$

$$\hat{\mathbf{x}}_m = \sum_{i=0}^{N-1} \mathbf{G}_i \hat{\mathbf{y}}_{m-i} . \tag{5.39}$$

In the absence of any processing, $\hat{\mathbf{y}}_m = \mathbf{y}_m$ and $\mathbf{F}(z)$ and $\mathbf{G}(z)$ are connected together back-to-back, so PR is possible if they are inverses of each other. Since the inverse of a causal FIR MIMO system may be noncausal, we can delay the entries of the inverse matrix to make it causal. Since the MIMO system's PTM is assumed to have order N (N is the overlap factor of the equivalent LT), PR requires that

$$\mathbf{G}(z)\mathbf{F}(z) = z^{-N+1}\mathbf{I}_M \quad \rightarrow \quad \mathbf{G}(z) = z^{-N+1}\mathbf{F}^{-1}(z) . \tag{5.40}$$

In this case, $\hat{\mathbf{x}}_m = \mathbf{x}_{m-N+1}$; the signal is perfectly reconstructed after a system's delay. Because of the delay chains combined with the block delay (system's order), the reconstructed signal delay is $\hat{x}(n) = x(n - NM + 1) = x(n - L - 1)$.

By combining Eqs. (5.38), (5.39), and (5.40), we can restate the PR conditions as

$$\sum_{i=0}^{N-1}\sum_{j=0}^{N-1} \mathbf{G}_i\mathbf{F}_j z^{-i-j} = z^{-N+1}\mathbf{I}_M , \tag{5.41}$$

which, by equating the powers of z, can be rewritten as

$$\sum_{k=0}^{N-1-m} \mathbf{G}_k\mathbf{F}_{k+m} = \sum_{k=0}^{N-1-m} \mathbf{G}_{k+m}\mathbf{F}_k = \delta(m)\mathbf{I}_M . \tag{5.42}$$

The reader should note the striking similarity of the above equation with Eq. (5.35). In fact, the simple comparison of the transformation process in the space domain notation Eq. (5.33) against the MIMO system notation in Eqs. (5.38) and (5.39) would reveal the following relations

$$\mathbf{F}_k = \mathbf{P}_{N-1-k}, \qquad \mathbf{G}_k = \mathbf{Q}_k^T \tag{5.43}$$

for $0 \le k < N$. In fact, the conditions imposed in Eqs. (5.34), (5.35), (5.40), and (5.42) are equivalent and each one implies the others. This is a powerful tool in the design of lapped transforms. As an LT, the matrix is nonsquare, but the entries are real. As a MIMO system, the matrix is square, but the entries are polynomials. One form may complement the other, facilitating tasks such as factorization, design, and implementation.

As mentioned earlier, paraunitary (lossless) systems belong to a class of MIMO systems of high interest. Let $\mathbf{E}(z)$ be a paraunitary PTM so that $\mathbf{E}^{-1}(z) = \mathbf{E}^T(z^{-1})$, and let

$$\mathbf{F}(z) = \mathbf{E}(z), \qquad \mathbf{G}(z) = z^{-(N-1)}\mathbf{E}^T\left(z^{-1}\right) . \tag{5.44}$$

As a result, the reader can verify that the equations above imply that $\mathbf{P}_i = \mathbf{Q}_i$, and that

$$\sum_{i=0}^{N-1-l} \mathbf{P}_i\mathbf{P}_{i+l}^T = \sum_{i=0}^{N-1-l} \mathbf{P}_i^T\mathbf{P}_{i+l} = \delta(l)\mathbf{I}_M , \tag{5.45}$$

$$\mathbf{H}\mathbf{H}^T = \mathbf{H}^T\mathbf{H} = \mathbf{I}_\infty . \tag{5.46}$$

In other words, if the system's PTM is paraunitary, then the corresponding LT ($\mathbf{H}$) is orthogonal and vice-versa.

5.4 Factorization of Lapped Transforms

There is an important result for paraunitary PTM which states that any paraunitary $\mathbf{E}(z)$ can be decomposed into a series of orthogonal matrices and delay stages [8, 64]. In this decomposition, there are N_z delay stages and $N_z + 1$ orthogonal matrices, where N_z is the McMillan degree of $\mathbf{E}(z)$ (the degree of the determinant of $\mathbf{E}(z)$). Then,

$$\mathbf{E}(z) = \mathbf{B}_0 \prod_{i=1}^{N_z} (\mathbf{\Upsilon}(z)\mathbf{B}_i) \tag{5.47}$$

where $\mathbf{\Upsilon}(z) = diag\{z^{-1}, 1, 1, \ldots, 1\}$, and $\mathbf{B}_i$ are orthogonal matrices. It is well known that an $M \times M$ orthogonal matrix can be expressed as a product of $M(M-1)/2$ plane rotations. However, in this case, only $\mathbf{B}_0$ is a general orthogonal matrix, while the matrices $\mathbf{B}_1$ through $\mathbf{B}_{N_z}$ have only $M - 1$ degrees of freedom [64].

This result states that it is possible to implement an orthogonal lapped transform using a sequence of delays and orthogonal matrices. It also defines the total number of degrees of freedom in a lapped transform; if one changes arbitrarily any of the plane rotations composing the orthogonal transforms, one will span all possible orthogonal lapped transforms, for given values of M and L. It is also possible to prove [35] that the (McMillan) degree of $\mathbf{E}(z)$ is bounded by $N_z \leq (L - M)/2$ with equality for a general structure to implement all orthogonal LTs whose bases have length up to $L = NM$, i.e., $\mathbf{E}(z)$ of order $N - 1$.

In fact Eq. (5.47) may be used to implement all lapped transforms (orthogonal or not) whose degree is N_z. To accomplish that, all of the multiplicative factors that compose the PTM must be invertible. Let us consider a more particular factorization:

$$\mathbf{F}(z) = \prod_{i=0}^{(N-1)/(K-1)} \mathbf{B}_i(z) \tag{5.48}$$

where $\mathbf{B}_i(z) = \sum_{k=0}^{K-1} \mathbf{B}_{ik} z^{-k}$ is a stage of order $K - 1$. If $\mathbf{F}(z)$ is paraunitary, then all $\mathbf{B}_i(z)$ must be paraunitary, so that perfect reconstruction is guaranteed if

$$\mathbf{G}(z) = z^{-N+1}\mathbf{F}^T\left(z^{-1}\right) = \prod_{i=(N-1)/(K-1)}^{0} \left(\sum_{k=0}^{K-1} \mathbf{B}_{ik}^T z^{-(K-1-k)}\right). \tag{5.49}$$

If PTM is not paraunitary, all factors have to be invertible in the unit circle for PR. More strongly put, there must be factors $\mathbf{C}_i(z)$ of order $K - 1$ such that

$$\mathbf{C}_i(z)\mathbf{B}_i(z) = z^{-K+1}\mathbf{I}_M . \tag{5.50}$$

Then the inverse PTM is simply given by

$$\mathbf{G}(z) = \prod_{i=(N-1)/(K-1)}^{0} \mathbf{C}_i(z) . \tag{5.51}$$

With this factorization, the design of $\mathbf{F}(z)$ is broken down to the design of $\mathbf{B}_i(z)$. Lower-order factors simplify the constraint analysis and facilitate the design of a useful transform, either paraunitary or invertible. It is even more desirable to factor the PTM as

$$\mathbf{F}(z) = \mathbf{B}_0 \prod_{i=0}^{N-1} \mathbf{\Lambda}(z)\mathbf{B}_i \qquad (5.52)$$

where $\mathbf{B}_i$ are square matrices and $\mathbf{\Lambda}(z)$ is a paraunitary matrix containing only entries 1 and z^{-1}. In this case, if the PTM is paraunitary

$$\mathbf{G}(z) = \left(\prod_{i=N-1}^{0} \mathbf{B}_i^T \tilde{\mathbf{\Lambda}}(z) \right) \mathbf{B}_0^T \qquad (5.53)$$

where $\tilde{\mathbf{\Lambda}}(z) = z^{-1}\mathbf{\Lambda}(1/z)$. If the PTM is not paraunitary, then

$$\mathbf{G}(z) = \left(\prod_{i=N-1}^{0} \mathbf{B}_i^{-1} \tilde{\mathbf{\Lambda}}(z) \right) \mathbf{B}_0^{-1} \; ; \qquad (5.54)$$

in other words, the design can be simplified by applying only invertible real matrices $\mathbf{B}_i$. This factorization approach is the basis for most useful LTs. It allows efficient implementation and design. We will discuss some useful LTs later on. For example, for M even, the symmetric delay factorization (SDF) is quite useful. In that,

$$\mathbf{\Lambda}(z) = \begin{bmatrix} z^{-1}\mathbf{I}_{M/2} & 0 \\ 0 & \mathbf{I}_{M/2} \end{bmatrix}, \qquad \tilde{\mathbf{\Lambda}}(z) = \begin{bmatrix} \mathbf{I}_{M/2} & 0 \\ 0 & z^{-1}\mathbf{I}_{M/2} \end{bmatrix}. \qquad (5.55)$$

The flow graph for implementing an LT which can be parameterized using SDF is shown in Fig. 5.9.

If we are given the SDF matrices instead of the basis coefficients, one can easily construct the LT matrix. For this, start with the last stage and recur the structure in Eq. (5.52) using Eq. (5.55). Let $\mathbf{P}^{(i)}$ be the partial reconstruction of $\mathbf{P}$ after including up to the i-th stage. Then,

$$\mathbf{P}^{(0)} = \mathbf{B}_{N-1} \qquad (5.56)$$

$$\mathbf{P}^{(i)} = \mathbf{B}_{N-1-i} \begin{bmatrix} \mathbf{I}_{M/2} & \mathbf{0}_{M/2} & \mathbf{0}_{M/2} & \mathbf{0}_{M/2} \\ \mathbf{0}_{M/2} & \mathbf{0}_{M/2} & \mathbf{0}_{M/2} & \mathbf{I}_{M/2} \end{bmatrix} \begin{bmatrix} \mathbf{P}^{(i-1)} & \mathbf{0}_M \\ \mathbf{0}_M & \mathbf{P}^{(i-1)} \end{bmatrix} \quad (5.57)$$

$$\mathbf{P} = \mathbf{P}^{(N-1)} \; . \qquad (5.58)$$

Similarly, one can find $\mathbf{Q}$ from the factors $\mathbf{B}_i^{-1}$.

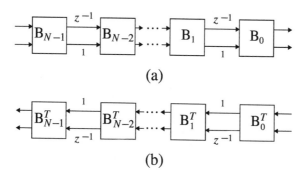

FIGURE 5.9
Flow graph for implementing an LT where $F(z)$ can be factorized using symmetric delays and N stages. Signals $\{x(n)\}$ and $\{y(n)\}$ are segmented and processed using blocks of M samples, all branches carry $M/2$ samples, and blocks B_i are $M \times M$ orthogonal or invertible matrices. (a) Forward transform section, (b) inverse transform section.

5.5 Hierarchical Connection of LTs: An Introduction

So far we have focused on the construction of a single LT resulting in M subband signals. What happens if we cascade LTs by connecting them hierarchically in such a way that a subband signal is the actual input for another LT? Also, what are the consequences of submitting only part of the subband signals to further stages of LTs? We will try to introduce the answers to these questions.

This subject has been intensively studied and many publications are available. Our intent, however, is to provide only a basic introduction, while leaving more detailed analysis to the references. Again, the relation between filter banks and discrete wavelets [53, 63, 65] is well known. Under conditions that are easily satisfied [63], an infinite cascade of filter banks will generate a set of continuous orthogonal wavelet bases. In general, if only the lowpass subband is connected to another filter bank, for a finite number of stages, we call the resulting filter bank a discrete wavelet transform (DWT) [63, 65]. A free cascading of filter banks, however, is better known as discrete wavelet packet (DWP) [7, 68, 34, 53]. As LTs and filter banks are generally equivalent, the same relations apply to LTs and wavelets. The system resulting from the hierarchical association of several LTs is called a hierarchical lapped transform (HLT) [23].

5.5.1 Time-Frequency Diagram

The cascaded connection of LTs is better described with the aid of simplifying diagrams. The first is the time-frequency (TF) diagram. It is based on the TF plane, which is well known from the fields of spectral and time-frequency analysis [31, 3, 4]. The time-frequency representation of signals is a well-known method (for example the time-dependent discrete Fourier transform — DFT — and the construction of spectrograms; see for example [31, 3, 4] for details on TF signal representation). The TF representation is obtained by expressing the signal $\{x(n)\}$ with respect to bases which are functions of both frequency and time. For example, the size-r DFT of a sequence extracted from $\{x(n)\}$ (from $x(n)$ to $x(n + r - 1)$) [31] can be

$$\alpha(k, n) = \sum_{i=0}^{r-1} x(i + n) \exp\left(-\frac{j2\pi ki}{r}\right) . \tag{5.59}$$

Using a sliding window $w(m)$ of length r which is nonzero only in the interval $n \le m \le n + r - 1$ (which in this case is rectangular), we can rewrite the last equation as

$$\alpha(k, n) = \sum_{i=-\infty}^{\infty} x(i) w(i) \exp\left(-\frac{jk(i - n)2\pi}{r}\right) . \tag{5.60}$$

For more general bases we may write

$$\alpha(k, n) = \sum_{i=-\infty}^{\infty} x(i) \phi(n - i, k) \tag{5.61}$$

where $\phi(n, k)$ represents the bases for the space of the signal, n represents the index where the base is located in time, and k is the frequency index.

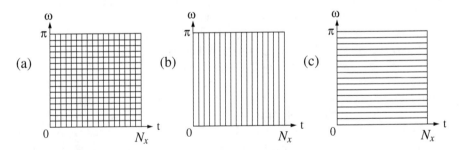

FIGURE 5.10
Examples of rectangular partitions of the time-frequency plane for a signal which has N_x samples. (a) Spectrogram with a N_x-length window, resulting in N_x^2 TF samples, (b) input signal, no processing, (c) a transform such as the DCT or DFT is applied to all N_x samples.

As the signal is assumed to have an infinite number of samples, consider a segment of N_x samples extracted from signal $\{x(n)\}$, which can be extended in any fashion in order to account for the overlap of the window of r samples outside the signal domain.

In such a segment we can construct a spectrogram with a resolution of r samples in the frequency axis and N_x samples in the time axis. Assuming a maximum frequency resolution, we can have a window with length up to $r = N_x$. The diagram for the spectrogram in this case is given in Fig. 5.10(a). We call such diagrams TF diagrams because they indicate only the number of samples used in the TF representation of the signal. Assuming an ideal partition of the TF plane (using filters with ideal frequency response and null transition regions), each TF coefficient represents a distinct region in a TF diagram. Note that in such representations, the signal is represented by N_x^2 TF coefficients. We are looking for a maximally decimated TF representation which is defined as a representation of the signal where the TF plane diagram would be partitioned into N_x regions; N_x TF coefficients will be generated. Also, we require that all N_x samples of $\{x(n)\}$ be able to be reconstructed from the N_x TF coefficients. If we use fewer than N_x samples in the TF plane, we clearly cannot reconstruct all possible combinations of samples in $\{x(n)\}$, from the TF coefficients, using solely linear relations.

Under these assumptions, Fig. 5.10(b) shows the TF diagram for the original signal (only resolution in the time axis) for $N_x = 16$. Also, for $N_x = 16$, Fig. 5.10(c) shows a TF diagram with maximum frequency resolution, which could be achieved by transforming the original N_x-sample sequence with an N_x-sample DCT or DFT.

5.5.2 Tree-Structured Hierarchical Lapped Transforms

The tree diagram is helpful in describing the hierarchical connection of filter banks. In this diagram we represent an M-band LT by nodes and branches of an M-ary tree. Fig. 5.11(a) shows an M-band LT, where all the M subband signals have sampling rates M times smaller than that of $\{x(n)\}$. Fig. 5.11(b) shows the equivalent notation for the LT in a tree diagram, i.e., a single-stage M-branch tree, which is called here a tree cell. Recalling Fig. 5.10, the equivalent TF diagram for an M-band LT is shown in Fig. 5.11(c), for a 16-sample signal and for $M = 4$. Note that the TF diagram of Fig. 5.11(c) resembles that of Fig. 5.10(a). This is because for each 4 samples in $\{x(n)\}$, there is a corresponding set of 4 transformed coefficients. So, the TF representation is maximally decimated. Compared to Fig. 5.10(b), Fig. 5.11(c) implies an exchange of resolution from time to frequency domain achieved by the LT.

The exchange of resolution in the TF diagram can be obtained from the LT. As we connect several LTs following the paths of a tree, each new set of branches (each new tree cell) connected to the tree will force the TF diagram to exchange from time to frequency resolution. We can achieve a more versatile TF representation by connecting cells in unbalanced ways. For example, Fig. 5.12 shows some examples of HLTs given by their tree diagrams and respective TF diagrams. Fig. 5.12(a) depicts the tree diagram for the 3-stages DWT. Note that only the lowpass subband is further processed. Also, as all stages are chosen to be 2-channel LTs, this HLT can be represented by a binary tree. Fig. 5.12(b) shows a more generic hierarchical connection of 2-channel LTs. First the signal is split into lowpass and highpass. Each output branch is further connected to another 2-channel LT. In the third stage, only the most

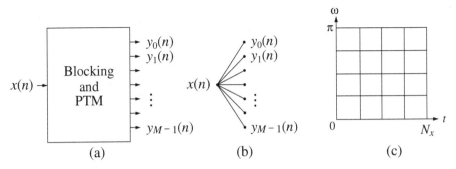

FIGURE 5.11
Representation of an M-channel LT as tree nodes and branches. (a) Forward section of an LT, including the blocking device. (b) Equivalent notation for (a) using an M-branch single-stage tree. (c) Equivalent TF diagram for (a) or (b) assuming $M = 4$ and $N_x = 16$.

lowpass subband signal is connected to another 2-channel LT. Fig. 5.12(c) shows a 2-stage HLT obtaining the same TF diagram as Fig. 5.12(b). Note that the succession of 2-channel LTs was substituted by a single stage 4-channel LT, i.e., the signal is split into four subbands and then one subband is connected to another LT. Fig. 5.12(d) shows the TF diagram corresponding to Fig. 5.12(a), while Fig. 5.12(e) shows the TF diagram corresponding to Figs. 5.12(b) and (c). The reader should note that, as the tree paths are unbalanced, we have irregular partitions of the TF plane. For example, in the DWT, low-frequency TF coefficients have poor time localization and good frequency resolution, while high-frequency ones have poor frequency resolution and better time localization.

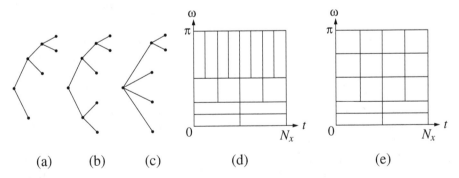

FIGURE 5.12
Tree and TF diagrams. (a) The 3-stage DWT binary-tree diagram, where only the lowpass subband is submitted to further LT stages. (b) A more generic 3-stage tree diagram. (c) A 2-stage tree-diagram resulting in the same TF diagram as (b). (d) TF diagram for (a). (e) TF diagram for (b) or (c).

To better understand how connecting an LT to the tree can achieve the exchange between time and frequency resolutions, see Fig. 5.13 which plots the basis functions resulting from two similar tree-structured HLTs.

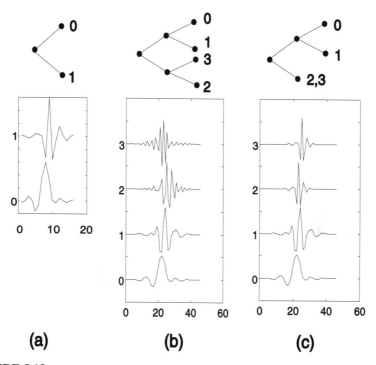

(a) **(b)** **(c)**

FIGURE 5.13

Two HLTs and resulting bases. (a) The 2-channel 16-tap-bases LT, showing low- and high-frequency bases, $f_0(n)$ and $f_1(n)$, respectively. (b) Resulting basis functions of a 2-stage HLT based on (a), given by $f_0(n)$ through $f_3(n)$. Its respective tree diagram is also shown. (c) Resulting HLT, by pruning one high-frequency branch in (b). Note that the two high-frequency basis functions are identical to the high-frequency basis function of (a), and, instead of having two distinct bases for high frequencies, occupying distinct spectral slots, the two bases are, now shifted in time. Thus, better time localization is attainable, at the expense of frequency resolution.

5.5.3 Variable-Length LTs

In the tree-structured method to cascade LTs, every time an LT is added to the structure, more subbands are created by further subdividing previous subbands, so the overall TF diagram of the decomposition is altered. There is a useful alternative to the tree structure in which the number of subbands does not change. See Fig. 5.14, where the "blocking" part of the diagram corresponds to the chain of delays and decimators (as in Fig. 5.8) that parallelizes the signal into polyphase components. System $\mathbf{A}(z)$ of M bases of length $N_A M$ is postprocessed by system $\mathbf{B}(z)$ of K bases

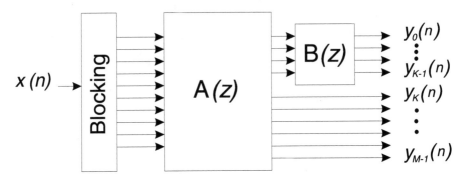

FIGURE 5.14
Cascade of PTMs $\mathbf{A}(z)$ of M channels and $\mathbf{B}(z)$ of K channels. The total number of subbands does not change; however, some of $\mathbf{A}(z)$ bases are increased in length and order.

of length $N_B K$. Clearly, entries in $\mathbf{A}(z)$ have order $N_A - 1$, and entries in $\mathbf{B}(z)$ have order $N_B - 1$. Without loss of generality we associate system $\mathbf{B}(z)$ to the first K output subbands of $\mathbf{A}(z)$. The overall PTM is given by

$$\mathbf{F}(z) = \begin{bmatrix} \mathbf{B}(z) & \mathbf{0} \\ \mathbf{0} & \mathbf{I}_{M-K} \end{bmatrix} \mathbf{A}(z) , \qquad (5.62)$$

where $\mathbf{F}(z)$ has K bases of order $N_A + N_B - 2$ and $M - K$ bases of order $N_A - 1$. As the resulting LT has M channels, the final orders dictate that the first K bases have length $(N_A + N_B - 1)M$ while the others still have length $N_A M$. In other words, the effect of cascading $\mathbf{A}(z)$ and $\mathbf{B}(z)$ was only to modify K bases, so the length of the modified bases is equal to or larger than the length of the initial bases. An example is shown in Fig. 5.15. We start with the bases corresponding to $\mathbf{A}(z)$, shown in Fig. 5.15(a). There are 8 bases of length 16 so that $\mathbf{A}(z)$ has order 1. $\mathbf{A}(z)$ is postprocessed by $\mathbf{B}(z)$ which is a 4×4 PTM of order 3, whose corresponding bases are shown in Fig. 5.15(b). The resulting LT is shown in Fig. 5.15(c). There are 4 bases of length 16, and 4 of length 40. The shorter ones are identical to those in Fig. 5.15(a), while the longer ones have order which is the sum of the orders of $\mathbf{A}(z)$ and $\mathbf{B}(z)$, i.e., order 4, and the shape of the longer bases in $\mathbf{F}(z)$ is very different than the corresponding ones in $\mathbf{A}(z)$.

The effect of postprocessing some bases is a means to construct a new LT with larger bases from an initial one. In fact it can be shown that variable length LTs can be factorized using postprocessing stages [60, 59]. A general factorization of LTs is depicted in Fig. 5.16. Assume a variable-length $\mathbf{F}(z)$ whose bases are arranged in

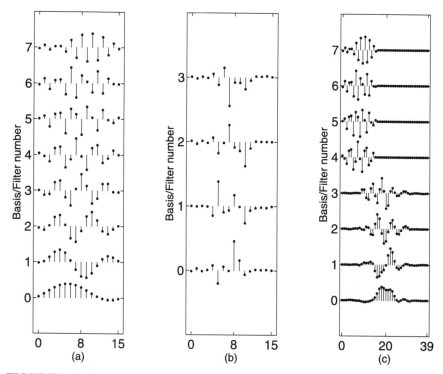

FIGURE 5.15

Example of constructing variable-length bases through cascading LTs. (a) The basis corresponding to A(z): an LT with 8 bases of length 16 (order 1). (b) The basis corresponding to B(z): an LT with 4 bases of length 16 (order 3). (c) The basis corresponding to F(z): 4 of the 8 bases have order 1, i.e., length 16, while the remaining 4 have order 4, i.e., length 40.

decreasing length order. Such a PTM can be factorized as

$$\mathbf{F}(z) = \prod_{i=0}^{M-2} \begin{bmatrix} \mathbf{B}_i(z) & \mathbf{0} \\ \mathbf{0} & \mathbf{I}_i \end{bmatrix} \quad (5.63)$$

where $\mathbf{I}_0$ is understood to be nonexisting and $\mathbf{B}_i(z)$ has size $(M - i) \times (M - i)$. The factors $\mathbf{B}_i$ can have individual orders K_i and can be factorized differently into factors $\mathbf{B}_{ik}(z)$ for $0 \le k < K_i$. Hence,

$$\mathbf{F}(z) = \prod_{i=0}^{M-2} \prod_{k=0}^{K_i-1} \begin{bmatrix} \mathbf{B}_{ik}(z) & \mathbf{0} \\ \mathbf{0} & \mathbf{I}_i \end{bmatrix}. \quad (5.64)$$

A later section presents a very useful LT based on the factorization principles of Eq. (5.64).

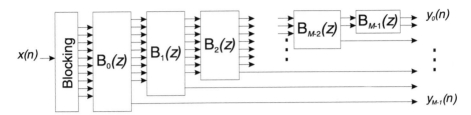

FIGURE 5.16
General factorization of a variable-length LT.

5.6 Practical Symmetric LTs

We have discussed LTs in a general sense as a function of several parameters, such as matrix entries, orthogonal, or invertible factors. The design of an LT suitable for a given application is the single most important step in the study of LTs. In order to do that, one may factorize the LT to facilitate optimization techniques.

An LT with symmetric bases is commonly used in image/video processing and compression applications. By symmetric bases we mean that the entries p_{ij} of $\mathbf{P}$ obey

$$p_{i,j} = (\pm 1)\, p_{i,L-1-j} \,. \tag{5.65}$$

The bases can be either symmetric or antisymmetric. In terms of the PTM, this constraint is given by

$$\mathbf{F}(z) = z^{-(N-1)}\mathbf{SF}\left(z^{-1}\right)\mathbf{J}_M \,, \tag{5.66}$$

where $\mathbf{S}$ is a diagonal matrix whose diagonal entries s_{ii} are ± 1, depending on whether the i-th basis is symmetric (+1) or anti-symmetric (-1). Note that we require that all bases share the same center of symmetry.

5.6.1 The Lapped Orthogonal Transform: LOT

The lapped orthogonal transform (LOT) [18, 19, 20] was the first useful LT with a well-defined factorization. Malvar developed the fast LOT based on the work by Cassereau [5] to provide not only a factorization, but a factorization based on the DCT. The DCT is attractive for many reasons, among them, fast implementation and near-optimal performance for block transform coding [45]. Also, since it is a popular transform, it has a reduced cost and is easily available in either software or hardware. The DCT matrix $\mathbf{D}$ is defined as having entries

$$d_{ij} = \sqrt{\frac{2}{M}}k_i \cos\left(\frac{(2j+1)i\pi}{2M}\right) \tag{5.67}$$

where $k_0 = 1$ and $k_i = 1/\sqrt{2}$, for $1 \leq i \leq M - 1$.

The LOT as defined by Malvar [20] is orthogonal. Then, according to our notation, $\mathbf{P} = \mathbf{Q}$ and $\mathbf{H}^{-1} = \mathbf{H}^T$. It is also a symmetric LT with M even. The LT matrix is given by

$$\mathbf{P}_{LOT} = \begin{bmatrix} \mathbf{I}_M & \mathbf{0} \\ \mathbf{0} & \mathbf{V}_R \end{bmatrix} \begin{bmatrix} \mathbf{D}_e - \mathbf{D}_o & \mathbf{J}_{M/2}\,(\mathbf{D}_e - \mathbf{D}_o) \\ \mathbf{D}_e - \mathbf{D}_o & -\mathbf{J}_{M/2}\,(\mathbf{D}_e - \mathbf{D}_o) \end{bmatrix} \qquad (5.68)$$

where $\mathbf{D}_e$ is the $M/2 \times M$ matrix with the even-symmetric basis functions of the DCT, and $\mathbf{D}_o$ is the matrix of the same size with the odd-symmetric ones. In our notation, $\mathbf{D}_e$ also corresponds to the even numbered rows of $\mathbf{D}$, and $\mathbf{D}_o$ corresponds to the odd numbered rows of $\mathbf{D}$. $\mathbf{V}_R$ is an $M/2 \times M/2$ orthogonal matrix, which according to Malvar and Staelin [20] and Malvar [26] should be approximated by $M/2 - 1$ plane rotations as

$$\mathbf{V}_R = \prod_{i=\frac{M}{2}-2}^{0} \Theta\,(i, i+1, \theta_i) \qquad (5.69)$$

where Θ is defined in Section 5.1.4. Suggestions of rotation angles that were designed to yield a good transform for image compression are [26]

$$\begin{aligned} M = 4 \quad &\rightarrow \quad \theta_0 = 0.1\pi & (5.70) \\ M = 8 \quad &\rightarrow \quad \{\theta_0, \theta_1, \theta_2\} = \{0.13, 0.16, 0.13\} \times \pi & (5.71) \\ M = 16 \quad &\rightarrow \quad \{\theta_0, \dots, \theta_7\} & \\ & \qquad = \{0.62, 0.53, 0.53, 0.50, 0.44, 0.35, 0.23, 0.11\} \times \pi\,. & (5.72) \end{aligned}$$

For $M \geq 16$ it is suggested to use

$$\mathbf{V}_R = \mathbf{D}_{IV}^T \mathbf{D}^T \qquad (5.73)$$

where $\mathbf{D}_{IV}$ is the DCT type IV matrix [45] whose entries are

$$d_{ij}^{IV} = \sqrt{\frac{2}{M}} \cos\left(\frac{(2j+1)(2i+1)\pi}{4M} \right)\,. \qquad (5.74)$$

A block diagram for the implementation of the LOT is shown in Fig. 5.17 for $M = 8$.

5.6.2   The Lapped Bi-Orthogonal Transform: LBT

The LOT is a great improvement over the DCT for image compression mainly because it reduces the so-called blocking effects. Although largely reduced, blocking is not eliminated because the format of the low frequency bases of LOT. In image compression, only a few bases are used to reconstruct the signal. From Fig. 5.4, one can see that the "tails" of the lower frequency bases of the LOT do not decay exactly

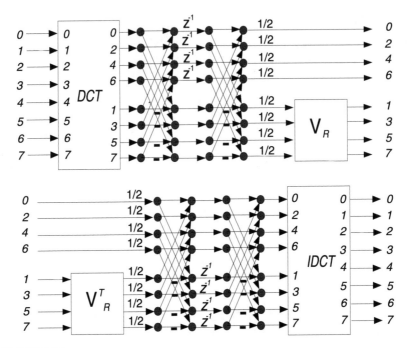

FIGURE 5.17
Implementation of the LOT for $M = 8$**. Top: forward transform; bottom:**
inverse transform.

to zero. For this reason, there is some blocking effect left in images compressed using
the LOT at lower bit rates.

To help resolve this problem, Malvar recently proposed to modify LOT, creating
the lapped bi-orthogonal transform (LBT) [28]. (Bi-orthogonal is a common term in
the filter banks community to designate PR transforms and filter banks that are not
orthogonal.) In any case, the factorization of the LBT is almost identical to that of
the LOT. However,

$$\mathbf{P}_{LBT} = \begin{bmatrix} \mathbf{I}_M & \mathbf{0} \\ \mathbf{0} & \mathbf{V}_R \end{bmatrix} \begin{bmatrix} \mathbf{D}_e - \mathbf{\Upsilon}\mathbf{D}_o & \mathbf{J}_{M/2}\,(\mathbf{D}_e - \mathbf{\Upsilon}\mathbf{D}_o) \\ \mathbf{D}_e - \mathbf{\Upsilon}\mathbf{D}_o & -\mathbf{J}_{M/2}\,(\mathbf{D}_e - \mathbf{\Upsilon}\mathbf{D}_o) \end{bmatrix} \quad (5.75)$$

where $\mathbf{\Upsilon}$ is the $M/2 \times M/2$ diagonal matrix given by $\mathbf{\Upsilon} = diag\{\sqrt{2}, 1, \ldots, 1\}$.
Note that it implies only that one of the DCT's output is multiplied by a constant.
The inverse is given by the LT $\mathbf{Q}_{LBT}$ which is found in an identical manner as in
Eq. (5.75) except that the multiplier is inverted; $\mathbf{\Upsilon} = diag\{1/\sqrt{2}, 1, \ldots, 1\}$. The
diagram for implementing an LBT for $M = 8$ is shown in Fig. 5.18.

Because of the multiplicative factor, the LT is no longer orthogonal. However, the
factor is very easily inverted. The result is a reduction of amplitude of lateral samples
of the first bases of LOT into the new bases of the forward LBT, as seen in Fig. 5.19. In
Fig. 5.19 the reader can note the reduction in the amplitude of the boundary samples

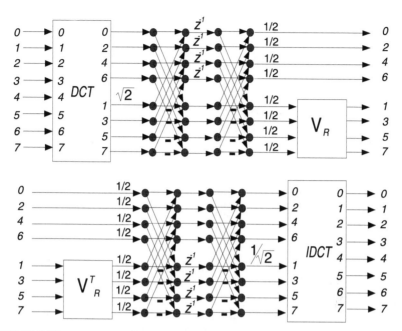

FIGURE 5.18

Implementation of the LBT for $M = 8$. Top: forward transform; bottom: inverse transform. Note that there is only one extra multiplication as compared to LOT.

of the inverse LBT and an enlargement of the same samples in the forward LBT. This simple "trick" noticeably improves the performance of the LOT/LBT for image compression at negligible overhead. Design of the other parameters of the LOT are not changed. It is recommended to use LBT instead of LOT whenever orthogonality is not a crucial constraint.

Another LBT with high practical value is LiftLT [62]. Instead of parameterizing the orthogonal matrix $\mathbf{V}_R$ in Eq. (5.69) by rotation angles, a series of dyadic lifting steps are used to construct $\mathbf{V}_R$, as shown in Fig. 5.20. The $\mathbf{V}_R$ matrix is still invertible, but no longer orthogonal. Hence, the LiftLT is a bi-orthogonal LT. The $\sqrt{2}$ factor in the Malvar's LBT can be replaced by a rational number to facilitate finite-precision implementations. A good rational scaling factor is $\frac{25}{16}$ for the forward transform, and $\frac{16}{25}$ for the inverse transform. The LiftLT offers a VLSI-friendly implementation using integer (even binary) arithmetic. Yet, it does not sacrifice anything in coding performance. It achieves 9.54 dB coding gain (a popular objective measure of transform performance to be described later), compared to LOT's 9.20 dB and LBT's 9.52 dB. It is the first step towards LTs that can map integers to integers and multiplierless LTs that can be implemented using only shift-and-add operations.

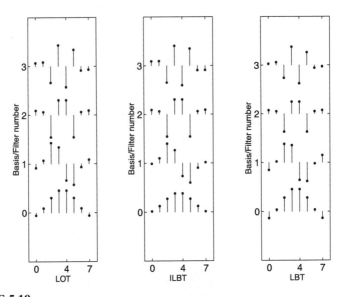

FIGURE 5.19
Comparison of bases for the LOT ($\mathbf{P}_{LOT}$), inverse LBT ($\mathbf{Q}_{LBT}$), and forward LBT ($\mathbf{P}_{LBT}$). The extreme samples of the lower frequency bases of the LOT are larger than those of the inverse LBT. This is an advantage for image compression.

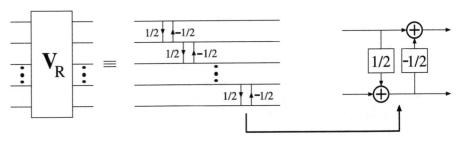

FIGURE 5.20
Parameterization of the $\mathbf{V}_R$ matrix using dyadic lifting steps in the LiftLT.

5.6.3   The Generalized LOT: GenLOT

The formulation for LOT [20] that is shown in Eq. (5.68) is not the most general there is for this kind of LT. In fact, it can be generalized to become

$$\mathbf{P} = \begin{bmatrix} \mathbf{U} & \mathbf{0} \\ \mathbf{0} & \mathbf{V} \end{bmatrix} \begin{bmatrix} \mathbf{D}_e - \mathbf{D}_o & \mathbf{J}_{M/2}\,(\mathbf{D}_e - \mathbf{D}_o) \\ \mathbf{D}_e - \mathbf{D}_o & -\mathbf{J}_{M/2}\,(\mathbf{D}_e - \mathbf{D}_o) \end{bmatrix}. \tag{5.76}$$

As long as **U** and **V** remain orthogonal matrices, LT is orthogonal. In terms of the PTM, $\mathbf{F}(z)$ can be similarly expressed. Let

$$\mathbf{W} = \frac{1}{\sqrt{2}} \begin{bmatrix} \mathbf{I}_{M/2} & \mathbf{I}_{M/2} \\ \mathbf{I}_{M/2} & -\mathbf{I}_{M/2} \end{bmatrix}, \tag{5.77}$$

$$\mathbf{\Phi}_i = \begin{bmatrix} \mathbf{U}_i & \mathbf{0}_{M/2} \\ \mathbf{0}_{M/2} & \mathbf{V}_i \end{bmatrix}, \tag{5.78}$$

$$\mathbf{\Lambda}(z) = \begin{bmatrix} \mathbf{I}_{M/2} & \mathbf{0}_{M/2} \\ \mathbf{0}_{M/2} & z^{-1}\mathbf{I}_{M/2} \end{bmatrix}, \tag{5.79}$$

and let **D** be the $M \times M$ DCT matrix. Then, for the general LOT,

$$\mathbf{F}(z) = \mathbf{\Phi}_1 \mathbf{W} \mathbf{\Lambda}(z) \mathbf{W} \mathbf{D}. \tag{5.80}$$

Where $\mathbf{U}_1 = \mathbf{U}$ and $\mathbf{V}_1 = -\mathbf{V}$. Note that the regular LOT is the case where $\mathbf{U}_1 = \mathbf{I}_{M/2}$ and $\mathbf{V}_1 = -\mathbf{V}_R$. The implementation diagram for $M = 8$ is shown in Fig. 5.21.

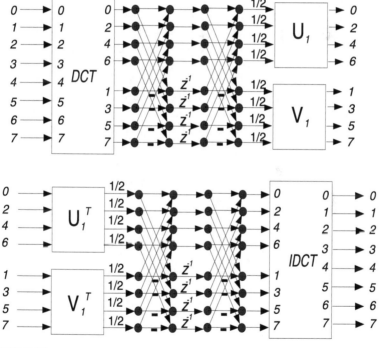

FIGURE 5.21
Implementation of a more general version of the LOT for $M = 8$. Top: forward transform; bottom: inverse transform.

From this formulation, along with other results, it was realized [40] that all orthogonal symmetric LTs can be expressed as

$$\mathbf{F}(z) = \mathbf{K}_{N-1}(z)\mathbf{K}_{N-2}(z) \cdots \mathbf{K}_1(z)\mathbf{K}_0 \tag{5.81}$$

where

$$\mathbf{K}_i(z) = \mathbf{\Phi}_i \mathbf{W} \mathbf{\Lambda}(z) \mathbf{W} , \tag{5.82}$$

and where $\mathbf{K}_0$ is any orthogonal symmetric matrix. The inverse is given by

$$\mathbf{G}(z) = \mathbf{K}_0^T \mathbf{K}_1'(z) \mathbf{K}_2'(z) \cdots \mathbf{K}_{N-1}'(z) \tag{5.83}$$

where

$$\mathbf{K}_i'(z) = z^{-1} \mathbf{W} \mathbf{\Lambda} \left(z^{-1} \right) \mathbf{W} \mathbf{\Phi}_i^T . \tag{5.84}$$

From this perspective, the generalized LOT (GenLOT) is defined as the orthogonal LT, as in Eq. (5.81) in which $\mathbf{K}_0 = \mathbf{D}$; i.e.,

$$\mathbf{F}(z) = \mathbf{K}_{N-1}(z) \cdots \mathbf{K}_1(z) \mathbf{D} . \tag{5.85}$$

A diagram for implementing a GenLOT for even M is shown in Fig. 5.22. In this diagram, the scaling parameters are $\beta = 2^{-(N-1)}$ and account for the terms $1/\sqrt{2}$ in the definition of $\mathbf{W}$.

The degrees of freedom of a GenLOT are the orthogonal matrices $\mathbf{U}_i$ and $\mathbf{V}_i$. There are $2(N-1)$ matrices to optimize, each of size $M/2 \times M/2$. From Section 5.1.4 we know that each one can be factorized into $M(M-2)/8$ rotations. Thus, the total number of rotations is $(L-M)(M-2)/4$, which is less than the initial number of degrees of freedom in a symmetric $M \times L$ matrix, $LM/2$. However, it is still a large number of parameters to design. In general, GenLOTs are designed through nonlinear unconstrained optimization. Rotation angles are searched to minimize some cost function. GenLOT examples are given elsewhere [40], and we present two examples, for $M = 8$, in Tables 5.1 and 5.2, which are also plotted in Fig. 5.23.

In the case when M is odd, the GenLOT is defined as

$$\mathbf{F}(z) = \mathbf{K}_{(N-1)/2}(z) \cdots \mathbf{K}_1(z) \mathbf{D} \tag{5.86}$$

where the stages $\mathbf{K}_i$ have necessarily order 2 as:

$$\mathbf{K}_i(z) = \mathbf{\Phi}_{2i}^o \mathbf{W}^o \mathbf{\Lambda}^{o1}(z) \mathbf{W}^o \mathbf{\Phi}_{2i-1}^o \mathbf{W}^o \mathbf{\Lambda}^{o2}(z) \mathbf{W}^o \tag{5.87}$$

and where

$$\mathbf{\Phi}_{2i}^o = \begin{bmatrix} \mathbf{U}_{2i} & \mathbf{0} \\ \mathbf{0} & \mathbf{V}_{2i} \end{bmatrix} , \tag{5.88}$$

$$\mathbf{\Phi}_{2i-1}^o = \begin{bmatrix} \mathbf{U}_{2i-1} & & \mathbf{0} \\ & 1 & \\ \mathbf{0} & & \mathbf{V}_{2i-1} \end{bmatrix} , \tag{5.89}$$

$$\mathbf{W}^o = \begin{bmatrix} \mathbf{I}_{(M-1)/2} & \mathbf{0}_{(M-1)/2 \times 1} & \mathbf{I}_{(M-1)/2} \\ \mathbf{0}_{1 \times (M-1)/2} & 1 & \mathbf{0}_{1 \times (M-1)/2} \\ \mathbf{I}_{(M-1)/2} & \mathbf{0}_{(M-1)/2 \times 1} & -\mathbf{I}_{(M-1)/2} \end{bmatrix} , \tag{5.90}$$

$$\mathbf{\Lambda}^{o1}(z) = diag \left\{ \underbrace{1, 1, \dots, 1}_{(M+1)/2 \times 1's}, \underbrace{z^{-1}, \dots, z^{-1}}_{(M-1)/2 \times z^{-1}} \right\} , \tag{5.91}$$

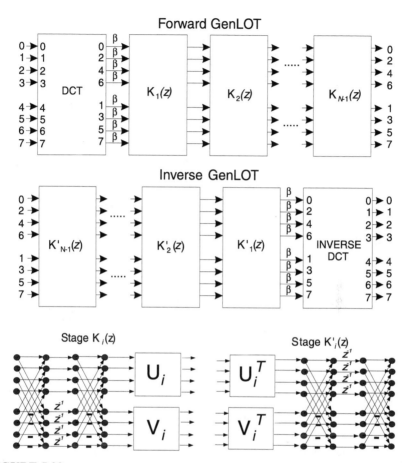

FIGURE 5.22

Implementation of a GenLOT for even M, ($M = 8$). Forward and inverse transforms are shown along with details of each stage. $\beta = 2^{-(N-1)}$ accounts for all terms of the form $1/\sqrt{2}$ which make the butterflies (W) orthogonal.

$$\Lambda^{o2}(z) = diag\left\{\underbrace{1, 1, \ldots, 1}_{(M-1)/2 \times 1's}, \underbrace{z^{-1}, \ldots, z^{-1}}_{(M+1)/2 \times z^{-1}}\right\}. \tag{5.92}$$

Although it may seem that the formulation of the odd-channel case is more complex than the one for the even-M case, the implementation is very similar in complexity as shown in Fig. 5.24. The main difference is that two stages have to be connected. The inverse transform is accomplished in the same way as for the even channel case:

$$G(z) = D^T K'_1(z) K'_2(z) \cdots K'_{N-1}(z) \tag{5.93}$$

Table 5.1 GenLOT Example for $N = 4$. The Even Bases are Symmetric while the Odd Ones are Anti-Symmetric, so Only their First Half is Shown

p_{0n}	p_{1n}	p_{2n}	p_{3n}	p_{4n}	p_{5n}	p_{6n}	p_{7n}
0.004799	0.004829	0.002915	−0.002945	0.000813	−0.000109	0.000211	0.000483
0.009320	−0.000069	−0.005744	−0.010439	0.001454	0.003206	0.000390	−0.001691
0.006394	−0.005997	−0.011121	−0.010146	0.000951	0.004317	0.000232	−0.002826
−0.011794	−0.007422	−0.001800	0.009462	−0.001945	−0.001342	−0.000531	0.000028
−0.032408	−0.009604	0.008083	0.031409	−0.005262	−0.007504	−0.001326	0.003163
−0.035122	−0.016486	0.001423	0.030980	−0.005715	−0.006029	−0.001554	0.001661
−0.017066	−0.031155	−0.027246	0.003473	−0.003043	0.005418	−0.000789	−0.005605
0.000288	−0.035674	−0.043266	−0.018132	−0.000459	0.013004	−0.000165	−0.010084
−0.012735	−0.053050	0.007163	−0.083325	0.047646	0.011562	0.048534	0.043066
−0.018272	−0.090207	0.131531	0.046926	0.072761	−0.130875	−0.089467	−0.028641
0.021269	−0.054379	0.109817	0.224818	−0.224522	0.136666	0.022488	−0.025219
0.126784	0.112040	−0.123484	−0.032818	−0.035078	0.107446	0.147727	0.109817
0.261703	0.333730	−0.358887	−0.379088	0.384874	−0.378415	−0.339368	−0.216652
0.357269	0.450401	−0.292453	−0.126901	−0.129558	0.344379	0.439129	0.317070
0.383512	0.369819	0.097014	0.418643	−0.419231	0.045807	−0.371449	−0.392556
0.370002	0.140761	0.478277	0.318691	0.316307	−0.433937	0.146036	0.427668

where the inverse factors are

$$\mathbf{K}'_i(z) = z^{-2}\mathbf{K}^T_i\left(z^{-1}\right) , \tag{5.94}$$

whose structure is evident from Fig. 5.24.

5.6.4   The General Factorization: GLBT

The general factorization for all symmetric LTs [61] can be viewed either as an extension of GenLOTs or as a generalization of the LBT. For M even, all LTs obeying Eq. (5.65) or Eq. (5.66) can be factorized as in Eq. (5.81), where the $\mathbf{K}_i(z)$ factors are given in Eq. (5.82) with the matrices $\mathbf{U}_i$ and $\mathbf{V}_i$ (which compose $\mathbf{\Phi}_i$) being required only to be general invertible matrices. From Section 5.1.4, each factor can be decomposed as

$$\mathbf{U}_i = \mathbf{U}_{iB}\mathbf{U}_{id}\mathbf{U}_{iA} \quad , \quad \mathbf{V}_i = \mathbf{V}_{iB}\mathbf{V}_{id}\mathbf{V}_{iA} , \tag{5.95}$$

where $\mathbf{U}_{iA}$, $\mathbf{U}_{iB}$, $\mathbf{V}_{iA}$, and $\mathbf{V}_{iB}$ are general $M/2 \times M/2$ orthogonal matrices, while $\mathbf{U}_{id}$ and $\mathbf{V}_{id}$ are diagonal matrices with nonzero diagonal entries.

The first factor $\mathbf{K}_0$ is given by

$$\mathbf{K}_0 = \mathbf{\Phi}_0\mathbf{W} , \tag{5.96}$$

where $\mathbf{\Phi}_i$ is given as in Eq. (5.78), and factors $\mathbf{U}_0$ and $\mathbf{V}_0$ are required only to be invertible. The general factorization can be viewed as a generalized LBT (GLBT) and its implementation flow graph for M is even shown in Fig. 5.25.

Table 5.2 GenLOT Example for $N = 6$. The Even Bases are Symmetric while the Odd Ones are Anti-Symmetric, so Only their First Half is Shown

p_{0n}	p_{1n}	p_{2n}	p_{3n}	p_{4n}	p_{5n}	p_{6n}	p_{7n}
−0.000137	−0.000225	0.000234	0.000058	−0.000196	−0.000253	0.000078	0.000017
−0.000222	−0.000228	0.000388	0.000471	0.000364	0.000163	−0.000220	−0.000283
0.001021	0.000187	0.002439	0.001211	−0.000853	−0.002360	0.000157	−0.000823
0.000536	0.000689	0.000029	0.000535	0.000572	0.000056	0.000633	0.000502
−0.001855	0.000515	−0.006584	−0.002809	0.003177	0.006838	−0.000886	0.001658
0.001429	0.001778	−0.000243	0.000834	0.000977	−0.000056	0.001687	0.001429
0.001440	0.001148	0.000698	0.000383	0.000109	−0.000561	−0.000751	−0.001165
0.001056	0.001893	0.002206	0.005386	0.005220	0.001676	0.001673	0.000792
0.009734	0.002899	0.018592	0.004888	−0.006600	−0.018889	−0.000261	−0.006713
−0.005196	−0.013699	−0.008359	−0.021094	−0.020406	−0.009059	−0.012368	−0.005263
−0.000137	−0.001344	−0.027993	−0.028046	0.026048	0.024169	−0.001643	−0.000402
−0.007109	−0.002130	0.002484	0.013289	0.013063	0.002655	−0.002180	−0.006836
−0.011238	−0.002219	0.033554	0.062616	−0.058899	−0.031538	−0.001404	0.004060
−0.020287	−0.006775	0.003214	0.019082	0.018132	0.004219	−0.006828	−0.019040
−0.028214	−0.018286	−0.059401	−0.023539	0.024407	0.056646	0.009849	0.021475
−0.034379	−0.055004	−0.048827	−0.052703	−0.051123	−0.048429	−0.049853	−0.031732
−0.029911	−0.106776	0.070612	−0.088796	0.086462	−0.066383	0.097006	0.031014
−0.004282	−0.107167	0.197524	0.049701	0.051188	0.193302	−0.104953	−0.006324
0.058553	−0.026759	0.144748	0.241758	−0.239193	−0.143627	0.020370	−0.048085
0.133701	0.147804	−0.123524	0.026563	0.025910	−0.125263	0.147501	0.130959
0.231898	0.330343	−0.376982	−0.365965	0.366426	0.377886	−0.332858	−0.228016
0.318102	0.430439	−0.312564	−0.174852	−0.174803	−0.314092	0.431705	0.317994
0.381693	0.368335	0.061832	0.393949	−0.395534	−0.060887	−0.369244	−0.384842
0.417648	0.144412	0.409688	0.318912	0.319987	0.411214	0.145256	0.419936

The inverse GLBT is similar to the GenLOT case, where

$$K_i'(z) = z^{-1}\mathbf{W}\mathbf{\Lambda}(z)\mathbf{W}\mathbf{\Phi}_i^{-1} \tag{5.97}$$

and

$$\mathbf{\Phi}_i^{-1} = \begin{bmatrix} \mathbf{U}_i^{-1} & \mathbf{0}_{M/2} \\ \mathbf{0}_{M/2} & \mathbf{V}_i^{-1} \end{bmatrix} = \begin{bmatrix} \mathbf{U}_{iA}^T\mathbf{U}_{id}^{-1}\mathbf{U}_{iB}^T & \mathbf{0}_{M/2} \\ \mathbf{0}_{M/2} & \mathbf{V}_{iA}^T\mathbf{V}_{id}^{-1}\mathbf{V}_{iB}^T \end{bmatrix} \tag{5.98}$$

while

$$\mathbf{K}_0^{-1} = \mathbf{W}\mathbf{\Phi}_0^{-1}. \tag{5.99}$$

The diagram for the implementation of the inverse stages of the GLBT is shown in Fig. 5.25. Examples of bases for the GLBT of particular interest to image compression are given in Tables 5.3 and 5.4.

For the odd case, the GLBT can be similarly defined. It follows the GenLOT factorization:

$$\mathbf{F}(z) = \mathbf{K}_{(N-1)/2}(z)\cdots\mathbf{K}_1(z)\mathbf{K}_0 \tag{5.100}$$

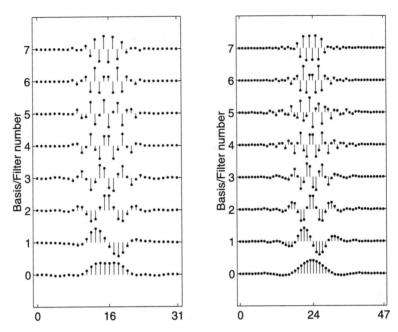

FIGURE 5.23
Example of optimized GenLOT bases for $M = 8$ and for $N = 4$ (left) and $N = 6$ (right).

Table 5.3 Forward GLBT Bases Example for $M = 8$ and $N = 2$. The Even Bases are Symmetric while the Odd Ones are Anti-Symmetric, so Only their First Half is Shown

p_{0n}	p_{1n}	p_{2n}	p_{3n}	p_{4n}	p_{5n}	p_{6n}	p_{7n}
−0.21192	−0.18197	0.00011	−0.09426	0.03860	−0.03493	0.04997	0.01956
−0.13962	−0.19662	0.16037	0.05334	0.09233	0.12468	−0.09240	−0.03134
−0.03387	−0.09540	0.17973	0.25598	−0.24358	−0.12311	0.01067	−0.01991
0.09360	0.10868	−0.06347	−0.01332	−0.05613	−0.10218	0.16423	0.11627
0.23114	0.34101	−0.36293	−0.39498	0.42912	0.36084	−0.35631	−0.22434
0.35832	0.46362	−0.35056	−0.16415	−0.13163	−0.31280	0.47723	0.31907
0.46619	0.42906	0.00731	0.42662	−0.45465	−0.07434	−0.40585	−0.38322
0.53813	0.22604	0.42944	0.36070	0.32595	0.43222	0.15246	0.39834

where the stages $\mathbf{K}_i$ are as in Eq. (5.87) with the following differences: (i) all factors $\mathbf{U}_i$ and $\mathbf{V}_i$ are only required to be invertible; (ii) the center element of $\mathbf{\Phi}_{2i-1}$ is a nonzero constant u_0 and not 1. Again $\mathbf{K}_0$ is a symmetric invertible matrix. Forward and inverse stages for the odd-channel case are illustrated in Fig. 5.26.

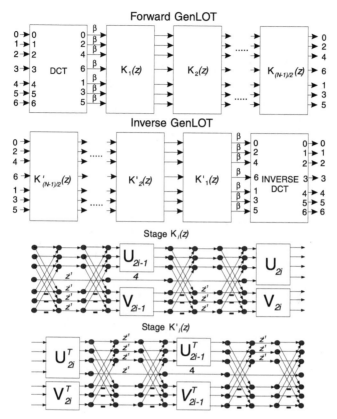

FIGURE 5.24
Implementation of a GenLOT for M odd. Forward and inverse transforms are shown along with details of each stage and $\beta = 2^{-(N-1)}$.

5.7 The Fast Lapped Transform: FLT

The motivation behind the fast lapped transform (FLT) is to design an LT with minimum possible complexity compared to a block transform, yet to provide some advantages over a block transform. For that we use the principles of Section 5.5.3 and define the FLT as the LT whose PTM is given by

$$\mathbf{F}(z) = \begin{bmatrix} \mathbf{E}(z) & \mathbf{0} \\ \mathbf{0} & \mathbf{I}_{M-K} \end{bmatrix} \mathbf{D}_M \qquad (5.101)$$

Stage $K_i(z)$

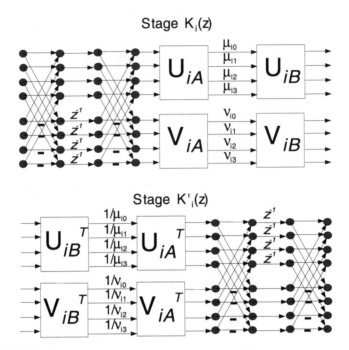

Stage $K'_i(z)$

FIGURE 5.25
Implementation of the factors of the general factorization (GLBT) for M even. Top: factor of the forward transform, $\mathbf{K}_i(z)$. Bottom: factor of the inverse transform, $\mathbf{K}'_i(z)$.

where $\mathbf{E}(z)$ is a $K \times K$ PTM and $\mathbf{D}_M$ is the $M \times M$ DCT matrix. The PTM for the inverse LT is given by

$$\mathbf{G}(z) = \mathbf{D}_M^T \begin{bmatrix} \mathbf{E}'(z) & \mathbf{0} \\ \mathbf{0} & \mathbf{I}_{M-K} \end{bmatrix}, \tag{5.102}$$

where $\mathbf{E}'(z)$ is the inverse of $\mathbf{E}(z)$.

The design of $\mathbf{E}(z)$ can be done in two basic ways. Firstly, one can use direct optimization. Secondly, one can design $\mathbf{E}(z)$ as

$$\mathbf{E}(z) = \mathbf{\Psi}(z)\mathbf{D}_K^T \tag{5.103}$$

where $\mathbf{\Psi}(z)$ is a known LT and $\mathbf{D}_K$ is the $K \times K$ DCT matrix; in other words, we perform an inverse DCT followed by a known LT. For example, if $\mathbf{\Psi}(z)$ is the LOT, GenLOT, or LBT of K channels, the first stage ($\mathbf{D}_K$) cancels the inverse DCT. Examples of FLT are given in Fig. 5.27. In that example, the first case where $K = 2$, direct optimization is recommended, for which the values $\{\alpha_{00}, \alpha_{01}, \alpha_{10}, \alpha_{11}, \alpha_{20}, \alpha_{21}\} = \{1.9965, 1.3193, 0.4388, 0.7136, 0.9385, 1.2878\}$ yield an excellent FLT for image compression. In the middle of Fig. 5.27 the case $K = 4$ can be optimized by optimizing two invertible matrices. In the case where we use the method in Eq. (5.103)

Table 5.4 Inverse GLBT Bases Example for $M = 8$ and $N = 2$. The Even Bases are Symmetric while the Odd Ones are Anti-Symmetric, so Only their First Half is Shown

p_{0n}	p_{1n}	p_{2n}	p_{3n}	p_{4n}	p_{5n}	p_{6n}	p_{7n}
0.01786	−0.01441	0.06132	0.01952	0.05243	0.05341	0.04608	0.08332
0.05692	−0.01681	0.16037	0.12407	0.04888	0.16065	−0.09042	−0.02194
0.10665	0.06575	0.12462	0.24092	−0.21793	−0.13556	0.02108	−0.00021
0.16256	0.20555	−0.12304	−0.03560	−0.02181	−0.08432	0.13397	0.12747
0.22148	0.34661	−0.38107	−0.35547	0.36530	0.39610	−0.30170	−0.23278
0.27739	0.40526	−0.32843	−0.12298	−0.12623	−0.35462	0.41231	0.34133
0.32711	0.33120	0.03939	0.38507	−0.38248	−0.08361	−0.35155	−0.40906
0.36617	0.13190	0.44324	0.30000	0.28191	0.45455	0.13232	0.41414

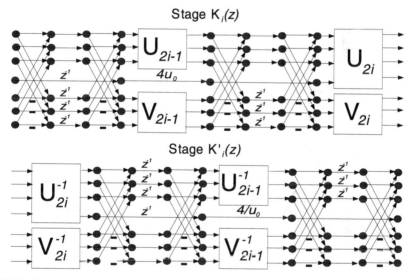

FIGURE 5.26

Implementation of the factors of the general factorization (GLBT) for M odd. Top: factor of the forward transform, $K_i(z)$. Bottom: factor of the inverse transform, $K'_i(z)$.

and the LBT as the K channel postprocessing stage, we can see that the LBT's DCT stage is canceled, yielding a very simple flow graph. The respective bases for forward and inverse transforms for the two FLTs ($K = 2$ with the given parameters, and $K = 4$ using the LBT) are shown in Fig. 5.28. Both bases are excellent for image coding, virtually eliminating ringing, despite the minimal complexity added to the DCT (which by itself can be implemented in a very fast manner [45]).

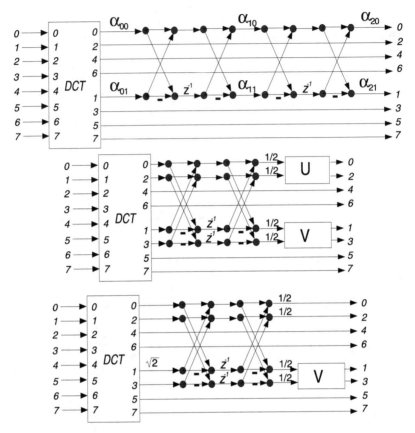

FIGURE 5.27
Implementation of examples of the FLT. On top, $K = 2$; middle, case $K = 4$; bottom, case $K = 4$ where $\Psi(z)$ is the LBT, thus having its DCT stage canceled.

5.8 Modulated LTs

Cosine modulated LTs or filter banks [63] use a lowpass prototype to modulate a cosine sequence. By a proper choice of the phase of the cosine sequence, Malvar developed the modulated lapped transform (MLT) [21], which led to the so-called extended lapped transforms (ELT) [24]–[27]. ELT allows several overlapping factors, generating a family of orthogonal cosine modulated LTs. Both designations (MLT and ELT) are frequently applied to this class of filter banks. Other cosine-modulation approaches have also been developed and the most significant difference among them is the lowpass prototype choice and the phase of the cosine sequence [17, 21, 25, 26, 30, 33, 48, 49, 55, 56, 63].

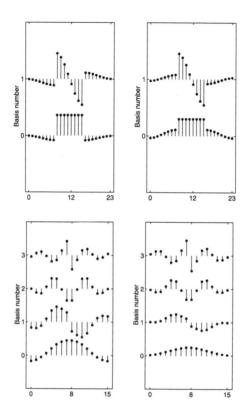

FIGURE 5.28
**Bases of the FLT in the case $M = 8$ for forward and inverse LTs. From left to
right: forward transform bases for the case $K = 2$; inverse transform bases for
the case $K = 2$; forward transform bases for the case $K = 4$; inverse transform
bases for the case $K = 4$. The remaining bases, not shown, are the regular bases
of the DCT and have length 8.**

In the ELTs, the filters' length L is basically an even multiple of the block size M,
as $L = NM = 2KM$. Thus, K is referred to as the *overlap factor* of the ELT. The
MLT-ELT class is defined by

$$p_{k,n} = h(n) \cos \left[\left(k + \frac{1}{2} \right) \left(\left(n - \frac{L-1}{2} \right) \frac{\pi}{M} + (N+1)\frac{\pi}{2} \right) \right] \qquad (5.104)$$

for $k = 0, 1 \ldots , M - 1$ and $n = 0, 1, \ldots , L - 1$. $h(n)$ is a symmetric window
modulating the cosine sequence and the impulse response of a lowpass prototype
(with cutoff frequency at $\pi/2M$), which is translated in frequency to M different
frequency slots in order to construct the LT. A very useful ELT is the one with $K = 2$,
which will be designated as ELT-2, while ELT with $K = 1$ will be referred to as the
MLT.

A major plus of the ELTs is its fast implementation algorithm. The algorithm is based on a factorization of the PTM into a series of plane rotation stages and delays and a DCT type IV [45] orthogonal transform in the last stage, which also has fast implementation algorithms. The lattice-style algorithm is shown in Fig. 5.29 for an ELT with generic overlap factor K. In Fig. 5.29, each branch carries $M/2$ samples, and both analysis (forward transform) and synthesis (inverse transform) flow graphs are shown. The plane rotation stages are of the form indicated in Fig. 5.30 and contain $M/2$ orthogonal butterflies to implement the $M/2$ plane rotations. The stages Θ_i contain the plane rotations and are defined by

$$\Theta_i = \begin{bmatrix} -\mathbf{C}_i & \mathbf{S}_i \mathbf{J}_{M/2} \\ \mathbf{J}_{M/2}\mathbf{S}_i & \mathbf{J}_{M/2}\mathbf{C}_i\mathbf{J}_{M/2} \end{bmatrix}, \qquad (5.105)$$

$$\mathbf{C}_i = diag\left\{\cos\left(\theta_{0,i}\right),\ \cos\left(\theta_{1,i}\right),\ldots,\cos\left(\theta_{\frac{M}{2}-1,i}\right)\right\},$$

$$\mathbf{S}_i = diag\left\{\sin\left(\theta_{0,i}\right),\ \sin\left(\theta_{1,i}\right),\ldots,\sin\left(\theta_{\frac{M}{2}-1,i}\right)\right\},$$

where $\theta_{i,j}$ are rotation angles. These angles are the free parameters in the design of an ELT because they define the modulating window $h(n)$. Note that there are KM angles, while $h(n)$ has $2KM$ samples; however, $h(n)$ is symmetric which brings the total number of degrees of freedom to KM.

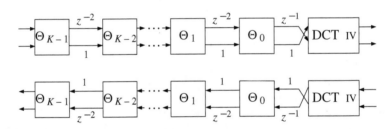

FIGURE 5.29
Flow graph for the direct (top) and inverse (bottom) ELT. Each branch carries $M/2$ samples.

In general, there is no simple relationship among the rotation angles and the window. Optimized angles for several values of M and K are presented in the extensive tables in [26]. In the ELT-2 case, however, one can use a parameterized design [25]–[27]. In this design, we have

$$\theta_{k,0} = -\frac{\pi}{2} + \mu_{M/2+k} \qquad (5.106)$$

$$\theta_{k,1} = -\frac{\pi}{2} + \mu_{M/2-1-k} \qquad (5.107)$$

where

$$\mu_i = \left[\left(\frac{1-\gamma}{2M}\right)(2k+1)+\gamma\right] \qquad (5.108)$$

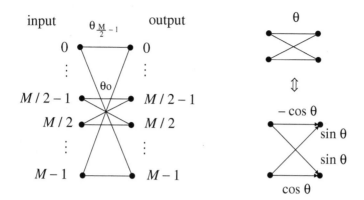

FIGURE 5.30
Implementation of the plane rotations stage showing the displacement of the
$M/2$ **butterflies.**

and γ is a control parameter, for $0 \le k \le (M/2) - 1$. In general, although suboptimal for individual applications, $\gamma = 0.5$ provides a balanced trade-off of stopband attenuation and transition range for the equivalent filters (which are the bases of the LT viewed as a filter bank). The equivalent modulating window $h(n)$ is related to the angles as

$$
\begin{aligned}
h(n) &= \cos(\theta_{n0}) \cos(\theta_{n1}) \\
h(M - 1 - n) &= \cos(\theta_{n0}) \sin(\theta_{n1}) \\
h(M + n) &= \sin(\theta_{n0}) \cos(\theta_{n1}) \\
h(2M - 1 - n) &= -\sin(\theta_{n0}) \sin(\theta_{n1})
\end{aligned}
\tag{5.109}
$$

for $0 \le n \le (M/2) - 1$. In the $K = 1$ case, some example angles are

$$
\theta_{k0} = \frac{\pi}{2} - \frac{\pi}{2M}\left(k + \frac{1}{2}\right)
\tag{5.110}
$$

for $0 \le k \le (M/2) - 1$. The corresponding modulating window $h(n)$ is

$$
\begin{aligned}
h(n) = h(2M - 1 - n) &= -\cos(\theta_{n0}) \\
h(M + n) = h(M - 1 - n) &= -\sin(\theta_{n0})
\end{aligned}
\tag{5.111}
$$

for $0 \le n \le (M/2) - 1$. The bases for the ELT using the suggested angles are shown in Fig. 5.31. In this figure, the 8-channel examples are for $N = 2$ ($K = 1$) and for $N = 4$ ($K = 2$).

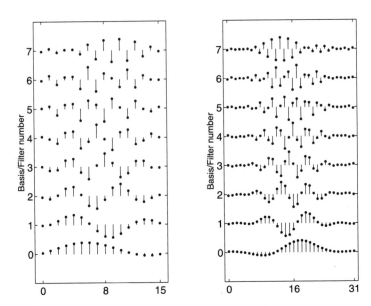

FIGURE 5.31
Example of ELT bases for the given angles design method for $M = 8$**. Left:**
$K = 1, N = 2$**, right:** $K = 2, N = 4$**.**

5.9   Finite-Length Signals

Since the LT matrices are not square, in order to obtain n transformed subband samples one has to process more than n samples of the input signal. For the same reason, n subband samples generate more than n signal samples after inverse transformation. Our analysis so far has assumed infinite-length signals. Processing finite-length signals, however, is not trivial. Without proper consideration, there will be a distortion in the reconstruction of the boundary samples of the signal. There are basically three methods to process finite-length signals with LTs:

• signal extension and windowing of subband coefficients;

• same as above but using different extensions for different bases;

• time-varying bases for the boundary regions.

We discuss only the first method. The second is applicable only to a few transforms and filter banks. The subject of time-varying LTs is very rich and provides solutions to several problems, including the processing of boundary samples; we do not cover it in this chapter, but the reader is referred to [34, 9, 38, 52, 35] and their references for further information on time-varying LTs.

5.9.1 Overall Transform

Here we assume the model of extension and windowing described in Fig. 5.32 [39]. The input vector $\mathbf{x}$ is assumed to have $N_x = N_B M$ samples and is divided into 3 sections: $\mathbf{x}^T = [\mathbf{x}_l^T, \mathbf{x}_c^T, \mathbf{x}_r^T]$, where $\mathbf{x}_l$ and $\mathbf{x}_r$ contain the first and last λ samples of $\mathbf{x}$, respectively. Following the signal extension model, $\mathbf{x}$ is extended into $\tilde{\mathbf{x}}$ as

$$\tilde{\mathbf{x}}^T = \left[\mathbf{x}_{e,l}^T, \mathbf{x}^T, \mathbf{x}_{e,r}^T \right] = \left[(\mathbf{R}_l x_l)^T, \mathbf{x}_l^T, \mathbf{x}_c^T, \mathbf{x}_r^T, (\mathbf{R}_r \mathbf{x}_r)^T \right]. \tag{5.112}$$

In other words, the extended sections are found by a linear transform of the boundary samples of $\mathbf{x}$, as shown in Fig. 5.33; i.e.,

$$\mathbf{x}_{e,l} = \mathbf{R}_l \mathbf{x}_l \quad , \quad \mathbf{x}_{e,r} = \mathbf{R}_r \mathbf{x}_r \tag{5.113}$$

and $\mathbf{R}_l$ and $\mathbf{R}_r$ are arbitrary $\lambda \times \lambda$ "extension" matrices. For example, $\mathbf{R}_l = \mathbf{R}_r = \mathbf{J}_\lambda$ yields a symmetric extension.

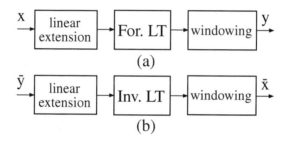

(a)

(b)

FIGURE 5.32
Extension and windowing in transformation of a finite-length signal using LTs.
(a) Overall forward transform section. (b) Overall inverse transform section.

The transformation from the $N_x + 2\lambda$ samples in $\tilde{\mathbf{x}}$ to vector $\mathbf{y}$ with $N_B M = N_x$ subband samples is achieved through the block-banded matrix $\tilde{\mathbf{P}}$:

$$\tilde{\mathbf{P}} = \begin{pmatrix} \ddots & & \ddots & & & & 0 \\ & \mathbf{P}_0 & \mathbf{P}_1 & \cdots & \mathbf{P}_{N-1} & & \\ & & \mathbf{P}_0 & \mathbf{P}_1 & \cdots & \mathbf{P}_{N-1} & \\ & & & \mathbf{P}_0 & \mathbf{P}_1 & \cdots & \mathbf{P}_{N-1} \\ 0 & & & & \ddots & & \ddots \end{pmatrix}. \tag{5.114}$$

Note that there are N_B block rows and that $\lambda = (N-1)M/2$. The difference between $\tilde{\mathbf{P}}$ and $\mathbf{H}$, defined in Eq. (5.21), is that $\mathbf{H}$ is assumed to be infinite and $\tilde{\mathbf{P}}$ is assumed to have only N_B block rows. We can use the same notation for $\tilde{\mathbf{Q}}$ with respect to $\mathbf{Q}_i$, so that, again, the difference between $\tilde{\mathbf{Q}}$ and $\mathbf{H}'$ defined in Eq. (5.32) is that $\mathbf{H}'$ is assumed to be infinite and $\tilde{\mathbf{Q}}$ is assumed to have only N_B block rows. The forward and inverse transform systems are given by

$$\tilde{\mathbf{y}} = \tilde{\mathbf{P}}\tilde{\mathbf{x}} \quad , \quad \bar{\tilde{\mathbf{x}}} = \tilde{\mathbf{Q}}^T \bar{\mathbf{y}}. \tag{5.115}$$

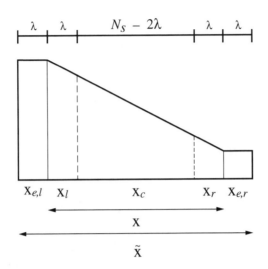

FIGURE 5.33

Illustration of signal extension of vector x into vector x̃. At each border, $\lambda = (L - M)/2$ samples outside the initial signal boundaries are found by linear relations applied to the λ boundary samples of x, i.e., $x_{e,l} = R_l x_l$ and $x_{e,r} = R_r x_r$. As only λ samples are affected across the signal boundaries, it is not necessary to use an infinite-length extension. Also, x_l and x_r contain the samples possibly affected by the border distortions after the inverse transformation.

In the absence of quantization or processing of the subband signals, $\tilde{y} = y$ and

$$\bar{\tilde{x}} = \tilde{Q}^T \tilde{y} = \tilde{Q}^T \tilde{P} \tilde{x} = \tilde{T} \tilde{x} \qquad (5.116)$$

where $\bar{\tilde{x}}$ is the reconstructed vector in the absence of quantization, and $\tilde{T} = \tilde{Q}^T \tilde{P}$ is the transform matrix between $\bar{\tilde{x}}$ and $\tilde{x}$. Note that $\tilde{T}$ has size $(N_x + \lambda) \times (N_x + \lambda)$ because it maps two extended signals. From Eq. (5.35) we can easily show that the transform matrix is

$$\tilde{T} = \tilde{Q}^T \tilde{P} = \begin{bmatrix} T_l & & 0 \\ \hline & I_{N_x - 2\lambda} & \\ \hline 0 & & T_r \end{bmatrix} \qquad (5.117)$$

where T_l and T_r are some $2\lambda \times 2\lambda$ matrices. Thus, distortion is incurred only to the λ boundary samples in each side of x (2λ samples in each side of $\tilde{x}$).

In another view of the process, regardless of the extension method, there is a transform T such that

$$y = Tx \quad , \quad \bar{x} = T^{-1} \bar{y} \qquad (5.118)$$

without resorting to signal extension. The key is to find T and to invert it. If T is made orthogonal, one can easily invert it by applying transposition. This is the concept behind the use of time-varying LTs for correcting boundary distortions. For example, the LT can be changed near the borders to ensure the orthogonality of $\mathbf{T}$ [38]. We will not use time-varying LTs here but rather use extended signals and transform matrices.

5.9.2 Recovering Distorted Samples

Let

$$
[\Phi_l|\Phi_r] = \begin{bmatrix} \mathbf{P}_0 & \mathbf{P}_1 & \cdots & \mathbf{P}_{N-2} & \mathbf{P}_{N-1} & & & \mathbf{0} \\ & \mathbf{P}_0 & \mathbf{P}_1 & \cdots & \mathbf{P}_{N-2} & \mathbf{P}_{N-1} & & \\ & & \ddots & \ddots & & \ddots & \ddots & \\ \mathbf{0} & & & \mathbf{P}_0 & \mathbf{P}_1 & \cdots & \mathbf{P}_{N-2} & \mathbf{P}_{N-1} \end{bmatrix},
$$
(5.119)

$$
[\Psi_l|\Psi_r] = \begin{bmatrix} \mathbf{Q}_0 & \mathbf{Q}_1 & \cdots & \mathbf{Q}_{N-2} & \mathbf{Q}_{N-1} & & & \mathbf{0} \\ & \mathbf{Q}_0 & \mathbf{Q}_1 & \cdots & \mathbf{Q}_{N-2} & \mathbf{Q}_{N-1} & & \\ & & \ddots & \ddots & & \ddots & \ddots & \\ \mathbf{0} & & & \mathbf{Q}_0 & \mathbf{Q}_1 & \cdots & \mathbf{Q}_{N-2} & \mathbf{Q}_{N-1} \end{bmatrix}.
$$
(5.120)

Hence,

$$
\mathbf{T}_l = \boldsymbol{\Psi}_l^T \boldsymbol{\Phi}_l , \quad \mathbf{T}_r = \boldsymbol{\Psi}_r^T \boldsymbol{\Phi}_r .
$$
(5.121)

If $\bar{\bar{\mathbf{x}}}$ is divided in the same manner as $\tilde{\mathbf{x}}$, i.e.,

$$
\bar{\bar{\mathbf{x}}} = \left[\bar{\mathbf{x}}_{e,l}^T, \bar{\mathbf{x}}_l^T, \bar{\mathbf{x}}_c^T, \bar{\mathbf{x}}_r^T, \bar{\mathbf{x}}_{e,r}^T \right],
$$
(5.122)

then,

$$
\begin{bmatrix} \bar{\mathbf{x}}_{c,l} \\ \bar{\mathbf{x}}_l \end{bmatrix} = \mathbf{T}_l \begin{bmatrix} \mathbf{x}_{e,l} \\ \mathbf{x}_l \end{bmatrix} = \mathbf{T}_l \begin{bmatrix} \mathbf{R}_l \mathbf{x}_l \\ \mathbf{x}_l \end{bmatrix} = \mathbf{T}_l \begin{bmatrix} \mathbf{R}_l \\ \mathbf{I}_\lambda \end{bmatrix} \mathbf{x}_l = \boldsymbol{\Gamma}_l \, \mathbf{x}_l \quad (5.123)
$$

where

$$
\boldsymbol{\Gamma}_l = \mathbf{T}_l \begin{bmatrix} \mathbf{R}_l \\ \mathbf{I}_\lambda \end{bmatrix}
$$
(5.124)

is a $2\lambda \times \lambda$ matrix. If and only if $\boldsymbol{\Gamma}_l$ has rank λ, then $\mathbf{x}_l$ can be recovered through the pseudo-inverse of $\boldsymbol{\Gamma}_l$ as

$$
\mathbf{x}_l = \boldsymbol{\Gamma}_l^+ \begin{bmatrix} \bar{\mathbf{x}}_{e,l} \\ \bar{\mathbf{x}}_l \end{bmatrix} = \left(\boldsymbol{\Gamma}_l^T \boldsymbol{\Gamma}_l \right)^{-1} \boldsymbol{\Gamma}_l^T \begin{bmatrix} \bar{\mathbf{x}}_{e,l} \\ \bar{\mathbf{x}}_l \end{bmatrix} .
$$
(5.125)

For the other ("right") border, the identical result is trivially found to be

$$
\mathbf{x}_r = \boldsymbol{\Gamma}_r^+ \begin{bmatrix} \bar{\mathbf{x}}_r \\ \bar{\mathbf{x}}_{e,r} \end{bmatrix} = \left(\boldsymbol{\Gamma}_r^T \boldsymbol{\Gamma}_r \right)^{-1} \boldsymbol{\Gamma}_r^T \begin{bmatrix} \bar{\mathbf{x}}_r \\ \bar{\mathbf{x}}_{e,r} \end{bmatrix} ,
$$
(5.126)

where

$$\Gamma_r = \mathbf{T}_r \begin{bmatrix} \mathbf{I}_\lambda \\ \mathbf{R}_r \end{bmatrix} \tag{5.127}$$

is also assumed to have rank λ. It is necessary, but not sufficient that Φ_l, Φ_r, Ψ_l, and Ψ_r have rank λ since rank can be reduced by the matrix products. It is also possible to express the conditions in more detail. However, without any useful analytical solution, numerical rank checking is the best approach.

To summarize, the steps to achieve PR for given $\mathbf{R}_l$ and $\mathbf{R}_r$ are

1. Select $\mathbf{P}$ and $\mathbf{Q}$ and identify their submatrices $\mathbf{P}_i$ and $\mathbf{Q}_i$.

2. Find $\Phi_l, \Phi_r, \Psi_l, \Psi_r$, from Eqs. (5.119) and (5.120).

3. Find $\mathbf{T}_l$ and $\mathbf{T}_r$ from Eq. (5.121).

4. Find Γ_l and Γ_r from Eqs. (5.124) and (5.127).

5. Test ranks of Γ_l and Γ_r.

6. If ranks are λ, obtain Γ_l^+, Γ_r^+ and reconstruct $\mathbf{x}_l$ and $\mathbf{x}_r$.

This is an extension of de Queiroz and Rao [39] to nonorthogonal LTs, with particular concern to test whether the pseudo-inverses exist.

The model in Fig. 5.32 and the proposed method are not applicable for some LTs. The notable classes of LTs include those whose bases have different lengths and different symmetries. Examples are (i) two-channel nonorthogonal LTs with odd-length (2-channel biorthogonal filter banks [55]); (ii) FLT; and (iii) other composite systems, i.e., cascaded systems such as those used in de Queiroz [41]. For the first example, it is trivial to use symmetric extensions but different symmetries for different bases [55]. The second example has the same reasoning. However, an FLT can be efficiently implemented by applying the method just described to each of the stages of the transformation (i.e., first apply the DCT and then use the method above for the second part). The reason for rank deficiency is that different filters would require different extensions during the forward transformation process; therefore, the model in Fig. 5.32 is not applicable.

The above method works very well for M-channel filter banks whose filters have the same length. The phase of the filters and the extensions can be arbitrary, and the method has been shown to be consistent for all uniform-length filter banks of interest tested.

5.9.3 Symmetric Extensions

In case the LT is symmetric and obeys Eqs. (5.65) and (5.66), there is a much simpler method to implement the LT over a finite-length signal of N_B blocks of M samples.

In the forward transform section, we perform symmetric extensions as described, applied to the last $\lambda = (L - M)/2$ samples on each border, resulting in a signal $\{\tilde{x}(n)\}$ with $N_x + 2\lambda = N_x + L - M$ samples,

$$x(\lambda - 1), \ldots, x(0), x(0), \ldots, x(N_x - 1), x(N_x - 1), \ldots, x(N_x - \lambda). \quad (5.128)$$

The signal is processed by the PTM $\mathbf{F}(z)$ as a clocked system, without concern for border locations. The internal states of the system $\mathbf{F}(z)$ can be initialized in any way. So, the $N_B + N - 1$ blocks of the extended signal are processed yielding an equal number of blocks of subband samples. Discarding the first $N - 1$ output blocks, obtain N_B transform-domain blocks corresponding to N_B samples of each subband.

The general strategy to achieve perfect reconstruction, without great increase in complexity or change in the implementation algorithm, is to extend the samples in the subbands, generating more blocks to be inverse transformed, in such a way that after inverse transformation (assuming no processing of the subband signals) the signal recovered is identical to the original at the borders. The extension of the k-th subband signal depends on the symmetry of the k-th basis. Let $p_{kn} = v_k p_{k,L-1-n}$ for $0 \leq k \leq M - 1$ and $0 \leq n \leq L - 1$, i.e., $v_k = 1$ if p_{kn} is symmetric and $v_k = -1$ if p_{kn} is anti-symmetric. Before inverse transformation, for each subband signal $\{\bar{y}_k(m)\}$ of N_B samples, fold the borders of $\{\bar{y}_k(m)\}$ (as in the analysis section) in order to find a signal $\{\hat{y}_k(m)\}$, and invert the sign of the extended samples if p_{kn} is anti-symmetric. For s samples reflected about the borders, the k-th subband signal will have samples

$$v_k \hat{y}_k(s - 1), \ldots, v_k \hat{y}_k(0), \hat{y}_k(0), \cdots \hat{y}_k(N_B - 1),$$
$$v_k \hat{y}_k(N_B - 1), \ldots, v_k \hat{y}_k(N_B - s).$$

The inverse transformation can be performed as

- N *odd* — Reflect $s = (N - 1)/2$ samples about each border, thus getting $N_B + N - 1$ blocks of subband samples to be processed. To obtain the inverse transformed samples $\{\hat{x}(n)\}$, initialize the internal states in any way, run the system $\mathbf{G}(z)$ over the $N_B + N - 1$ blocks, and discard the first $N - 1$ reconstructed blocks, retaining the $N_x = N_B M$ remaining samples.

- N *even* — Reflect $s = N/2$ samples about each border, thus getting $N_B + N$ blocks to be processed. To obtain the inverse transformed samples $\{\hat{x}(n)\}$, initialize the internal states in any way and run the system $\mathbf{G}(z)$ over the $N_B + N$ blocks. Discard the first $N - 1$ reconstructed blocks and the first $M/2$ samples of the N-th block. Include in the reconstructed signal the last $M/2$ samples of the N-th block and the subsequent $(N_B - 1)M$ samples. In the last block, include the first $M/2$ samples in the reconstructed signal and discard the rest.

This approach will assure the perfect reconstruction property and orthogonality of the overall transformation if the LT is orthogonal [38]. The price paid is running the algorithm over extra N or $N - 1$ blocks. As it is common to have $N_B \gg N$, the computational increase is only marginal.

5.10 Design Issues for Compression

Block transform coding and subband coding have been two dominant techniques in existing image compression standards and implementations. Both methods actually exhibit many similarities: relying on a certain transform to convert the input image to a more decorrelated representation, then utilizing the same basic building blocks such as bit allocator, quantizer, and entropy coder to achieve compression.

Block transform coders enjoyed success due to their low complexity in implementation and their reasonable performance. The most popular block transform coder led to the current image compression standard JPEG [32] which utilizes the 8×8 DCT at its transformation stage. At high bit rates, JPEG offers almost visually lossless reconstruction image quality. However, when more compression is needed (i.e., at lower bit rates), annoying blocking artifacts show up for two reasons: (i) the DCT bases are short, nonoverlapped, and have discontinuities at the ends; and (ii) JPEG processes each image block independently. So, interblock correlation is not taken into account.

The development of lapped transforms helps solve the blocking problem by borrowing pixels from the adjacent blocks to produce the transform coefficients of the current block. In other words, lapped transforms are block transforms with overlapping basis functions. Compared to the traditional block transforms such as DCT, DST, and the Walsh-Hadamard transform [46], lapped transforms offer two main advantages: (i) from the analysis viewpoint, they take into account interblock correlation and, hence, provide better energy compaction that leads to more efficient entropy coding of the coefficients; and (ii) from the synthesis viewpoint, their basis functions decay asymptotically to zero at the ends, reducing blocking discontinuities drastically.

All of the lapped transforms presented in the previous sections are designed to have high practical value. They all have perfect reconstruction. Some of them even have real and symmetric basis functions. However, for the transforms to achieve high coding performance, several other properties are also needed. Transforms can be obtained using unconstrained nonlinear optimization where some of the popular cost criteria are coding gain, DC leakage, attenuation around mirror frequencies, and stopband attenuation. In the particular field of image compression, all of these criteria are well-known desired properties in yielding the best reconstructed image quality [44, 55]. The cost function in the optimization process can be a weighted linear combination of these measures as follows

$$C_{\text{overall}} = \alpha_1 \, C_{\text{coding gain}} + \alpha_2 \, C_{\text{DC}} + \alpha_3 \, C_{\text{mirror}}$$
$$+ \, \alpha_4 \, C_{\text{analysis stopband}} + \alpha_5 \, C_{\text{synthesis stopband}} \cdot \qquad (5.129)$$

Coding Gain

The coding gain of a transform is defined as the reduction in transform coding mean-square error over pulse-code modulation (PCM) which simply quantizes the samples of the signal with the desired number of bits per sample. Define σ_x^2 as the variance of the input signal, $\sigma_{x_i}^2$ as the variance of the i-th subband, and $||q_i||^2$ as the $\mathcal{L}^2$ norm of the i-th bases of the inverse LT $\mathbf{Q}$. With several assumptions including scalar quantization and a sufficiently large bit rate, the generalized coding gain can be formulated [12, 15, 28] as

$$
C_{\text{coding gain}} = 10 \log_{10} \frac{\sigma_x^2}{\left(\displaystyle\prod_{i=0}^{M-1} \sigma_{x_i}^2 \, ||q_i||^2 \right)^{\frac{1}{M}}} . \tag{5.130}
$$

The coding gain can be thought of as an approximate measure of the transform's energy compaction capability. Among the listed criteria, higher coding gain correlates most consistently with higher objective performance (measured in MSE or PSNR). Transforms with higher coding gain compact more signal energy into a fewer number of coefficients, leading to more efficient entropy coding.

Low DC Leakage

The DC leakage cost function measures the amount of DC energy that leaks out to the bandpass and highpass subbands. The main idea is to concentrate all signal energy at DC into the DC coefficients, which proves to be advantageous in both signal decorrelation and in the prevention of discontinuities in the reconstructed signals. Low DC leakage can prevent the annoying checkerboard artifact that usually occurs when high frequency bands are severely quantized [55]. The DC cost function can be defined as

$$
C_{\text{DC}} = \sum_{i=1}^{M-1} \sum_{n=0}^{L-1} p_{in} , \tag{5.131}
$$

where $\{p_{in}\}$ are entries of the LT matrix $\mathbf{P}$. The reader should note that all antisymmetric filters have a zero at DC (zero frequency). Therefore, the above formula needs to apply only to symmetric bases to reduce the complexity of the optimization process. It is interesting to note that the zero leakage condition is equivalent to having one vanishing moment — a necessary condition in the construction of wavelets.

Attenuation at Mirror Frequencies

Viewing the transform bases as filters, we can generalize C_{DC} to also encompass mirror frequency points. The concern is now at all aliasing frequencies $\omega_m = \frac{2\pi m}{M}$, $m \in \mathcal{Z}$, $1 \leq m \leq \frac{M}{2}$. Ramstad, Aase, and Husoy [44] have shown that frequency attenuation at mirror frequencies is very important in the further reduction of blocking artifacts; the filter (basis function) response should be small at these mirror frequencies as well. The corresponding cost function is

$$
C_{\text{mirror}} = \sum_{i=0}^{M-2} \left| P_i \left(e^{j\omega_m} \right) \right|^2, \qquad \omega_m = \frac{2\pi m}{M}, \qquad 1 \leq m \leq \frac{M}{2} , \tag{5.132}
$$

where $P_i(e^{j\omega})$ is the Fourier transform of $\{p_{in}\}$. Low DC leakage and high attenuation near the mirror frequencies are not as essential to the coder's objective performance as coding gain. However, they do improve the visual quality of the reconstructed image significantly.

Stopband Attenuation

Stopband attenuation is a classical performance criterion in filter design. In the forward transform, the stopband attenuation cost helps in improving the signal decorrelation and decreasing the amount of aliasing. In meaningful images, we know a priori that most of the energy is concentrated in a low frequency region. Hence, high stopband attenuation in this part of the frequency spectrum becomes extremely desirable. In the inverse transform, the synthesis filters (basis functions) covering low-frequency bands need to have high stopband attenuation near and/or at $\omega = \pi$ to enhance their smoothness. The biased weighting can be enforced using two simple functions $W_i^a(e^{j\omega})$ and $W_i^s(e^{j\omega})$. For our purposes, the stopband attenuation criterion measures the sum of all of the filters' energy outside the designated passbands:

$$C_{\text{analysis stopband}} = \sum_{i=0}^{M-1} \int_{\omega \in \Omega_{stopband}} W_i^a\left(e^{j\omega}\right) \left|P_i\left(e^{j\omega}\right)\right|^2 d\omega$$

$$C_{\text{synthesis stopband}} = \sum_{i=0}^{M-1} \int_{\omega \in \Omega_{stopband}} W_i^s\left(e^{j\omega}\right) \left|Q_i\left(e^{j\omega}\right)\right|^2 d\omega , \quad (5.133)$$

where $Q_i(e^{j\omega})$ is the Fourier transform of $\{q_{in}\}$, which are the entries of $\mathbf{Q}$.

5.11   Transform-Based Image Compression Systems

Transform coding is the single-most popular approach for image compression. The basic building blocks of a transform coder are illustrated in Fig. 5.34. The entropy coder is the step that actually performs any compression. The entropy of the symbols to be compressed is reduced by the quantizer which is the only building block which is not reversible; it is a lossy operator. The transform step neither causes losses nor performs compression, but it is the core of the compression system. It enables compression by compacting the energy into few coefficients, thus reducing the distortion caused by the quantization step.

A separable transformation is one where all rows are transformed, and then all columns are transformed. Let $\mathbf{X}$ be the matrix containing the image pixels and let $\mathbf{Y}$ be the transformed image; then from Eq. (5.33) we have

$$\mathbf{Y} = \mathbf{H}_F \mathbf{X} \mathbf{H}_F^T \qquad \mathbf{X} = \mathbf{H}_I^T \mathbf{Y} \mathbf{H}_I . \qquad (5.134)$$

FIGURE 5.34
Basic building blocks of a transform coder. The image is transformed and quantized in order to submit the data to a lossless (entropy) coder.

$\mathbf{Y}$ is composed by $M \times M$ blocks $\mathbf{Y}_{ij}$, each block having a full set of transform coefficients (subband samples). Each of the blocks $\mathbf{Y}_{ij}$ is then quantized and encoded. In the case of a block transform, each transformed block $\mathbf{Y}_{ij}$ is related to only one image block of $M \times M$ pixels. For LTs, of course, each block $\mathbf{Y}_{ij}$ is related to several pixel blocks. We will discuss only the performance of LTs in the context of two popular image coders.

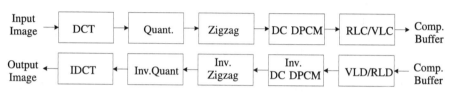

FIGURE 5.35
JPEG building blocks. The image is broken into blocks and each block is transformed using the DCT, quantized, and encoded. The decoder performs the inverse operations.

5.11.1   JPEG

Transform coding is the framework employed by the Joint Photographic Experts Group (JPEG) still image compression standard. The reader is referred to Pennebacker and Mitchell [32] for a detailed description of the JPEG image compression system; the publication not only covers JPEG well but it also includes a copy of the standard draft! The steps for JPEG compression are shown in Fig. 5.35. The image is broken into blocks of 8×8 pixels, which are transformed using a separable 8-channel DCT. The transformed block is quantized (effectively divided by an integer number and rounded) and then encoded. The quantized transformed samples in a block are scanned into a vector following a zigzag pattern starting from the lowest frequency band to the highest. The lowest frequency sample of a block is known as the DC coefficient (DCC). Before encoding, the quantized DCC of a block is actually replaced by the difference of itself and the DCC of a previous block (this is referred to as DPCM). Finally the scanned vector is fed into an entropy coder which uses a combination of run-length coding (RLC) and variable-length coding (VLC) to compress the data. The decoder runs the inverse of all the encoding steps in reverse order to reconstruct the image block from the compressed data.

For our purposes, we compare the performance of LTs against the DCT. LTs can be very easily incorporated into JPEG by simply replacing the DCT. Even though the bases overlap, the subband samples are arranged in a block just like the DCT. The transformed block is fed to the rest of the JPEG coder (which does not care if the samples were found through the DCT) to be quantized and encoded. All results here for JPEG are found by merely replacing the DCT by an 8-channel ($M = 8$) LT while maintaining all the other parameters and settings unchanged.

5.11.2 Embedded Zerotree Coding

Embedded Zerotree Coding (EZC) is often associated with the dyadic wavelet transform. The multiresolution characteristics of the wavelet transform have created an intuitive foundation on which simple, yet sophisticated, methods of encoding the transform coefficients are developed. Exploiting the relationship between the parent and the offspring coefficients in a wavelet tree, the original Embedded Zerotree Wavelet (EZW) coder [50] and its variations [47, 71] can effectively order the coefficients by bit planes and transmit the more significant bits first. This coding scheme results in an embedded bit stream along with many other advantages, such as exact bit rate control and near-idempotency (perfect idempotency is obtained when the transform maps integers to integers). In these subband coders, global information is taken into account fully. In this section, we confirm that the embedded zerotree framework is not limited only to the wavelet transform but it can be utilized with various LTs as well. In fact, the combination of a LT and several levels of wavelet decomposition of the DC band can provide much finer frequency spectrum partitioning, leading to significant improvement over current embedded wavelet coders [58, 61].

The EZC approach relies on the fundamental idea that the most important information (defined here as what decreases a certain distortion measure the most) should be transmitted first. Assuming that the distortion measure is the mean-squared error (MSE), the transform is orthogonal, and transform coefficients C_{ij} are transmitted one by one, it is well known that the MSE decreases by $\frac{1}{N}|C_{ij}|^2$, where N is the total number of pixels. Therefore, large magnitude coefficients should always be transmitted first. If one bit is transmitted at a time, this approach can be generalized to ranking the coefficients by bit planes, and the most significant bits are transmitted first [43] as demonstrated in Fig. 5.36. The progressive transmission scheme results in an embedded bit stream (i.e., it can be truncated at any point by the decoder to yield the best corresponding reconstructed image). The algorithm can be thought of as an elegant combination of a scalar quantizer with power-of-two stepsizes and an entropy coder to encode wavelet coefficients.

Embedded algorithm relies on the hierarchical coefficients' tree structure called a *wavelet tree* — a set of wavelet coefficients from different scales that belong in the same spatial locality. This is demonstrated in Fig. 5.37, where the tree in the vertical direction is circled. All of the coefficients in the lowest frequency band make up the *DC band* or the *reference signal* (located at the upper left corner). Besides these DC coefficients, in a wavelet tree of a particular direction, each lower frequency

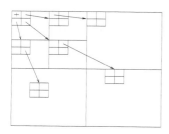

		sign	s	s	s	s	s	s	s	s	s	s	s	s
msb	5	1	1	0	0	0	0	0	0	0	0	0	0	0
	4	>	>	1	1	0	0	0	0	0	0	0	0	0
	3	>	>	>	>	1	1	1	1	0	0	0	0	0
	2	>	>	>	>	>	>	>	>	1	1	1	1	1
	1	>	>	>	>	>	>	>	>	>	>	>	>	>
lsb	0	>	>	>	>	>	>	>	>	>	>	>	>	>

FIGURE 5.36
Embedded zerotree coding as a bit-plane refinement scheme.

parent node has four corresponding higher frequency *offspring nodes*. All coefficients below a parent node in the same spatial locality is defined as its *descendants*. Define a coefficient C_{ij} to be *significant* with respect to a given threshold T if $|C_{ij}| \geq T$, and *insignificant* otherwise. Meaningful image statistics have shown that if a coefficient is insignificant, it is very likely that its offspring and descendants are insignificant as well. Exploiting this fact, sophisticated embedded wavelet coders can output a single marker to represent very efficiently a large, smooth image area (an insignificant tree). For more details on the algorithm, the reader is referred to references [47, 50], and [71].

LTs obtain a uniform spectrum partitioning whereas the wavelet transform has an octave-band signal decomposition. All LT subbands have the same size. A parent node would not have four offspring nodes as in the case of the wavelet representation. In order to use embedded zerotree algorithms to encode the LT coefficients, we have to modify the data structure. Investigating the analogy between the wavelet and the LT, as in Fig. 5.37, reveals that the parent, offspring, and descendants in a wavelet tree cover the same spatial locality, and so do the coefficients of a block of LT coefficients. In fact, a wavelet tree in an L-level decomposition is analogous to a 2^L-band LT's coefficient block. The difference lies at the bases that generate these coefficients.

Let $\mathcal{O}(i, j)$ be the set of coordinates of all offsprings of the node (i, j) in an M-band LT $(0 \leq i, j \leq M - 1)$; then $\mathcal{O}(i, j)$ can be represented as follows:

$$\mathcal{O}(i, j) = \{(2i, 2j), (2i, 2j + 1), (2i + 1, 2j), (2i + 1, 2j + 1)\}. \qquad (5.135)$$

All $(0, 0)$ coefficients from all transform blocks form the DC band, which is similar to the wavelet transform's reference signal, and each of these nodes has only three offspring: $(0, 1)$, $(1, 0)$, and $(1, 1)$. The complete tree is now available locally; we do not have to search for the offspring across the subbands anymore. The only requirement here is that the number of bands that M has be a power of two. Fig. 5.38 demonstrates, through a simple rearrangement of the LT coefficients, that the redefined tree structure in Eq. (5.135) does possess a wavelet-like multiscale representation. To decorrelate the DC band even more, several levels of wavelet decomposition can be used depending on the input image size. Besides the obvious increase in the coding efficiency of DC coefficients, thanks to deeper coefficient trees, wavelets provide variably longer bases for the signal's DC component, leading to smoother reconstructed images; in other words, blocking artifacts are further reduced. LT

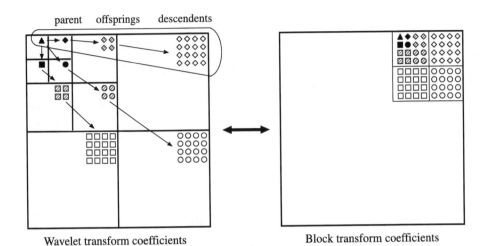

Wavelet transform coefficients   Block transform coefficients

FIGURE 5.37
Wavelet and block transform analogy.

iteration (resulting in the HLTs or M-band wavelets) can be applied to the DC band as well.

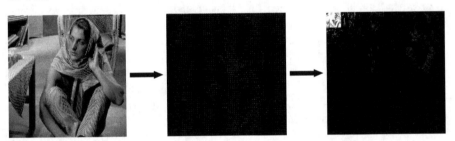

FIGURE 5.38
A demonstration of the analogy between block transform and wavelet representation. Reproduced by Special Permission of *Playboy* magazine. Copyright ©1972, 2000 by Playboy.

5.11.3 Other Coders

Although we use JPEG and EZC only as examples, LTs can be used to replace other transforms in a variety of coders. If the coder treats the subbands independently, encoding of the LT's output is not different from encoding the output of any other transform. If the subbands are not encoded independently, but the coder was designed for a nonhierarchical M-channel transform (e.g., DCT), then the LT can be used to immediately replace the transform, just like we do for JPEG. If the subbands are not encoded independently, and the coder was designed for a hierarchical transform (like wavelets), then one can use the same approach as in the previous section to incorporate the LTs. It is straightforward to utilize LTs in coders such as the new standard (JPEG 2000) [11] or other efficient coders based on optimized classification of subband samples [13]. Furthermore, LTs can be used to replace the wavelet transform in several efficient coders [1, 51, 67, 69]. Adaptive LTs have also been applied to image compression [16].

5.12 Performance Analysis

The performance of any compression system is measured by computing the distortion achieved by compressing a particular image at a certain compression ratio. For every compression attempt for a particular image and coding settings, one can compute the rate achieved R, often expressed in bits per pixel (or bpp), and the distortion is some measure of the difference between the original image and its reconstructed approximation after decompression. We use the peak signal-to-noise ratio measure which is given in decibels (dB) and is defined as

$$
PSNR = 10 \log_{10} \left(\frac{(255)^2}{\dfrac{1}{N_{\text{pixels}}} \displaystyle\sum_{ij} [x_o(i, j) - x_r(i, j)]^2} \right), \qquad (5.136)
$$

where $x_o(i, j)$ and $x_r(i, j)$ are the original and reconstructed pixels, respectively.

We used four 512×512-pixel test images shown in Fig. 5.39 for benchmarking the LTs in image compression. The image, "Barbara" has large detailed areas with high-frequency patterns, while "Lena" has mainly smooth areas with occasional edges and textures. "Goldhill" is a typical landscape image with many details, and the "text" image is basically composed of high contrast edges.

5.12.1 JPEG

For JPEG, we compared the following transforms for $M = 8$: ELT-2, MLT, DCT, LOT, GenLOT ($L = 48$), GLBT ($L = 16$), and FLT (two 24-tap bases along with six 8-tap DCT bases). For each image and transform, several rate-distortion points are obtained by compressing the image using JPEG's default (example) quantizer table [32] scaled by a multiplicative factor. Instead of providing the actual PSNR obtained with every experiment we compared every result to the PSNR obtained by compressing the same image at the same bit rate using the DCT. Objective results are shown in Fig. 5.40. The curves can be viewed as incremental PSNR plots, in which for each transform, image, and bit rate, it indicates the gain in PSNR (dB) obtained by replacing the DCT by a given LT. Note that the performance of the LTs is far superior to that of the DCT except for the "text" image. This image is not suitable

ling gain [33] as

$$G(n) = \left(\frac{\dfrac{1}{M} \displaystyle\sum_{i=0}^{M-1} \hat{\sigma}_i^2}{\left(\displaystyle\prod_{i=0}^{M-1} \hat{\sigma}_i^2\right)^{1/M}} \right)$$

threshold g in order to

FIGURE 5.39
Test images of 512×512 pixels. From left to right and top to bottom: "Barbara," "Lena," "Goldhill," and "text." Reproduced by Special Permission of *Playboy* magazine. Copyright ©1972, 2000 by Playboy.

for transform compression because it is mainly composed of sharp edges. Hence, the short bases of the DCT concentrate the artifacts (caused by compressing the sharp edges) into small areas. In any case, the GLBT and the GenLOT are always good performers.

Objective comparisons are not always the best. The FLT, for example, seems to have similar performance to DCT in terms of PSNR. However, it produces noticeably better images. It is virtually free of the ringing and blocking artifacts which are the main drawbacks of using block transforms like DCT. Reconstructed images are shown in Fig. 5.41.

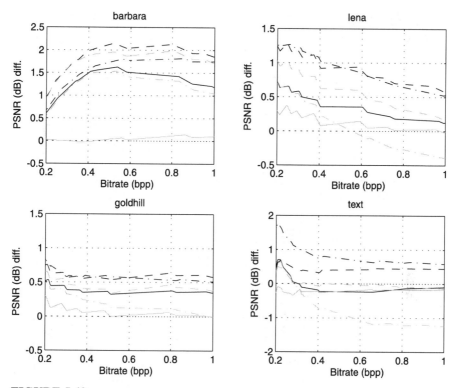

FIGURE 5.40

Comparison among transforms in JPEG for several images and bit rates. Incremental PSNR plots are shown, indicating the gain in PSNR (dB) obtained by replacing the DCT by the given transform. Curve line styles: black solid — LOT; black dashed — GenLOT ($L = 48$); black dash-dot — GLBT ($L = 16$); gray solid — FLT ($2 \times L = 24$ and $6 \times L = 8$); gray dashed — ELT-2 ($L = 32$); gray dash-dot — MLT ($L = 16$).

5.12.2   Embedded Zerotree Coding

The objective coding results (PSNR in dB) for the four test images ("Barbara," "Lena," "Goldhill," and "text") are tabulated in Table 5.5. The transforms in comparison are the ELT-2, MLT, DCT, LOT, GenLOT ($L = 40$), GLBT ($L = 16$), FLT (all with $M = 8$), and the 9/7-tap bi-orthogonal wavelet [2]. In the LT cases, we use three additional levels of 9/7 wavelet decomposition on the DC bands. All computed PSNR quotes in dB are obtained from a real compressed bit stream with all overheads included. Incremental rate-distortion plots are shown in Fig. 5.42 where the 9/7 wavelet serves as the performing benchmark.

The coding results clearly confirm the potential of LTs. For a smooth image such as "Lena," which the wavelet transform can sufficiently decorrelate, the 9/7 wavelet offers a comparable performance. However, for a highly-textured image such as

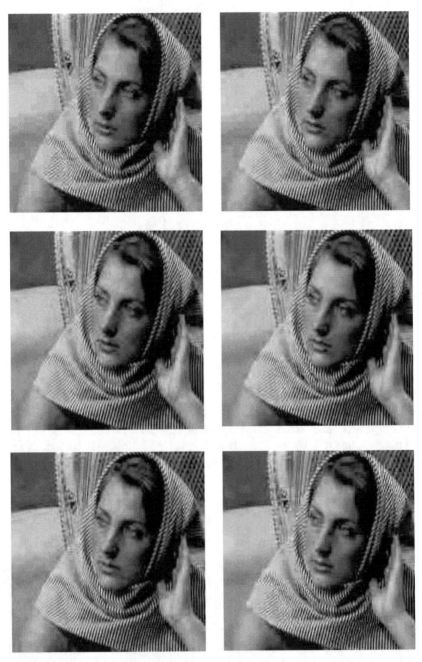

FIGURE 5.41
Enlarged portion of reconstructed images using JPEG at 0.3 bpp. Top left:
DCT (25.67); top right: LOT (26.94); middle left: GenLOT, $L = 48$, (27.19);
middle right: GLBT, $L = 16$, (26.88); bottom left: ELT-2 (27.24); bottom right:
FLT, 2×24 and 6×8, (25.63). Number in parentheses indicates PSNR in dB.
Reproduced by Special Permission of *Playboy* magazine. Copyright ©1972, 2000
by Playboy.

Table 5.5 Objective Coding Results (PSNR in dB). (a) Lena. (b) Goldhill. (c) Barbara. (d) Text

Lena	Transform							
Comp. Ratio	9 / 7 wavelet	8 x 8 DCT	8 x 16 LOT	8 x 40 GenLOT	2x24 6x8 FLT	8 x 16 GLBT	8 x 32 ELT	8 x 16 MLT
8:1	40.41	39.91	40.05	40.43	39.89	40.35	40.01	39.20
16:1	37.21	36.38	36.72	37.32	36.51	37.28	36.93	36.27
32:1	34.11	32.90	33.56	34.23	33.25	34.14	33.94	33.42
64:1	31.10	29.67	30.48	31.16	30.15	31.04	30.93	30.44
100:1	29.35	27.80	28.62	29.31	28.31	29.14	29.14	28.61

(a)

Goldhill	Transform							
Comp. Ratio	9 / 7 wavelet	8 x 8 DCT	8 x 16 LOT	8 x 40 GenLOT	2x24 6x8 FLT	8 x 16 GLBT	8 x 32 ELT	8 x 16 MLT
8:1	36.55	36.25	36.63	36.80	36.22	36.69	36.46	36.19
16:1	33.13	32.76	33.18	33.36	32.76	33.31	33.14	32.93
32:1	30.56	30.07	30.56	30.79	30.25	30.70	30.65	30.48
64:1	28.48	27.93	28.36	28.60	28.17	28.58	28.52	28.33
100:1	27.38	26.65	27.09	27.40	27.06	27.33	27.33	27.10

(b)

Barbara	Transform							
Comp. Ratio	9 / 7 wavelet	8 x 8 DCT	8 x 16 LOT	8 x 40 GenLOT	2x24 6x8 FLT	8 x 16 GLBT	8 x 32 ELT	8 x 16 MLT
8:1	36.41	36.31	37.43	38.08	36.22	37.84	37.53	37.07
16:1	31.40	31.11	32.70	33.47	31.12	33.02	33.12	32.59
32:1	27.58	27.28	28.80	29.53	27.42	29.04	29.33	28.74
64:1	24.86	24.58	25.70	26.37	24.86	26.00	26.30	25.65
100:1	23.76	23.42	24.34	24.95	23.74	24.55	24.90	24.28

(c)

Text	Transform							
Comp. Ratio	9 / 7 wavelet	8 x 8 DCT	8 x 16 LOT	8 x 40 GenLOT	2x24 6x8 FLT	8 x 16 GLBT	8 x 32 ELT	8 x 16 MLT
8:1	40.49	40.24	40.36	40.41	40.05	40.57	39.89	38.80
16:1	37.14	36.81	36.85	36.89	36.60	37.24	36.38	35.34
32:1	33.11	32.93	33.36	33.41	32.75	33.84	32.89	32.60
64:1	29.14	28.01	29.10	29.09	28.01	29.51	28.77	28.61
100:1	25.89	25.08	26.50	26.47	25.36	26.73	26.15	26.00

(d)

"Barbara," the 8×40 GenLOT and the 8×16 GLBT can provide a PSNR gain of more than 1.5 dB over a wide range of bit rates. Fig. 5.43 demonstrates the high level of reconstructed image quality as well. The ELT, GLBT, and GenLOT can completely eliminate blocking artifacts.

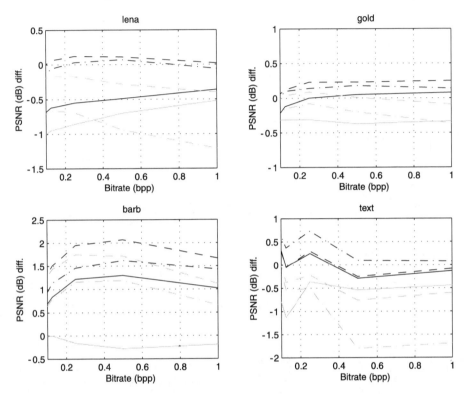

FIGURE 5.42

Comparison among transforms in EZC for several images and bit rates. Incremental PSNR plots are shown, indicating the gain (or loss) in PSNR (dB) obtained by replacing the 9/7 bi-orthogonal wavelet by the given transform. Curve lines: black solid — LOT; black dashed — GenLOT ($L = 40$); black dash-dot — GLBT ($L = 16$); gray solid — FLT ($2 \times L = 24$ and $6 \times L = 8$); gray dashed — ELT-2 ($L = 32$); gray dash-dot — MLT ($L = 16$).

5.13   Conclusions

We hope that this chapter serves as an eye-catching introduction to lapped transforms and their potentials in image/video compression. As for the theory of LTs, this

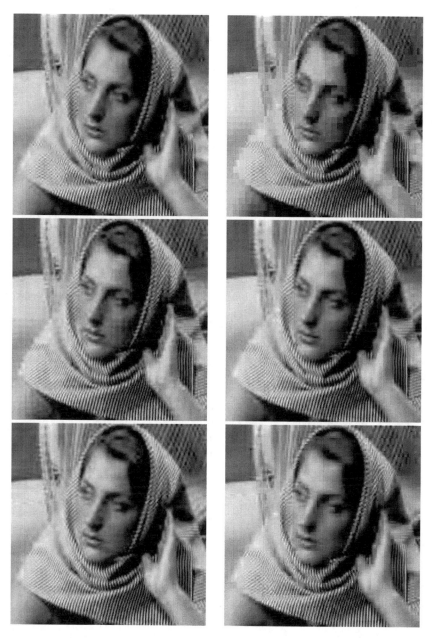

FIGURE 5.43
Perceptual coding comparison using EZC at a bit-rate of 0.25 bpp. The 9/7 wavelet transform is compared to various LTs and enlarged portions of the reconstructed "Barbara" image are shown. Top left: 9/7 bi-orthogonal wavelet (27.58); top right: DCT (27.28); middle left: LOT (28.80); middle right: GLBT, $L = 16$, (29.04); bottom left: GenLOT, $L = 40$, (29.53); bottom right: 2×24 and 6×8 FLT (27.42). Number in parentheses indicates PSNR in dB. Reproduced by Special Permission of *Playboy* magazine. Copyright ©1972, 2000 by Playboy.

chapter should be viewed as a first step, whereas the references and the references therein should give a more detailed treatment of the subject.

It was shown how lapped transforms can replace block transforms allowing an overlap of the basis functions. It was also shown that the analysis of MIMO systems, mainly their factorization, are invaluable tools for the design of useful lapped transforms. That was the case of transforms such as LOT, LBT, GenLOT, GLBT, FLT, MLT, and ELT. We presented these practical LTs by not only describing the general factorization, but also by plotting bases and discussing in detail how to construct at least a good design example. We made an effort in tabulating the bases entries or providing all parameters necessary to construct the bases. Even if the examples are not ideal for a particular application that the reader might have in mind, they may provide the basics upon which the reader can build by exploring the references and performing customized optimization. Invariably, the design examples presented here were tuned for image compression applications.

Some image compression methods were briefly described to serve as a comparison framework in which the LTs are applied for compression of typical images. Several LTs were compared to transforms such as DCT and wavelets, showing how truly promising LTs are for image compression. FLT and LBT with lifting steps require minimal computation apart from the DCT computation and are very attractive replacements for DCT, rivalling wavelet transforms at a lower implementation complexity. Buffering is also reduced since the transforms are not implemented hierarchically. Parallel computation and region-of-interest coding/decoding are also greatly facilitated.

It is worth noting that we intentionally avoided viewing the transforms as filter banks, so the bases were not discussed as impulse responses of filters, and their frequency response was not analyzed. Nevertheless, the framework is the same as is the analysis of MIMO systems. We trust that this chapter will give some insight into this vast field which is based on the study of multirate systems.

References

[1] Andrew, J., Simple and efficient hierarchical image coder, *Proc. of IEEE International Conference on Image Processing,* 3, 658–661, Santa Barbara, CA, Oct. 1997.

[2] Antonini, M., Barlaud, M., Mathieu, P., and Daubechies, I., Image coding using the wavelet transform, *IEEE Trans. on Image Processing,* 1, 205–220, 1992.

[3] Boashash, B., Ed., *Time-Frequency Signal Analysis,* John Wiley & Sons, New York, 1992.

[4] Bordreaux-Bartels, G.F., Mixed time-frequency signal transformations, *The Transforms and Applications Handbook,* Poularikas, A., Ed., CRC Press, Boca Raton, FL, 1996.

[5] Cassereau, P., *A New Class of Optimal Unitary Transforms for Image Processing,* Master's thesis, M.I.T., Cambridge, MA, May 1985.

[6] Clarke, R.J., *Transform Coding of Images,* Academic Press, Orlando, FL, 1985.

[7] Coifman, R., Meier, Y., Quaker, D., and Wickerhauser, V., Signal processing and compression using wavelet packets, Technical Report, Dept. of Mathematics, Yale Univ., New Haven, CT, 1991.

[8] Doğanata, Z., Vaidyanathan, P.P., and Nguyen, T.Q., General synthesis procedures for FIR lossless transfer matrices, for perfect reconstruction multirate filter banks applications, *IEEE Trans. Acoust., Speech, Signal Processing,* 36(10), 1561–1574, 1988.

[9] Herley, C., Kovacevic, J., Ramchandran, K., and Vetterli, M., Tilings of the time-frequency plane: construction of arbitrary orthogonal bases and fast tiling algorithms, *IEEE Trans. on Signal Processing,* 41, 3341–3359, 1993.

[10] Hohn, F.E., *Elementary Matrix Algebra,* 2nd ed., MacMillan, New York,1964.

[11] ISO/IEC JTC1/SC29/WG1, JPEG 2000 Committee, Working Draft 2.0, June 25, 1999.

[12] Jayant, N.S. and Noll, P., *Digital Coding of Waveforms,* Prentice-Hall, Englewood Cliffs, NJ, 1984.

[13] Joshi, R.L., Crump, V.J., and Fisher, T.R., Image subband coding using arithmetic coded trellis coded quantization, *IEEE Trans. Circuits and Systems for Video Technology,* 5, 515–523, 1995.

[14] Jozawa, H. and Watanabe, H., Intrafield/Interfield adaptive lapped transform for compatible HDTV coding, *4th International Workshop on HDTV and Beyond,* Torino, Italy, Sept. 4–6, 1991.

[15] Katto, J. and Yasuda, Y., Performance evaluation of subband coding and optimization of its filter coefficients, *SPIE Proc. Visual Comm. and Image Proc.,* 1991.

[16] Klausutis, T.J. and Madisetti, V.K., Variable block size adaptive lapped transform-based image coding, *Proc. of IEEE International Conference on Image Processing,* 3, 686–689, Santa Barbara, CA, Oct. 1997.

[17] Koilpillai, R.D. and Vaidyanathan, P.P., Cosine modulated FIR filter banks satisfying perfect reconstruction, *IEEE Trans. Signal Processing,* 40, 770–783, 1992.

[18] Malvar, H.S., *Optimal pre- and post-filtering in noisy sampled-data systems,* Ph.D. dissertation, M.I.T., Cambridge, MA, Aug. 1986.

[19] Malvar, H.S., Reduction of blocking effects in image coding with a lapped orthogonal transform, *Proc. of Intl. Conf. on Acoust., Speech, Signal Processing,* Glasgow, Scotland, 781–784, Apr. 1988.

[20] Malvar, H.S. and Staelin, D.H., The LOT: transform coding without blocking effects, *IEEE Trans. Acoust., Speech, Signal Processing,* ASSP-37, 553–559, 1989.

[21] Malvar, H.S., Lapped transforms for efficient transform/subband coding, *IEEE Trans. Acoust., Speech, Signal Processing,* ASSP-38, 969–978, 1990.

[22] Malvar, H.S., The LOT: a link between block transform coding and multirate filter banks, *Proc. Intl. Symp. Circuits and Systems,* Espoo, Finland, 835–838, June 1988.

[23] Malvar, H.S., Efficient signal coding with hierarchical lapped transforms, *Proc. of Intl. Conf. on Acoust., Speech, Signal Processing,* Albuquerque, NM, 761–764, 1990.

[24] Malvar, H.S., Modulated QMF filter banks with perfect reconstruction, *Elect. Letters,* 26, 906–907, 1990.

[25] Malvar, H.S., Extended lapped transform: fast algorithms and applications, *Proc. of Intl. Conf. on Acoust., Speech, Signal Processing,* Toronto, Canada, 1797–1800, 1991.

[26] Malvar, H.S., *Signal Processing with Lapped Transforms,* Artech House, Norwood, MA, 1992.

[27] Malvar, H.S., Extended lapped transforms: properties, applications and fast algorithms, *IEEE Trans. Signal Processing,* 40, 2703–2714, 1992.

[28] Malvar, H.S., Biorthogonal and nonuniform lapped transforms for transform coding with reduced blocking and ringing artifacts, *IEEE Trans. on Signal Processing,* 46, 1043–1053, 1998.

[29] Nayebi, K., Barnwell, T.P., and Smith, M.J., The time domain filter bank analysis: a new design theory, *IEEE Trans. on Signal Processing,* 40, 1412–1429, 1992.

[30] Nguyen, T.Q. and Koilpillai, R.D., Theory and design of arbitrary-length cosine-modulated filter banks and wavelets satisfying perfect reconstruction, *IEEE Trans. on Signal Processing,* 44, 473–483, 1996.

[31] Oppenheim, A.V. and Schafer, R.W., *Discrete-Time Signal Processing,* Prentice-Hall, Englewood Cliffs, NJ, 1989.

[32] Pennebaker, W.B. and Mitchell, J.L., *JPEG: Still Image Compression Standard,* Van Nostrand Reinhold, New York, 1993.

[33] Princen, J.P. and Bradley, A.B., Analysis/synthesis filter bank design based on time domain aliasing cancellation, *IEEE Trans. Acoust., Speech, Signal Processing,* ASSP-34, 1153–1161, 1986.

[34] de Queiroz, R.L. and Rao, K.R., Time-varying lapped transforms and wavelet packets, *IEEE Trans. on Signal Processing,* 41, 3293–3305, 1993.

[35] de Queiroz, R.L., *On Lapped Transforms,* Ph.D. dissertation, The University of Texas at Arlington, August 1996.

[36] de Queiroz, R.L. and Rao, K.R., The extended lapped transform for image coding, *IEEE Trans. on Image Processing,* 4, 828–832, 1995.

[37] de Queiroz, R.L., Nguyen, T.Q., and Rao, K.R., The generalized lapped orthogonal transforms, *Electronics Letters,* 30, 107–107, 1994.

[38] de Queiroz, R.L. and Rao, K.R., On orthogonal transforms of images using paraunitary filter banks, *J. Visual Communications and Image Representation,* 6(2), 142–153, 1995.

[39] de Queiroz, R.L. and Rao, K.R., On reconstruction methods for processing finite-length signals with paraunitary filter banks, *IEEE Trans. on Signal Processing,* 43, 2407–2410, 1995.

[40] de Queiroz, R.L., Nguyen, T.Q., and Rao, K.R., The GenLOT: generalized linear-phase lapped orthogonal transform, *IEEE Trans. on Signal Processing,* 44, 497–507, 1996.

[41] de Queiroz, R.L., Uniform filter banks with non-uniform bands: post-processing design, *Proc. of Intl. Conf. Acoust. Speech. Signal Proc.,* Seattle, WA, III, 1341–1344, May 1997.

[42] de Queiroz, R.L. and Eschbach, R., Fast downscaled inverses for images compressed with M-channel lapped transforms, *IEEE Trans. on Image Processing,* 6, 794–807, 1997.

[43] Rabbani, M. and Jones, P.W., *Digital Image Compression Techniques,* SPIE Optical Engineering Press, Bellingham, WA, 1991.

[44] Ramstad, T.A., Aase, S.O., and Husoy, J.H., *Subband Compression of Images: Principles and Examples,* Elsevier, New York, 1995.

[45] Rao, K.R. and Yip, P., *Discrete Cosine Transform: Algorithms, Advantages, Applications,* Academic Press, San Diego, CA, 1990.

[46] Rao, K.R., Ed., *Discrete Transforms and Their Applications,* Van Nostrand Reinhold, New York,1985.

[47] Said, A. and Pearlman, W.A., A new fast and efficient image codec based on set partitioning in hierarchical trees, *IEEE Trans. on Circuits Syst. Video Tech.,* 6, 243–250, 1996.

[48] Schiller, H., Overlapping block transform for image coding preserving equal number of samples and coefficients, *Proc. SPIE, Visual Communications and Image Processing,* 1001, 834–839, 1988.

[49] Schuller, G.D.T. and Smith, M.J.T., New framework for modulated perfect reconstruction filter banks, *IEEE Trans. on Signal Processing,* 44, 1941–1954, 1996.

[50] Shapiro, J.M., Embedded image coding using zerotrees of wavelet coefficients, *IEEE Trans. on Signal Processing,* 41, 3445–3462, 1993.

[51] Silva, E.A.B., Sampson, D.G., and Ghanbari, M., A successive approximation vector quantizer for wavelet transform image coding, *IEEE Trans. Image Processing,* 5, 299–310, 1996.

[52] Sodagar, I., Nayebi, K., and Barnwell, T.P., A class of time-varying wavelet transforms, *Proc. of Intl. Conf. on Acoust., Speech, Signal Processing,* Minneapolis, MN, III, 201–204, Apr. 1993.

[53] Soman, A.K. and Vaidyanathan, P.P., Paraunitary filter banks and wavelet packets, *Proc. of Intl. Conf. on Acoust., Speech, Signal Processing,* IV, 397–400, 1992.

[54] Soman, A.K., Vaidyanathan, P.P., and Nguyen, T.Q., Linear-phase paraunitary filter banks: theory, factorizations and applications, *IEEE Trans. on Signal Processing,* 41, 3480–3496, 1993.

[55] Strang, G. and Nguyen, T., *Wavelets and Filter Banks,* Wellesley-Cambridge, Wellesley, MA, 1996.

[56] Temerinac, M. and Edler, B., A unified approach to lapped orthogonal transforms, *IEEE Trans. Image Processing,* 1, 111–116, 1992.

[57] Tran, T.D. and Nguyen, T.Q., On M-channel linear-phase FIR filter banks and application in image compression, *IEEE Trans. on Signal Processing,* 45, 2175–2187, 1997.

[58] Tran, T.D. and Nguyen, T.Q., A progressive transmission image coder using linear phase uniform filter banks as block transforms, *IEEE Trans. on Image Processing,* in press.

[59] Tran, T.D., *Linear Phase Perfect Reconstruction Filter Banks: Theory, Structure, Design, and Application in Image Compression,* Ph.D. thesis, University of Wisconsin, Madison, WI, May 1998.

[60] Tran, T.D., de Queiroz, R.L., and Nguyen, T.Q., The variable-length generalized lapped biorthogonal transform, *Proc. Intl. Conf. Image Processing,* Chicago, IL, III, 697–701, 1998.

[61] Tran, T.D., de Queiroz, R.L., and Nguyen, T.Q., The generalized lapped biorthogonal transform, *Proc. Intl. Conf. Acoust., Speech Signal Proc.,* Seattle, III, 1441–1444, May 1998.

[62] Tran, T.D., The LiftLT: fast lapped transforms via lifting steps, *Proc. IEEE Int. Conf. on Image Processing*, Kobe, Japan, Oct. 1999.

[63] Vaidyanathan, P.P., *Multirate Systems and Filter Banks.* Prentice-Hall, Englewood Cliffs, NJ, 1993.

[64] Vaidyanathan, P.P. and Hoang, P., Lattice structures for optimal design and robust implementation of 2-channel PR-QMF banks, *IEEE Trans. Acoust., Speech, Signal Processing*, ASSP-36, 81–94, 1988.

[65] Vetterli, M. and Herley, C., Wavelets and filter banks: theory and design, *IEEE Trans. Signal Processing*, 40, 2207–2232, 1992.

[66] Vetterli, M. and Kovacevic, J., *Wavelets and Subband Coding*, Prentice-Hall, Englewood Cliffs, NJ, 1995.

[67] Wang, H. and Kuo, C.-C.J., A multi-threshold wavelet coder (MTWC) for high fidelity image compression, *Proc. of IEEE Int. Conf. on Image Processing*, 1, 652–655, Santa Barbara, CA, Oct. 1997.

[68] Wickerhauser, M.V., Acoustical signal compression using wavelet packets, in *Wavelets: A Tutorial in Theory and Applications*, Chui, C.K., Ed., Academic Press, San Diego, CA, 1992.

[69] Xiong, Z., Ramchandran, K., and Orchard, M.T., Space frequency quantization for wavelet image coding, *IEEE Trans. Image Processing*, 6, 677–693, 1997.

[70] Young, R.W. and Kingsbury, N.G., Frequency domain estimation using a complex lapped transform, *IEEE Trans. Image Processing*, 2, 2–17, 1993.

[71] Zandi, A., Allen, J., Schwartz, E., and Boliek, M., CREW: compression with reversible embedded wavelets, *Proc. IEEE Data Compression Conf.*, Snowbird, UT, 212–221, 1995.

Chapter 6

Wavelet-Based Image Compression

James S. Walker
University of Wisconsin-Eau Claire

Truong Q. Nguyen
Boston University

6.1 Introduction

One of the most successful applications of wavelet methods is transform-based image compression (also called coding). Such a coder [depicted in Fig. 6.1(a)] operates by transforming the data to remove redundancy, then quantizing the transform coefficients (a lossy step), and finally entropy coding the quantizer output. Because of their superior energy compaction properties and correspondence with the human visual system, wavelet compression methods have produced superior objective and subjective results [5]. Since a wavelet basis consists of functions with both short support (for high frequencies) and long support (for low frequencies), large smooth areas of an image may be represented with very few bits, and detail added where it is needed.

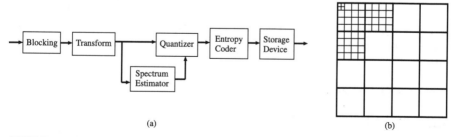

(a)   (b)

FIGURE 6.1

(a) Transform-based coder. (b) Subband decomposition used in the FBI finger-print compression standard.

Both orthogonal [73] and bi-orthogonal [1, 70] wavelets have been used for image compression. The recent FBI fingerprint compression standard [70] uses symmetric dyadic wavelets and significantly outperforms the JPEG (Joint Picture Expert Group) standard [38] at compression ratios above 10:1. Fig. 6.1(b) shows the subband decomposition used in the FBI fingerprint compression standard. Interestingly, the wavelet tree used in the FBI specification is a predominantly 4-channel decomposition achieved by cascading 2-channel filter banks.

Most high-quality algorithms today use some form of transform coder. One widely used standard is the JPEG compression algorithm, based on the discrete cosine transform (DCT) [38]. The image is partitioned into 8×8 blocks, each of which is then transformed via a tensor product of two 8-point DCTs. The transform coefficients are then arranged into 64 subbands, scalar-quantized, and adaptively Huffman coded. The JPEG algorithm yields good results for compression ratios of 10:1 and below (on 8-bit gray-scale images), but at higher compression ratios the underlying block nature of the transform begins to show through the compressed image. By the time compression ratios have reached 24:1, only the DC (lowest frequency) coefficient is getting any bits allocated to it, and the input image has been approximated by a set of 8×8 blocks. Consequently, the decompressed image has substantial blocking artifacts for medium and high compression ratios.

Researchers have applied subband coding to images for over a decade [69, 60]; their results reached a new level with the advent of the wavelet transform. Wavelet methods involve *overlapping* transforms with varying-length basis functions. The overlapping nature of the transform (each pixel contributes to several output points) alleviates blocking artifacts, while the multiresolution character of the wavelet decomposition leads to superior energy compaction and perceptual quality of the decompressed image. Furthermore, the multiresolution transform domain means that wavelet compression methods degrade much more gracefully than block-DCT methods as the compression ratio increases. One wavelet algorithm, the embedded zerotree wavelet (EZW) coder, yields acceptable compression at a ratio of 100:1 [48]. The EZW coder is described in detail in Section 6.3.2.

Section 6.2 briefly reviews the concepts of dyadic wavelet transform and multiresolution representation and their design and implementation using two-channel filter banks. Further details can be found in Mallat [26] and Strang and Nguyen [53]. Readers familiar with wavelet theory could skip this section and proceed to Section 6.3 where several coding schemes based on zerotree wavelet coding are described and compared to JPEG.

6.2   Dyadic Wavelet Transform

The dyadic wavelet transform is an octave-band representation for signals; the discrete wavelet transform may be obtained by iterating a two-channel filter bank on its lowpass output. This multiresolution decomposition of a signal into its coarse and detail components is useful for data compression, feature extraction, and denoising. In the case of images, the wavelet representation is well-matched to psychovisual models, and compression systems based on the wavelet transform yield perceptual quality superior to other methods at medium and high compression ratios. Furthermore, the multiresolution nature of the wavelet transform enables fast browsing of image databases (the user may decompress only the coarsest scale representation of an image to decide whether he or she wants to examine it at a finer resolution).

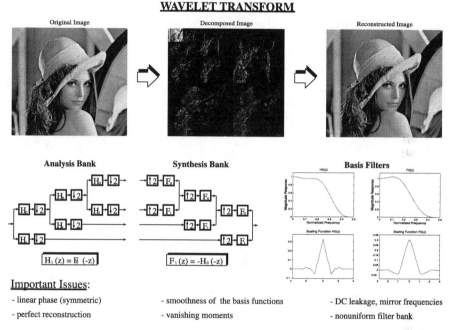

FIGURE 6.2

Dyadic wavelet transform, multiresolution representation, implementation using two-channel filter bank and filter characteristics. Reproduced by Special Permission of *Playboy* magazine. Copyright ©1972, 2000 by Playboy.

Fig. 6.2 shows an original image, its wavelet representation, and the reconstructed image without coefficient quantization. Since the wavelet transform is invertible, the reconstructed image is exactly the same as the original image. The decomposed image (wavelet representation) shows a coarse approximation image in the upper left corner and several detail images at various scales. As the scale changes, the subimage size changes. This is multiresolution and is enabled by the downsampling operation in the structure shown in the bottom left portion of Fig. 6.2. The coarse approximation and detail images are computed by first filtering the original image by lowpass and highpass filters $H_0(z)$ and $H_1(z)$, respectively. The filtered images are then downsampled by a factor of 2 to preserve the total image size. This is reflected in the structure as a two-channel filter bank.

This filter bank is repeated on the coarse approximation image since it still has large energy content (*coarse approximation image* is also referred to as the *all-lowpass subband* in Section 6.3). The structure shows a three-level filter bank and the corresponding decomposed image shows a three-level wavelet decomposition. The above process is repeated column-wise and row-wise. Throughout this chapter, we use the following terminology: *horizontal subband* to denote lowpass filtering on rows and highpass filtering on columns, *vertical subband* to denote highpass filtering on rows and lowpass filtering on columns, and *diagonal subband* to denote highpass filtering on both rows and columns. As shown, most of the energy is concentrated in the coarse approximation image, which is $\frac{1}{64}$ the original image size. The detailed images have small coefficients, as observed from the dark regions in the detailed images. The resulting multiresolution representation enables the user to treat each subband independently; for example, he or she may selectively allocate bits depending on the energy content (variance) of each subband, and the subsequent perceptual or algorithmic processing. Using the above multiresolution representation for image compression, one needs to develop an efficient coding algorithm for the locations of these small coefficients in the detailed images. This topic is discussed in detail in Section 6.3.

The bottom right portion of Fig. 6.2 shows the magnitude frequency responses of the lowpass filters $H_0(z)$ and $F_0(z)$ and their corresponding scaling functions. These are the Daub 9/7 filters used in the FBI fingerprint standard [70] as well as in the JPEG2000 standard. These filters are designed appropriately such that the whole filter bank is invertible and the corresponding basis functions are smooth. The perceptual quality of the reconstructed image depends on both the basis functions and the coding algorithm. Note that both lowpass filters have zeros at frequency π (in fact, they both have four zeros at π in this example). In general, filters with more zeros at π yields smoother basis functions [53].

6.2.1 Two-Channel Perfect-Reconstruction Filter Bank

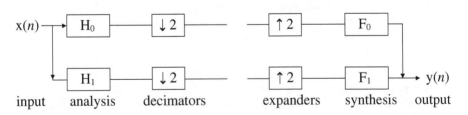

The figure above shows a two-channel filter bank where H_0 and H_1 are analysis filters used in the decomposition process, and F_0 and F_1 are synthesis filters, used in the reconstruction process. The boxes with down and up arrows denote downsampling and upsampling operations, respectively [54]. The objective is to design these filters such that the overall filter bank has perfect reconstruction; the output is a delayed version of the input. It is clear that a two-channel perfect-reconstruction filter bank yields an invertible discrete dyadic wavelet transform.

Since these filters are used in image compression, only filters with symmetric and finite impulse responses (FIR) are considered. When applying wavelet and filter bank transforms to finite-length signals, symmetry of the filters becomes an important consideration. This is because one must provide special treatment at signal boundaries (e.g., edges of an image). A simple periodic wrap of the signal (circular convolution) will work but can lead to unpleasant artifacts if the signal intensities at opposite boundaries differ significantly. When the filters of the filter bank/wavelet transform are linear-phase (symmetric or antisymmetric), it has been shown [51, 2, 21, 3] that one may *symmetrically extend* the input signal (by reflecting it), and that the subband outputs will also be symmetric, leading to a critically sampled, perfectly invertible representation that behaves smoothly at signal boundaries.

Another advantage of symmetric filters is that they maintain the correct spatial and time positioning of events. In a wavelet representation, if the filters are linear-phase and odd-length (whole-sample-symmetric), then signal details remain centered on signal samples under iteration of the filtering and decimation operation. This is important both for frame-to-frame correlation schemes used in video processing, and for event localization in geophysical signal processing.

There are several design methods for two-channel filter banks and dyadic wavelets. They are based on spectral factorization [32, 50], lattice structure [55, 34], time-domain optimization [33], and quadratic-constrained least-squares (QCLS) [35]. The design method based on spectral factorization is outlined below.

Using z-transform analysis, one obtains the following conditions on the filters such that the overall two-channel filter bank is perfectly reconstructed [54, 53]:

$$\begin{cases} H_0(z)F_0(z) - H_0(-z)F_0(-z) = 2z^{-(2L+1)} \\ H_1(z) = F_0(-z), \quad F_1(z) = -H_0(-z) . \end{cases} \tag{6.1}$$

Defining $P_0(z) = H_0(z)F_0(z)$, the first condition above is equivalent to finding a polynomial $P_0(z)$ such that

$$P_0(z) - P_0(-z) = 2z^{-(2L+1)} . \tag{6.2}$$

A $P_0(z)$ that satisfies the above condition is a halfband filter with length $(4L + 3)$ [54, 53]. The design procedure is as follows:

1. Design a symmetric halfband filter $P_0(z)$ with length $(4L + 3)$.

2. Factorize $P_0(z)$ into $H_0(z)$ and $F_0(z)$ such that they are symmetric filters.

3. The highpass filters can be obtained from $H_1(z) = F_0(-z)$, $F_1(z) = -H_0(-z)$.

This design procedure yields a two-channel perfect reconstruction filter bank. Recall that one also needs the lowpass filters to have zeros at frequency π so that the resulting basis functions are smooth. This condition implies that the halfband filter $P_0(z)$ also has zeros at frequency π. Daubechies [12] discusses a design method to obtain a halfband filter $P_0(z)$ with the maximum number of zeros at π. For $P_0(z)$

with length $(4L + 3)$, the maximum number of zeros at π is $(2L + 2)$. The Daub 9/7 filter comes from a halfband filter with length 15 and with 8 zeros at π. Each lowpass filter in this case has 4 zeros at π.

6.2.2 Dyadic Wavelet Transform, Multiresolution Representation

A *wavelet* decomposition arises from iteration of the lowpass filtering and decimation steps of a multirate filter bank. For a dyadic wavelet decomposition, one iterates on the lowpass output only, whereas for a wavelet-packet decomposition one may iterate on any output [26, 53]. A finite number of iterations will lead to a discrete-time multiresolution analysis with lowpass frequency response $\prod_{k=1}^{n} H_0 \left(\frac{\omega}{2^k} \right)$. If the lowpass filter h_0 satisfies the orthonormality constraint, $\sum_k h_0[k] = \frac{1}{\sqrt{2}}$, and has one vanishing moment ($\sum_k k h_0[k] = 0$), then the infinite product $\lim_{n \to \infty} \prod_{k=1}^{n} H_0 \left(\frac{\omega}{2^k} \right)$ converges to a function $\phi(\omega)$, whose inverse Fourier transform is the continuous time function $\phi(t)$ called the scaling function [12, 26, 53]. The scaling function $\phi(t)$ is the solution to the dilation equation

$$\phi(t) = 2 \sum_k h_0[k]\phi(2t - k) , \tag{6.3}$$

and it is orthogonal to its integer translates. If the filter $h_0[n]$ is FIR, then $\phi(t)$ has compact support. The scaling function determines the wavelet $w(t)$ by means of the highpass filter h_1:

$$w(t) = 2 \sum_k h_1[k]\phi(2t - k) . \tag{6.4}$$

The set of dilates and translates $\{w(2^k t - l)\}_{k,l \in \mathbf{Z}}$ forms a tight frame (and in most cases an orthonormal basis) for $L^2(\mathbf{R})$ [8, 23]. The functional relations Eqs. (6.3) and (6.4) introduce an entirely new set of relationships between discrete and continuous-time signal processing, unique to wavelet representations.

The span of integer translates of the scaling function $\phi(t)$ is the "lowpass" space V_0, the set of *scale-limited signals* [17]. Any continuous-time function $f(t)$ in V_0 can be expanded as a linear combination $f(t) = \sum_n v_n^0 \phi(t - n)$. The superscript 0 denotes an expansion "at scale level 0." $f(t)$ is completely described by the sequence $\{v_n^0\}$. Given such a sequence, its coarse approximation [component in V_1, where V_1 is the signal space with basis function $\phi(2t - n)$] is computed with the lowpass filter of the wavelet filter bank:

$$v_n^1 = \left(\left(v^0 * h_0 \right) \downarrow 2 \right) [n] .$$

This is essentially implemented as lowpass filtering followed by downsampling in the two-channel filter bank structure. Analogously, the details [component in W_1, where W_1 is the signal space with basis function $w(2t - n)$] are computed with the highpass filter $h_1[n]$. Hence, if we take a discrete sequence v_n to be the coefficients of a signal $f(t)$ at some fixed scale, the discrete wavelet transform of v_n will decompose the underlying signal f into a coarse-scale component and detail at several intermediate scales, as follows:

$$V_0 = V_1 \oplus W_1 = [V_2 \oplus W_2] \oplus W_1 = [[V_3 \oplus W_3] \oplus W_2] \oplus W_1 = \ldots = V_J \oplus \sum_{j=1}^{J} W_j .$$

In summary, the signal is represented in terms of its coarse approximation at scale J [with basis function $\phi(2^J t - n)$], and the J details [with basis functions $w(2^j t - n)$, $1 \le j \le J$]. This transform matches multiresolution models of human and computer vision [27] and has proven very effective for high-quality image compression. It also allows multiscale access to information, for applications such as image browsing and selective decoding of individual channels in a multicarrier system.

6.2.3 Wavelet Smoothness

As with any signal processing structure, one must consider the performance of the filters involved. In the case of the wavelet transform, one is concerned with the smoothness of the iterated lowpass filter. When using wavelets for lossy transform-based image coding, any quantization noise will appear in the decompressed image as linear combinations of the wavelet transform basis functions $\phi(t)$ and $w(t)$. If these basis functions (which are derived from the iterated discrete filter) are not smooth, then perceptually unacceptable artifacts will result. In fact, for all commonly used wavelets, the cascade converges fast enough so that the smoothness of the infinite limit is visually comparable to that of a six-level iterate [41]. Five- and six-level iterates are common in commercial implementations of wavelet-transform-based image compression [73, 10].

The smoothness of continuous-time wavelet systems has been the object of intensive study [12, 13, 15, 18, 64, 65]. Because the wavelet $w(t)$ is determined from the scaling function by means of the highpass filter taps Eq. (6.4), the smoothness of the scaling function (infinitely iterated lowpass filter) determines the smoothness of the overall wavelet system. In the two-band case, Daubechies' construction [12] imposed N vanishing wavelet moments $\int t^k w(t) dt = 0$, $0 \le k \le N - 1$ as a means of ensuring smoothness; this condition is equivalent to an N-th order zero at π for the lowpass filter: $\sum_n (-1)^n n^k h_0[n] = 0$, $0 \le k \le N - 1$. This condition was motivated by a theorem [11] stating that if an orthonormal system of dilates and translates $\{2^{j/2} w(2^j t - k)\}$ is made up of N times continuously differentiable functions, then the generating wavelet $w(t)$ must have N vanishing moments.

Vanishing moments are also associated with polynomial interpolation properties of the lowpass filter [52]. If a wavelet system has N vanishing moments, then polynomials of degree less than N may be represented as a linear combination of translates of the scaling function. In the setting of digital filter banks, this means that any locally polynomial component of a signal (of degree less than N) is preserved by the lowpass filter and zeroed out by the highpass filter — so long as the wavelet system has N

vanishing moments. These smoothness-under-iteration and polynomial approximation properties help explain why wavelet filters with vanishing moments perform so
well in image compression.

6.3   Wavelet-Based Image Compression

There are two types of image compression: *lossless* and *lossy*. With lossless compression, the original image is recovered exactly after decompression. Unfortunately,
with images of natural scenes it is rarely possible to obtain error-free compression
at a rate beyond 2:1. Much higher compression ratios can be obtained if some error,
which is usually difficult to perceive, is allowed between the decompressed image
and the original image. This is lossy compression. In many cases, it is not necessary
or even desirable that there be error-free reproduction of the original image. For
example, if some noise is present, then the error due to that noise will usually be
significantly reduced via some denoising method. In such a case, the small amount
of error introduced by lossy compression may be acceptable. Lossy compression is
also acceptable in fast transmission of still images over the Internet.

We concentrate on wavelet-based lossy compression of gray-level still images.
When there are 256 levels of possible intensity for each pixel, then we shall call these
images 8 bpp (bits per pixel) images. Images with 4096 gray-levels are referred to
as 12 bpp. Some brief comments on color images are also given, and we also briefly
describe some wavelet-based lossless compression methods.

6.3.1   Lossy Compression

We concentrate on the following methods of lossy compression: EZW (embedded
zerotree wavelet) algorithm, SPIHT (set partitioning in hierarchical trees) algorithm,
WDR (wavelet difference reduction) algorithm, and ASWDR (adaptively scanned
wavelet difference reduction) algorithm. These are relatively recent algorithms which
achieve some of the lowest errors per compression rate and highest perceptual quality
yet reported. After describing these algorithms in detail, we shall list some of the
other algorithms that are available.

Before we examine the algorithms listed above, we shall outline the basic steps that
are common to all wavelet-based image compression algorithms. The five stages of
compression and decompression are shown in Figs. 6.3 and 6.4. All of the steps shown
in the compression diagram are invertible, hence lossless, except for the *quantize* step.
Quantizing refers to a reduction of the precision of the floating point values of the
wavelet transform, which are typically either 32- or 64-bit floating point numbers.
To use less bits in the compressed transform — which is necessary if compression of
8 bpp or 12 bpp images is to be achieved — these transform values must be expressed
with less bits for each value. This leads to rounding error. These approximate,

quantized, wavelet transforms will produce approximations to the images when an inverse transform is performed. Thus creating the error inherent in lossy compression.

FIGURE 6.3
Compression of an image.

FIGURE 6.4
Decompression of an image.

The relationship between the quantize and encode steps, shown in Fig. 6.3, is the crucial aspect of wavelet transform compression. Each of the algorithms described below takes a different approach to this relationship.

The purpose served by the wavelet transform is that it produces a large number of values having zero, or near zero, magnitudes. For example, consider the image shown in Fig. 6.5(a), which is called "Lena." Fig. 6.5(b), shows a 7-level Daub 9/7 wavelet transform of the "Lena" image. This transform has been thresholded, using a threshold of 8. That is, all values with magnitudes less than 8 have been set equal to 0; they appear as a uniformly gray background in the image in Fig. 6.5(b). These large areas of gray background indicate that there is a large number of zero values in the thresholded transform. If an inverse wavelet transform is performed on this thresholded transform, then the image in Fig. 6.5(c) results (after rounding to integer values between 0 and 255). It is difficult to detect any difference between the images in Figs. 6.5(a) and (c).

The image in Fig. 6.5(c) was produced using only the 32,498 nonzero values of the thresholded transform, instead of all 262,144 values of the original transform. This represents an 8:1 compression ratio. We are, of course, ignoring difficult problems such as how to transmit concisely the positions of the nonzero values in the thresholded transform, and how to encode these nonzero values with as few bits as possible. Solutions to these problems are described below, when the various compression algorithms are discussed.

Two commonly used measures for quantifying the error between images are *mean square error* (MSE) and *peak signal to noise ratio* (PSNR). The MSE between two images f and g is defined by

$$\text{MSE} = \frac{1}{N} \sum_{j,k} (f[j,k] - g[j,k])^2 \tag{6.5}$$

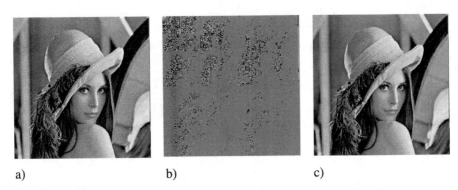

a)   b)   c)

FIGURE 6.5
(a) "Lena" image, 8 bpp. (b) Wavelet transform of image, threshold = 8. (c) Inverse of thresholded wavelet transform, PSNR = 39.14 dB. Reproduced by Special Permission of *Playboy* magazine. Copyright ©1972, 2000 by Playboy.

where the sum over j, k denotes the sum over all pixels in the images, and N is the number of pixels in each image. For the images in Figs. 6.5(a) and (c), the MSE is 7.921. The PSNR between two (8 bpp) images is, in decibels,

$$\text{PSNR} = 10 \log_{10} \left(\frac{255^2}{\text{MSE}} \right). \tag{6.6}$$

PSNR tends to be cited more often since it is a logarithmic measure, and our brains seem to respond logarithmically to intensity. Increasing PSNR represents increasing fidelity of compression. For the images in Figs. 6.5(a) and (c), the PSNR is 39.14 dB. Generally, when the PSNR is 40 dB or larger, then the two images are virtually indistinguishable by human observers. In this case, we can see that 8:1 compression should yield an image almost identical to the original. The methods described below do in fact produce such results with even greater PSNR than we have just achieved with our crude approach.

Before we begin our treatment of various "state of the art" algorithms, it may be helpful to briefly outline a baseline compression algorithm of the kind described in Davis and Nosratinia [14] and Mallat [26]. This algorithm has two main parts.

First, the positions of the significant transform values — the ones having larger magnitudes than the threshold T — are determined by scanning through the transform as shown in Fig. 6.6. The positions of the significant values are then encoded using a runlength method. To be precise, it is necessary to store the values of the significance map:

$$s(m) = \begin{cases} 0 & \text{if } |w(m)| < T \\ 1 & \text{if } |w(m)| \geq T, \end{cases} \tag{6.7}$$

where m is the scanning index, and $w(m)$ is the wavelet transform value at index m. From Fig. 6.5(b) we can see that there will be long runs of $s(m) = 0$. If the scan

1	2	5	8	17	24	25	32
3	4	6	7	18	23	26	31
9	10	13	14	19	22	27	30
12	11	15	16	20	21	28	29
33	34	35	36	49	50	54	55
40	39	38	37	51	53	56	61
41	42	43	44	52	57	60	62
48	47	46	45	58	59	63	64

(a) 2-level

1	2	5	8	17	24	25	32
3	4	6	7	18	23	26	31
9	10	13	14	19	22	27	30
12	11	15	16	20	21	28	29
33	34	35	36	49	50	54	55
40	39	38	37	51	53	56	61
41	42	43	44	52	57	60	62
48	47	46	45	58	59	63	64

(b) 3-level

FIGURE 6.6
Scanning for wavelet transforms: zigzag through all-lowpass subband, column scan through vertical subbands, row scan through horizontal subbands, zigzag through diagonal subbands. (a) and (b): Order of scanned elements for 2-level and 3-level transforms of 8 by 8 image.

order illustrated in Fig. 6.6 is used, then there will also be long runs of $s(m) = 1$. The positions of significant values can then be concisely encoded by recording sequences of 6 bits according to the following pattern:

$$0\,abcde : \quad \text{run of } 0 \text{ of length } (abcde)_2$$
$$1\,abcde : \quad \text{run of } 1 \text{ of length } (abcde)_2 \,.$$

A lossless compression, such as Huffman or arithmetic compression, of these data is also performed for a further reduction in bits.

Second, the significant values of the transform are encoded. This can be done by dividing the range of transform values into subintervals (*bins*) and rounding each transform value into the midpoint of the bin in which it lies. Fig. 6.7 shows the histogram of the frequencies of significant transform values lying in 512 bins for the 7-level Daub 9/7 transform of "Lena" shown in Fig. 6.5(b). The extremely rapid drop in the frequencies of occurrence of higher transform magnitudes implies that the very low magnitude values, which occur much more frequently, should be encoded using shorter length bit sequences. This is typically done with either Huffman encoding or arithmetic coding. If arithmetic coding is used, then the average number of bits needed to encode each significant value in this case is about 1 bit.

We have only briefly sketched the steps in this baseline compression algorithm. More details can be found in Davis and Nosratinia [14] and Mallat [26].

Our purpose in discussing the baseline compression algorithm is to introduce some basic concepts, such as scan order and thresholding, which are needed for our examination of the algorithms to follow. The baseline algorithm was one of the first to be

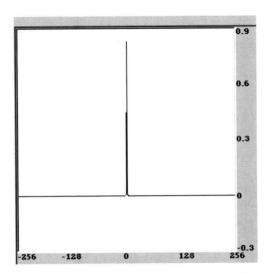

FIGURE 6.7
Histogram for 512 bins for thresholded transform of "Lena."

proposed using wavelet methods [1]. It suffers from some defects which later algorithms have remedied. For instance, with the baseline algorithm it is very difficult, if not impossible, to specify in advance the exact compression rate or the exact error to be achieved. This is a serious defect. Another problem with the baseline method is that it does not allow for progressive transmission. In other words, it is not possible to send successive data packets (over the Internet, for instance) which produce successively increasing resolution for the received image. Progressive transmission is vital for applications that include some level of interaction with the receiver.

Let us now turn to these improved wavelet image compression algorithms. The algorithms to be discussed are the EZW, SPIHT, WDR, and ASWDR algorithms.

6.3.2   EZW Algorithm

The EZW algorithm was one of the first algorithms to show the full power of wavelet-based image compression. It was introduced in the groundbreaking paper of Shapiro [48]. We shall describe EZW in some detail because a solid understanding of it will make it much easier to comprehend the other algorithms we shall be discussing. These other algorithms build upon the fundamental concepts that were first introduced with EZW.

Our discussion of EZW will be focused on the fundamental ideas underlying it. We will not use it to compress any images because it has been superceded by a far superior algorithm, SPIHT. Since SPIHT is only a highly refined version of EZW, it makes sense to first describe EZW.

EZW stands for *embedded zerotree wavelet*. An embedded coding is a process of encoding the transform magnitudes that allows for progressive transmission of the compressed image. Zerotrees allow for a concise encoding of the positions of significant values that result during the embedded coding process. We shall first discuss embedded coding, and then examine the notion of zerotrees.

The embedding process used by EZW is called *bit-plane encoding*. It consists of the following five-step process:

Bit-plane encoding —

Step 1: *Initialize.* Choose initial threshold, $T = T_0$, such that *all* transform values satisfy $|w(m)| < T_0$ and at least one transform value satisfies $|w(m)| \geq T_0/2$.

Step 2: *Update threshold.* Let $T_k = T_{k-1}/2$.

Step 3: *Significance pass.* Scan through insignificant values using baseline algorithm scan order. Test each value $w(m)$ as follows:

```
If |w(m)| ≥ Tₖ, then
    Output sign of w(m)
    Set wQ(m) = Tₖ
Else if |w(m)| < Tₖ then
    Let wQ(m) retain its initial value of 0 .
```

Step 4: *Refinement pass.* Scan through significant values found with higher threshold values T_j, for $j < k$ (if $k = 1$ skip this step). For each significant value $w(m)$, do the following:

```
If |w(m)| ∈ [wQ(m), wQ(m) + Tₖ), then
    Output bit 0
Else if |w(m)| ∈ [wQ(m) + Tₖ, wQ(m) + 2Tₖ), then
    Output bit 1
    Replace value of wQ(m) by wQ(m) + Tₖ .
```

Step 5: *Loop.* Repeat steps 2 through 4.

This bit-plane encoding procedure can be continued for as long as necessary to obtain quantized transform magnitudes $w_Q(m)$ which are as close as desired to the transform magnitudes $|w(m)|$. During decoding, the signs and the bits output by this method can be used to construct an approximate wavelet transform to any desired degree of accuracy. If instead, a given compression ratio is desired, then it can be achieved by stopping the bit-plane encoding as soon as a given number of bits (a *bit budget*) is exhausted. In either case, the execution of the bit-plane encoding procedure can terminate at any point (not just at the end of one of the loops).

As a simple example of bit-plane encoding, suppose that we just have two transform values $w(1) = -9.5$ and $w(2) = 42$. For an initial threshold, we set $T_0 = 64$. During the first loop, when $T_1 = 32$, the output is the sign of $w(2)$, and the quantized transform magnitudes are $w_Q(1) = 0$ and $w_Q(2) = 32$. For the second loop, $T_2 = 16$, and there is no output from the significance pass. The refinement pass produces the bit 0 because $w(2) \in [32, 32 + 16)$. The quantized transform magnitudes are $w_Q(1) = 0$ and $w_Q(2) = 32$. During the third loop, when $T_3 = 8$, the significance pass outputs the sign of $w(1)$. The refinement pass outputs the bit 1 because $w(2) \in [32+8, 32+16)$. The quantized transform magnitudes are $w_Q(1) = 8$ and $w_Q(2) = 40$.

It is not hard to see that *after n loops, the maximum error between the transform values and their quantized counterparts is less than $T_0/2^n$*. It follows that we can reduce the error to as small a value as we wish by performing a large enough number of loops. For instance, in the simple example just described, with seven loops the error is reduced to zero. The output from these seven loops, arranged to correspond to $w(1)$ and $w(2)$, is

$$w(1): \qquad\qquad -\quad 0\quad 0\quad 1\quad 1$$
$$w(2): \qquad +\quad 0\quad 1\quad 0\quad 1\quad 0\quad 0.$$

Notice that $w(2)$ requires seven symbols, but $w(1)$ requires only five.

Bit-plane encoding consists simply of computing binary expansions — using T_0 as unit — for the transform values and recording *in magnitude order* only the significant bits in these expansions. Because the first significant bit is always 1, it is *not* encoded. Instead, the sign of the transform value is encoded first. This coherent ordering of encoding, with highest magnitude bits encoded first, is what allows for progressive transmission.

Wavelet transforms are particularly well-adapted for bit-plane encoding[1] because wavelet transforms of images of natural scenes often have relatively few high-magnitude values, which are mostly found in the highest level subbands. These high-magnitude values are first coarsely approximated during the initial loops of the bit-plane encoding, thereby producing a low-resolution, but often recognizable, version of the image. Subsequent loops encode lower magnitude values and refine the high magnitude values, adding further details to the image and refining existing details. Thus, progressive transmission is possible, and encoding/decoding can cease once a given bit budget is exhausted or a given error target is achieved.

Now that we have described the embedded coding of wavelet transform values, we will describe the zerotree method by which EZW transmits the positions of significant transform values. The zerotree method gives an implicit, very compact, description of the location of significant values by creating a highly compressed description of the location of insignificant values. For many images of natural scenes, such as the "Lena" image for example, insignificant values at a given threshold T are organized in zerotrees.

To define a zerotree we first define a *quadtree* — a tree of locations in the wavelet transform with a root $[i, j]$ and its *children* located at $[2i, 2j], [2i+1, 2j], [2i, 2j+1]$, and $[2i+1, 2j+1]$, and each of their children, and so on. These *descendants* of the root reach all the way back to the 1st level of the wavelet transform. For example, Fig. 6.8(a) shows two quadtrees (enclosed in dashed boxes). One quadtree has root at index 12 and children at indices $\{41, 42, 47, 48\}$. This quadtree has two levels. We denote it by $\{12 \mid 41, 42, 47, 48\}$. The other quadtree, which has three levels, has its

---

[1] Although other transforms, such as the block discrete cosine transform, can also be bit-plane encoded.

root at index 4, the children of this root at indices {13, 14, 15, 16}, and their children at indices {49, 50, ... , 64}. It is denoted by {4 | 13, ... , 16 | 49, ... , 64}.

1	2	5	8	17	24	25	32
3	4	6	7	18	23	26	31
9	10	13	14	19	22	27	30
12	11	15	16	20	21	28	29
33	34	35	36	49	50	54	55
40	39	38	37	51	53	56	61
41	42	43	44	52	57	60	62
48	47	46	45	58	59	63	64

(a) Scan order, with two quadtrees

63	-34	49	10	5	18	-12	7
-31	23	14	-13	3	4	6	-1
-25	-7	-14	8	5	-7	3	9
-9	14	3	-12	4	-2	3	2
-5	9	-1	47	4	6	-2	2
3	0	-3	2	3	-2	0	4
2	-3	6	-4	3	6	3	6
5	11	5	6	0	3	-4	4

(b) Wavelet transform

+	-	+	R	I	I	•	•
I	R	R	R	I	I	•	•
R	I	•	•	•	•	•	•
R	R	•	•	•	•	•	•
•	•	I	+	•	•	•	•
•	•	I	I	•	•	•	•
•	•	•	•	•	•	•	•
•	•	•	•	•	•	•	•

(c) Threshold = 32

+	-	+	R	I	+	•	•
-	+	R	R	I	I	•	•
-	R	R	R	•	•	•	•
R	R	R	R	•	•	•	•
I	I	•	+	•	•	•	•
I	I	•	•	•	•	•	•
•	•	•	•	•	•	•	•
•	•	•	•	•	•	•	•

(d) Threshold = 16

FIGURE 6.8
First two stages of EZW. (a) 3-level scan order. (b) 3-level wavelet transform. (c) Stage 1, threshold = 32. (d) Stage 2, threshold = 16.

Now that we have defined a quadtree, we can give a simple definition of a zerotree. *A zerotree is a quadtree which, for a given threshold T, has insignificant wavelet transform values at each of its locations.* For example, if the threshold is $T = 32$, then each of the quadtrees shown in Fig. 6.8(a) is a zerotree for the wavelet transform in Fig. 6.8(b). But if the threshold is $T = 16$, then $\{12 \mid 41, 42, 47, 48\}$ remains a zerotree, but $\{4 \mid 13, \ldots, 16 \mid 49, \ldots, 64\}$ is no longer a zerotree because its root value is no longer insignificant.

Zerotrees can provide very compact descriptions of the locations of insignificant values because it is only necessary to encode one symbol, such as R, to mark the root location. The decoder can infer that all other locations in the zerotree have insignificant values, so their locations are not encoded. For the threshold $T = 32$, in the example just discussed, two R symbols are enough to specify all 26 locations in the two zerotrees.

Zerotrees can be useful only if they occur frequently. Fortunately, with wavelet transforms of natural scenes, the multiresolution structure of the wavelet transform does produce many zerotrees (especially at higher thresholds). For example, consider the images shown in Fig. 6.9. Fig. 6.9(a) shows the second all-lowpass subband of a Daub 9/7 transform of the "Lena" image. Image 6.9(b), on its right, is the third vertical subband produced from this all-lowpass subband, with a threshold of 16. Notice that there are large patches of gray pixels in this image. These represent insignificant transform values for the threshold of 16 which correspond to regions of nearly constant, or nearly linearly graded, intensities in the image in 6.9(a). Such intensities are nearly orthogonal to the analyzing Daub 9/7 wavelets. Zerotrees arise for the threshold of 16 because in image 6.9(c) — the second all-lowpass subband — there are similar regions of constant or linearly graded intensities. In fact, it was precisely these regions that were smoothed and downsampled to create the corresponding regions in image 6.9(a). These regions in image 6.9(c) produce insignificant values *in the same relative locations* (the child locations) in the second vertical subband shown in image 6.9(d).

Likewise, there are uniformly gray regions in the same relative locations in the first vertical subband [see Fig. 6.9(f)]. Because the second vertical subband in Fig. 6.9(d) is magnified by a factor of two in each dimension, and the third vertical subband in Fig. 6.9(b) is magnified by a factor of four in each dimension, it follows that the common regions of gray background shown in these three vertical subbands are all zerotrees. Similar images could be shown for horizontal and diagonal subbands, and they would also indicate a large number of zerotrees.

The "Lena" image is typical of many images of natural scenes, and the above discussion gives some background for understanding how zerotrees arise in wavelet transforms. A more rigorous, statistical discussion can be found in Shapiro [48].

Now that we have laid the foundations of zerotree encoding, we can complete our discussion of the EZW algorithm. The EZW algorithm consists simply of replacing the significance pass in the Bit-plane encoding procedure with the following step:

EZW Step 3: *Significance pass.* Scan through insignificant values using baseline

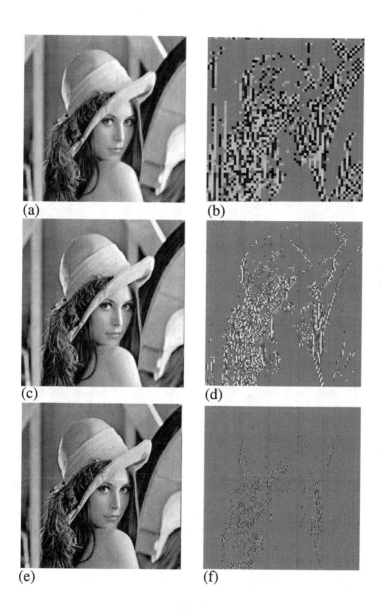

FIGURE 6.9

(a) Second all-lowpass subband. (b) Third vertical subband. (c) First all-lowpass subband. (d) Second vertical subband. (e) Original "Lena." (f) First vertical subband. Reproduced by Special Permission of *Playboy* magazine. Copyright ©1972, 2000 by Playboy.

algorithm scan order. Test each value $w(m)$ as follows:

```
If |w(m)| ≥ Tₖ, then
      Output the sign of w(m)
      Set wQ(m) = Tₖ
Else if |w(m)| < Tₖ then
      Let wQ(m) remain equal to 0
      If m is at 1st level, then
            Output I
      Else
            Search through quadtree having root m
            If this quadtree is a zerotree, then
                  Output R
            Else
                  Output I .
```

During a search through a quadtree, values that were found to be significant at higher thresholds are treated as zeros. All descendants of a root of a zerotree are skipped in the rest of the scanning at this threshold.

As an example of the EZW method, consider the wavelet transform shown in Fig. 6.8(b), which will be scanned through using the scan order shown in Fig. 6.8(a). Suppose that the initial threshold is $T_0 = 64$. In the first loop, the threshold is $T_1 = 32$. The results of the first significance pass are shown in Fig. 6.8(c). The coder output after this first loop would be

$$+ - I R + R R R R I R R I I I I I + I I \qquad (6.8)$$

corresponding to a quantized transform having only two values: $\pm 32 - +32$ at each location marked by a plus sign in Fig. 6.8(c), -32 at each location marked by a minus sign, and 0 at all other locations. In the second loop, with threshold $T_2 = 16$, the results of the significance pass are indicated in Fig. 6.8(d). Notice, in particular, that the symbol R is at the position 10 in the scan order because the plus sign which lies at a child location is from the previous loop, so it is treated as zero. Hence, position 10 is at the root of a zerotree. There is also a refinement pass done in this second loop. The output from this second loop is then

$$- + R R R - R R R R R R R I I I + I I I I 1 0 1 0 \qquad (6.9)$$

with corresponding quantized wavelet transform shown in Fig. 6.10(a). The MSE between this quantized transform and the original transform is 48.6875. This is a 78% reduction in error from the start of the method (when the quantized transform has all zero values).

A couple of final remarks are in order concerning the EZW method. First, it should be clear from the discussion above that the decoder, whose structure is outlined in

Fig. 6.4 above, can reverse each of the steps of the coder and produce the quantized wavelet transform. It is standard practice for the decoder to then round the quantized values to the midpoints of the intervals that they were last found to belong to during the encoding process (i.e., add half of the last threshold used to their magnitudes). This generally reduces MSE. For instance, in the example just considered, if this rounding is done to the quantized transform in Fig. 6.10(a), then the result is shown in Fig. 6.10(b). The MSE is then 39.6875, a reduction of more than 18%. A good discussion of the theoretical justification for this rounding technique can be found in Mallat [26]. *This rounding method will be employed by all of the other algorithms that we shall discuss.*

48	-32	48	0	0	16	0	0
-16	16	0	0	0	0	0	0
-16	0	0	0	0	0	0	0
0	0	0	0	0	0	0	0
0	0	0	32	0	0	0	0
0	0	0	0	0	0	0	0
0	0	0	0	0	0	0	0
0	0	0	0	0	0	0	0

(a)

56	-40	56	0	0	24	0	0
-24	24	0	0	0	0	0	0
-24	0	0	0	0	0	0	0
0	0	0	0	0	0	0	0
0	0	0	40	0	0	0	0
0	0	0	0	0	0	0	0
0	0	0	0	0	0	0	0
0	0	0	0	0	0	0	0

(b)

FIGURE 6.10

(a) Quantization at end of second stage, MSE = 48.6875. (b) After rounding to midpoints, MSE = 39.6875, reduction by more than 18%.

Second, since we live in a digital world, it is usually necessary to transmit just bits. A simple encoding of the symbols of the EZW algorithm into bits would be to use a code such as $+ = 01$, $- = 00$, $R = 10$, and $I = 11$. Since the decoder can always infer precisely when the encoding of these symbols ends (the significance pass is complete), the encoding of refinement bits can simply be as single bits 0 and 1. This form of encoding is the fastest to perform, but it does not achieve the greatest compression. In Shapiro [48], a lossless form of arithmetic coding was recommended in order to further compress the bit stream from the encoder.

6.3.3   SPIHT Algorithm

The SPIHT algorithm is a highly refined version of the EZW algorithm. It was introduced in Said and Pearlman [44, 45]. Some of the best results — highest PSNR values for given compression ratios — for a wide variety of images have been obtained with SPIHT. Consequently, it is probably the most widely used wavelet-based algorithm for image compression, providing a basic standard of comparison for all subsequent algorithms.

SPIHT stands for set partitioning in hierarchical trees. The term *hierarchical trees* refers to the quadtrees that we defined in our discussion of EZW. *Set partitioning* refers to the way these quadtrees partition the wavelet transform values at a given threshold. By a careful analysis of this partitioning of transform values, Said and Pearlman were able to greatly improve the EZW algorithm, significantly increasing its compressive power.

Our discussion of SPIHT will consist of three parts. First, we describe a modified version of the algorithm introduced in Said and Pearlman [44]. We refer to it as the *spatial-orientation tree wavelet* (STW) algorithm. STW is essentially the SPIHT algorithm; the only difference is that SPIHT is slightly more careful in its organization of coding output. Second, we describe the SPIHT algorithm. It is easier to explain SPIHT using the concepts underlying STW. Third, we see how well SPIHT compresses images.

The only difference between STW and EZW is that STW uses a different approach to encoding the zerotree information. STW uses a *state transition model.* From one threshold to the next, the locations of transform values undergo state transitions. This model allows STW to reduce the number of bits needed for encoding. Instead of code for the symbols R and I output by EZW to mark locations, the STW algorithm uses states I_R, I_V, S_R, and S_V and outputs code for state-transitions such as $I_R \rightarrow I_V$, $S_R \rightarrow S_V$, etc. To define the states involved, some preliminary definitions are needed.

For a given index m in the baseline scan order, define the set $D(m)$ as follows. If m is either at the first level or at the all-lowpass level, then $D(m)$ is the empty set $\emptyset$. Otherwise, if m is at the jth level for $j > 1$, then

$$D(m) = \{\text{Descendents of index } m \text{ in quadtree with root } m\} \,.$$

The significance function $\mathcal{S}$ is defined by

$$\mathcal{S}(m) = \begin{cases} \max_{n \in D(m)} |w(n)|, & \text{if } D(m) \neq \emptyset \\ \infty, & \text{if } D(m) = \emptyset \,. \end{cases}$$

With these preliminary definitions in hand, we can now define the states. For a given threshold T, the states I_R, I_V, S_R, and S_V are defined by

$$m \in I_R \quad \text{if and only if} \quad |w(m)| < T, \ \mathcal{S}(m) < T \tag{6.10}$$

$$m \in I_V \quad \text{if and only if} \quad |w(m)| < T, \ \mathcal{S}(m) \geq T \tag{6.11}$$

$$m \in S_R \quad \text{if and only if} \quad |w(m)| \geq T, \ \mathcal{S}(m) < T \tag{6.12}$$

$$m \in S_V \quad \text{if and only if} \quad |w(m)| \geq T, \ \mathcal{S}(m) \geq T \,. \tag{6.13}$$

Fig. 6.11 shows the state transition diagram for these states when a threshold is decreased from T to $T' < T$. Note that once a location m arrives in state S_V, it will remain in that state. Furthermore, there are only two transitions from each of the

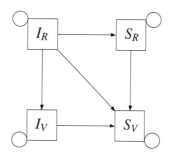

FIGURE 6.11
State transition diagram for STW.

Table 6.1 Code for State
Transitions, • Indicates that
$S_V \to S_V$ Transition is Certain
(Hence no Encoding Needed)

Old\New	I_R	I_V	S_R	S_V
I_R	00	01	10	11
I_V		0		1
S_R			0	1
S_V				•

states I_V and S_R, so those transitions can be coded with one bit each. A simple binary coding for these state transitions is shown in Table 6.1.

Now that we have laid the groundwork for the STW algorithm, we can give its full description.

STW encoding —

Step 1: *Initialize.* Choose initial threshold, $T = T_0$, such that *all* transform values satisfy $|w(m)| < T_0$ and at least one transform value satisfies $|w(m)| \geq T_0/2$. Assign all indices for the Lth level, where L is the number of levels in the wavelet transform, to the *dominant list* (this includes all locations in the all-lowpass subband as well as the horizontal, vertical, and diagonal subbands at the Lth level). Set the *refinement list* of indices equal to the empty set.

Step 2: *Update threshold.* Let $T_k = T_{k-1}/2$.

Step 3: *Dominant pass.* Use the following procedure to scan through indices in the dominant list (which can change as the procedure is executed).

```
Do
    Get next index m in dominant list
    Save old state S_old = S(m, T_{k-1})
    Find new state S_new = S(m, T_k) using Eqs. (6.10)--(6.13)
    Output code for state transition S_old → S_new
    If S_new ≠ S_old then do the following
        If S_old ≠ S_R and S_new ≠ I_V then
            Append index m to refinement list
                Output sign of w(m) and set w_Q(m) = T_k
        If S_old ≠ I_V and S_new ≠ S_R then
            Append child indices of m to dominant list
        If S_new = S_V then
            Remove index m from dominant list
    Loop until end of dominant list
```

Step 4: *Refinement pass.* Scan through indices m in the refinement list found with higher threshold values T_j, for $j < k$ (if $k = 1$ skip this step). For each value $w(m)$, do the following:

```
If |w(m)| ∈ [w_Q(m), w_Q(m) + T_k), then
    Output bit 0
Else if |w(m)| ∈ [w_Q(m) + T_k, w_Q(m) + 2T_k), then
    Output bit 1
    Replace value of w_Q(m) by w_Q(m) + T_k .
```

Step 5: *Loop.* Repeat steps 2 through 4.

To see how STW works — and how it improves the EZW method — it helps to reconsider the example shown in Fig. 6.8. In Fig. 6.12, we show STW states for the wavelet transform in Fig. 6.8(b) using the same two thresholds we used previously with EZW. It is important to compare the three quadtrees enclosed in the dashed boxes in Fig. 6.12 with the corresponding quadtrees in Figs. 6.8(c) and (d). There is a large savings in coding output for STW represented by these quadtrees. The EZW symbols for these three quadtrees are $+ I I I I$, $- I I I I$, and $+ R R R R$. For STW, however, they are described by the symbols $+ S_R$, $- S_R$, and $+ S_R$, which is a substantial reduction in the information that STW needs to encode.

There is not much difference between STW and SPIHT. The one thing that SPIHT does differently is to carefully organize the output of bits in the encoding of state transitions in Table 6.1, *so that only one bit is output at a time.* For instance, for the transition $I_R \to S_R$, which is coded as 1 0 in Table 6.1, SPIHT outputs a 1 first and then (after further processing) outputs a 0. Even if the bit budget is exhausted before the second bit can be output, the first bit of 1 indicates that there is a new significant value.

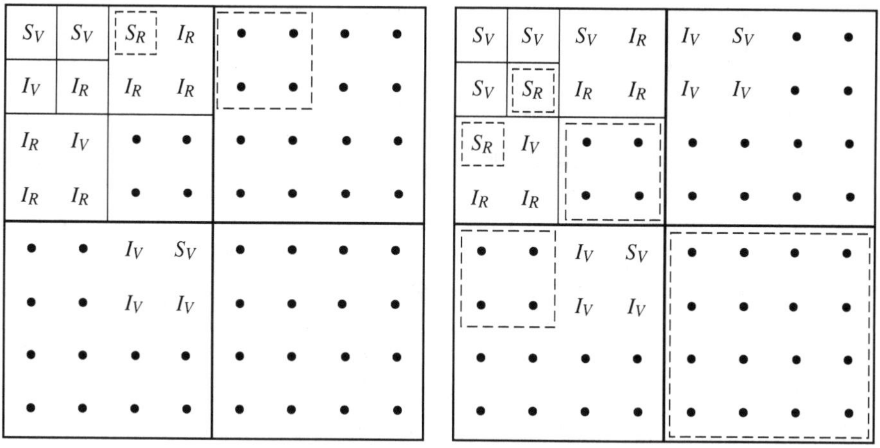

(a) Threshold = 32   (b) Threshold = 16

FIGURE 6.12
First two stages of STW for wavelet transform in Fig. 6.8.

The SPIHT encoding process, as described in Said and Pearlman [45], is phrased in terms of pixel locations $[i, j]$ rather than indices m in a scan order. To avoid introducing new notation, and to highlight the connections between SPIHT and the other algorithms, EZW and STW, we rephrase the description of SPIHT from Said and Pearlman [45] in terms of scanning indices. We also slightly modify their notation in the interests of clarity.

First, we need some preliminary definitions. For a given set I of indices in the baseline scan order, the significance $\mathcal{S}_T[\mathtt{I}]$ of I relative to a threshold T is defined by

$$
\mathcal{S}_T[\mathtt{I}] = \begin{cases} 1, & \text{if } \max_{n \in \mathtt{I}} |w(n)| \geq T \\ 0, & \text{if } \max_{n \in \mathtt{I}} |w(n)| < T . \end{cases}
\tag{6.14}
$$

It is important to note that, for the initial threshold T_0, we have $\mathcal{S}_{T_0}[\mathtt{I}] = 0$ for all sets of indices. If I is a set containing just a single index m, then for convenience we write $\mathcal{S}_T[m]$ instead of $\mathcal{S}_T[\{m\}]$.

For a succinct presentation of the method, we need the following definitions of sets of indices:

$$
\begin{aligned}
\mathtt{D}(m) &= \{\text{Descendent indices of the index } m\} \\
\mathtt{C}(m) &= \{\text{Child indices of the index } m\} \\
\mathtt{G}(m) &= \mathtt{D}(m) - \mathtt{C}(m) \\
&= \{\text{Grandchildren of } m, \text{ i.e., descendants which are not children}\} .
\end{aligned}
$$

In addition, the set H consists of indices for the Lth level, where L is the number of levels in the wavelet transform (this includes all locations in the all-lowpass subband as well as the horizontal, vertical, and diagonal subbands at the Lth level). It is important to remember that the indices in the all-lowpass subband have no descendants. If m marks a location in the all-lowpass subband, then $D(m) = \emptyset$.

SPIHT keeps track of the states of sets of indices by means of three lists. They are the *list of insignificant sets* (LIS), the *list of insignificant pixels* (LIP), and the *list of significant pixels* (LSP). For each list a set is identified by a single index, in the LIP and LSP these indices represent the singleton sets $\{m\}$ where m is the identifying index. An index m is called either significant or insignificant, depending on whether the transform value $w(m)$ is significant or insignificant with respect to a given threshold. For the LIS, the index m denotes either $D(m)$ or $G(m)$. In the former case, the index m is said to be of type D and, in the latter case, of type G.

The following is the pseudocode for the SPIHT algorithm. For simplicity, we write the significance function $\mathcal{S}_{T_k}$ as $\mathcal{S}_k$.

SPIHT encoding —

Step 1: *Initialize.* Choose initial threshold T_0 such that *all* transform values satisfy $|w(m)| < T_0$ and at least one value satisfies $|w(m)| \geq T_0/2$. Set LIP equal to H, set LSP equal to $\emptyset$, and set LIS equal to all the indices in H that have descendants (assigning them all type D).

Step 2: *Update threshold.* Let $T_k = T_{k-1}/2$.

Step 3: *Sorting pass.* Proceed as follows:

```
For each m in LIP do:
    Output Sₖ[m]
    If Sₖ[m] = 1 then
        Move m to end of LSP
        Output sign of w(m); set w_Q(m) = Tₖ
Continue until end of LIP
For each m in LIS do:
    If m is of type D then
        Output Sₖ[D(m)]
        If Sₖ[D(m)] = 1 then
```

```
For each n ∈ C(m) do:
    Output 𝒮ₖ[n]
    If 𝒮ₖ[n] = 1 then
        Append n to LSP
        Output sign of w(n); set w_Q(n) = Tₖ
    Else If 𝒮ₖ[n] = 0 then
        Append n to LIP
    If G(m) ≠ ∅ then
        Move m to end of LIS as type G
    Else
        Remove m from LIS
Else If m is of type G then
    Output 𝒮ₖ[G(m)]
    If 𝒮ₖ[G(m)] = 1 then
        Append C(m) to LIS, all type D indices
        Remove m from LIS
Continue until end of LIS
```

Notice that the set LIS can undergo many changes during this procedure, it typically does not remain fixed throughout.

Step 4: *Refinement pass.* Scan through indices m in LSP found with higher threshold values T_j, for $j < k$ (if $k = 1$ skip this step). For each value $w(m)$, do the following:

```
If |w(m)| ∈ [w_Q(m), w_Q(m) + Tₖ), then
    Output bit 0
Else if |w(m)| ∈ [w_Q(m) + Tₖ, w_Q(m) + 2Tₖ), then
    Output bit 1
    Replace value of w_Q(m) by w_Q(m) + Tₖ .
```

Step 5: *Loop.* Repeat steps 2 through 4.

It helps to carry out this procedure on the wavelet transform shown in Fig. 6.8. Then one can see that SPIHT simply performs STW with the binary code for the states in Table 6.1 being output one bit at a time.

Now comes the payoff. We shall see how well SPIHT performs in compressing images. To do these compressions we used the public domain SPIHT programs that can be downloaded from the Internet [46]. In Fig. 6.13 we show several SPIHT compressions of the "Lena" image. The original "Lena" image is shown in Fig. 6.13(f). Five SPIHT compressions are shown with compression ratios of 128:1, 64:1, 32:1, 16:1, and 8:1.

Several things are worth noting about these compressed images. First, they were all produced from one file containing the 1 bpp compression of the "Lena image."

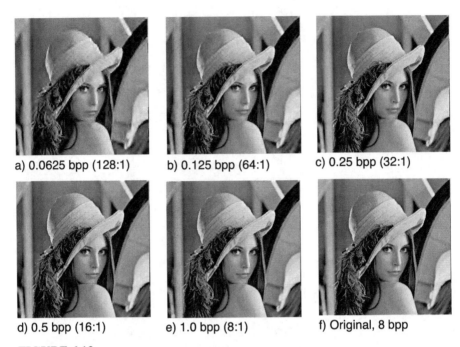

a) 0.0625 bpp (128:1)   b) 0.125 bpp (64:1)   c) 0.25 bpp (32:1)

d) 0.5 bpp (16:1)   e) 1.0 bpp (8:1)   f) Original, 8 bpp

FIGURE 6.13
SPIHT compressions of "Lena" image. PSNR values: (a) 27.96 dB. (b) 30.85
dB. (c) 33.93 dB. (d) 37.09 dB. (e) 40.32 dB. Reproduced by Special Permission
of *Playboy* magazine. Copyright ©1972, 2000 by Playboy.

By specifying a bit budget, a certain bpp value up to 1, the SPIHT decompression program will stop decoding the 1 bpp compressed file once the bit budget is exhausted. This illustrates the embedded nature of SPIHT.

Second, the rapid convergence of the compressed images to the original is nothing short of astonishing. Even the 64:1 compression in Fig. 6.13(b) is almost indistinguishable from the original. A close examination of the two images is needed in order to see some differences, e.g., the blurring of details in the top of Lena's hat. The image in (b) would be quite acceptable for some applications, such as the first image in a sequence of video telephone images or as a thumbnail display within a large archive.

Third, notice that the 1 bpp image has a 40.32 dB PSNR value and is virtually indistinguishable — even under very close examination — from the original. Here we find that SPIHT is able to exceed the simple thresholding compression we first discussed (see Fig. 6.5). For reasons of space, we cannot show SPIHT compressions of many test images, so in Table 6.2 we give PSNR values for several test images [19]. These data show that SPIHT produces higher PSNR values than the two other algorithms that we shall describe below. SPIHT is well-known for its superior performance when PSNR is used as the error measure. High PSNR values, however, are not the sole criterion for the performance of lossy compression algorithms. We discuss other criteria below.

Table 6.2 PSNR Values, *With* Arithmetic Compression

Image/Method	SPIHT	WDR	ASWDR
Lena, 0.5 bpp	**37.09**	36.45	36.67
Lena, 0.25 bpp	**33.85**	33.39	33.64
Lena, 0.125 bpp	**30.85**	30.42	30.61
Goldhill, 0.5 bpp	**33.10**	32.70	32.85
Goldhill, 0.25 bpp	**30.49**	30.33	30.34
Goldhill, 0.125 bpp	**28.39**	28.25	28.23
Barbara, 0.5 bpp	**31.29**	30.68	30.87
Barbara, 0.25 bpp	**27.47**	26.87	27.03
Barbara, 0.125 bpp	**24.77**	24.30	24.52
Airfield, 0.5 bpp	**28.57**	28.12	28.36
Airfield, 0.25 bpp	**25.90**	25.49	25.64
Airfield, 0.125 bpp	**23.68**	23.32	23.50

Fourth, these SPIHT compressed images were obtained using SPIHT's arithmetic compression option. The method that SPIHT uses for arithmetic compression is quite involved and space does not permit a discussion of the details here. Some details are provided in Said and Pearlman [47].

Finally, it is interesting to compare SPIHT compressions with compressions obtained with the JPEG method[2]. The JPEG method is a sophisticated implementation of block discrete cosine transform encoding [67, 38]. It is used extensively for compression of images, especially for transmission over the Internet. In Fig. 6.14, we compare compressions of the "Lena" image obtained with JPEG and with SPIHT at three different compression ratios. (JPEG does not allow for specifying the bpp value in advance; the 59:1 compression was the closest we could get to 64:1.) It is clear from these images that SPIHT is far superior to JPEG. It is better both in perceptual quality and in terms of PSNR. Notice, in particular, that the 59:1 JPEG compression is very distorted (exhibiting blocking artifacts stemming from coarse quantization within the blocks making up the block DCT used by JPEG). The SPIHT compression, even at the slightly higher ratio of 64:1, exhibits none of these objectionable features. In fact, for quick transmission of a thumbnail image (say, as part of a much larger webpage), this SPIHT compression would be quite acceptable. The 32:1 JPEG image might be

---

[2]JPEG stands for Joint Photographic Experts Group, a group of engineers who developed this compression method.

acceptable for some applications, but it also contains some blocking artifacts. The 32:1 SPIHT compression is almost indistinguishable (at these image sizes) from the original "Lena" image. The 16:1 compressions for both methods are nearly indistinguishable. In fact, they are both nearly indistinguishable from the original "Lena" image.

Although we have compared JPEG with SPIHT using only one image, the results we have found are generally valid. SPIHT compressions are superior to JPEG compressions both in perceptual quality and in PSNR values. In fact, all of the wavelet-based image compression techniques that we discuss here are superior to JPEG. Hence, we will not make any further comparisons with the JPEG method.

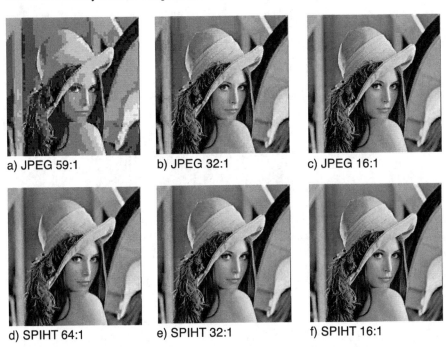

a) JPEG 59:1   b) JPEG 32:1   c) JPEG 16:1

d) SPIHT 64:1   e) SPIHT 32:1   f) SPIHT 16:1

FIGURE 6.14
Comparison of JPEG and SPIHT compressions of "Lena" image. PSNR values:
(a) 24.16 dB. (b) 30.11 dB. (c) 34.12 dB. (d) 30.85 dB. (e) 33.93 dB. (f) 37.09 dB.
Reproduced by Special Permission of *Playboy* magazine. Copyright ©1972, 2000
by Playboy.

6.3.4   WDR Algorithm

One of the defects of SPIHT is that it only *implicitly* locates the position of significant coefficients. This makes it difficult to perform operations which depend on the exact position of significant transform values, such as region selection on compressed data. By *region selection,* also known as *region of interest* (ROI), we mean selecting a portion of a compressed image that requires increased resolution. This can occur, for example, with a portion of a low resolution medical image that has been sent at a low bpp rate in order to arrive quickly.

Such compressed data operations are possible with the wavelet difference reduction (WDR) algorithm of Tian and Wells [56]–[58]. The term *difference reduction* refers to the way in which WDR encodes the locations of significant wavelet transform values, which we describe below. Although WDR will not typically produce higher PSNR values than SPIHT (see Table 6.2), we will see that WDR can produce perceptually superior images, especially at high compression ratios.

The only difference between WDR and the bit-plane encoding described above is in the significance pass. In WDR, the output from the significance pass consists of the signs of significant values along with sequences of bits which concisely describe the precise locations of significant values. The best way to see how this is done is to consider a simple example.

Suppose that the significant values are $w(2) = +34.2$, $w(3) = -33.5$, $w(7) = +48.2$, $w(12) = +40.34$, and $w(34) = -54.36$. The indices for these significant values are 2, 3, 7, 12, and 34. Rather than working with these values, WDR works with their successive differences: 2, 1, 4, 5, 22. In this latter list, the first number is the *starting index*, and each successive number is the *number of steps* needed to reach the next index. The binary expansions of these successive differences are $(10)_2$, $(1)_2$, $(100)_2$, $(101)_2$, and $(10110)_2$. Since the most significant bit for each of these expansions is always 1, this bit can be dropped and the signs of the significant transform values can be used instead as separators in the symbol stream. The resulting symbol stream for this example is then $+0 - +00 + 01 - 0110$.

When this most significant bit is dropped, we will refer to the binary expansion that remains as the reduced binary expansion. Notice, in particular, that the reduced binary expansion of 1 is empty. The reduced binary expansion of 2 is just the 0 bit, the reduced binary expansion of 3 is just the 1 bit, and so on.

The WDR algorithm simply consists of replacing the significance pass in the bit-plane encoding procedure with the following step:

WDR Step 3: *Significance pass.* Perform the following procedure on the insignificant indices in the baseline scan order:

```
Initialize step-counter C = 0
Let  C_old = 0
Do
      Get next insignificant index m
      Increment step-counter C by 1
```

```
If |w(m)| ≥ T_k then
      Output sign w(m) and set w_Q(m) = T_k
      Move m to end of sequence of significant indices
      Let n = C − C_old
      Set C_old = C
      If n > 1 then
            Output reduced binary expansion of n
   Else if |w(m)| < T_k then
      Let w_Q(m) retain its initial value of 0.
Loop until end of insignificant indices
Output end-marker
```

The output for the end-marker is a plus sign, followed by the reduced binary expansion of $n = C + 1 - C_{old}$, and a final plus sign.

It is not hard to see that WDR is of no greater computational complexity than SPIHT. For one thing, WDR does not need to search through quadtrees as SPIHT does. The calculations of the reduced binary expansions adds some complexity to WDR, but they can be done rapidly with bit-shift operations. As explained in Tian and Wells [56]–[58], the output of the WDR encoding can be arithmetically compressed. The method that they describe is based on the elementary arithmetic coding algorithm described in Witten, Neal, and Cleary [68]. This form of arithmetic coding is substantially less complex (at the price of poorer performance) than the arithmetic coding employed by SPIHT.

As an example of the WDR algorithm, consider the scan order and wavelet transform shown in Fig. 6.8. For the threshold $T_1 = 32$, the significant values are $w(1) = 63$, $w(2) = -34$, $w(5) = 49$, and $w(36) = 47$. The output of the WDR significance pass will then be the following string of symbols:

$$+ - +1 + 1111 + 1101+$$

which compares favorably with the EZW output in Eq. (6.8). The last six symbols are the code for the end-marker. For the threshold $T_2 = 16$, the new significant values are $w(3) = -31$, $w(4) = 23$, $w(9) = -25$, and $w(24) = 18$. Since the previous indices 1, 2, 5, and 36, are removed from the sequence of insignificant indices, the values of n in the WDR significance pass will be 1, 1, 4, and 15. In this case, the value of n for the end-marker is 40. Adding on the four refinement bits, which are the same as in Eq. (6.9), the WDR output for this second threshold is

$$- + -00 + 111 + 01000 + 1010$$

which is also a smaller output than the corresponding EZW output. It is also clear that, for this simple case, WDR does *not* produce as compact an output as STW does.

As an example of WDR performance for a natural image, Fig. 6.15 shows several compressions of the "Lena" image. These compressions were produced with free software [16].

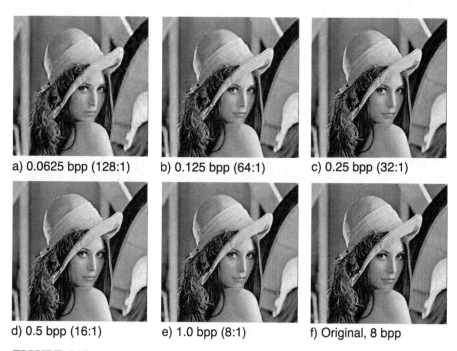

a) 0.0625 bpp (128:1) b) 0.125 bpp (64:1) c) 0.25 bpp (32:1)

d) 0.5 bpp (16:1) e) 1.0 bpp (8:1) f) Original, 8 bpp

FIGURE 6.15
**WDR compressions of "Lena" image. PSNR values: (a) 27.63 dB. (b) 30.42 dB.
(c) 33.39 dB. (d) 36.45 dB. (e) 39.62 dB. Reproduced by Special Permission of
Playboy magazine. Copyright ©1972, 2000 by Playboy.**

There are a couple things to observe about these compressions. First, the PSNR values are lower than for SPIHT. This is typically the case. In Table 6.2 we compare PSNR values for WDR and SPIHT on several images at various compression ratios. In every case, SPIHT has higher PSNR values.

Second, at high compression ratios, the visual quality of WDR compressions of "Lena" are superior to those of SPIHT. For example, the 0.0625 bpp and 0.125 bpp compressions have higher resolution with WDR. This is easier to see if the images are magnified as in Fig. 6.16. At 0.0625 bpp, the WDR compression does a better job in preserving the shape of Lena's nose and in retaining some of the striping in the band around her hat. Similar remarks apply to the 0.125 bpp compressions. SPIHT, however, does a better job in preserving parts of Lena's eyes. These observations point to the need for an objective, quantitative measure of image quality.

There is no universally accepted objective measure for image quality. We shall now describe a simple measure that we have found useful. There is some evidence that the visual system of humans concentrates on analyzing edges in images [30, 40].

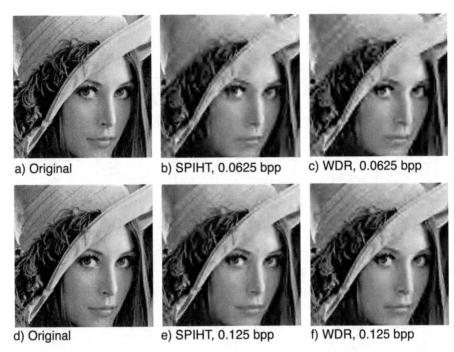

| a) Original | b) SPIHT, 0.0625 bpp | c) WDR, 0.0625 bpp |

| d) Original | e) SPIHT, 0.125 bpp | f) WDR, 0.125 bpp |

FIGURE 6.16
SPIHT and WDR compressions of "Lena" at low bpp. Reproduced by Special Permission of *Playboy* magazine. Copyright ©1972, 2000 by Playboy.

To produce an image that retains only edges, we proceed as follows. First, a 3-level Daub 9/7 transform of an image f is created. Second, the all-lowpass subband is subtracted away from this transform. Third, an inverse transform is performed on the remaining part of the transform. This produces a highpass filtered image, which exhibits edges from the image f. A similar highpass filtered image is created from the compressed image. Both of these highpass filtered images have mean values that are approximately zero. We define the *edge correlation* γ_3 by

$$\gamma_3 = \frac{\sigma_c}{\sigma_o}$$

where σ_c denotes the standard deviation of the values of the highpass filtered version of the compressed image, and σ_o denotes the standard deviation of the values of the highpass filtered version of the original image. Thus γ_3 measures how well the compressed image captures the variation of edge details in the original image.

Using this edge correlation measure, we obtained the results shown in Table 6.3. In every case, the WDR compressions exhibit higher edge correlations than the SPIHT compressions. These numerical results are also consistent with the increased preservation of details within WDR images, and with the informal reports of human observers.

Table 6.3 Edge Correlations, *With* Arithmetic
Compression

Image/Method	SPIHT	WDR	ASWDR
Lena, 0.5 bpp	.966	.976	**.978**
Lena, 0.25 bpp	.931	.946	**.951**
Lena, 0.125 bpp	.863	.885	**.894**
Goldhill, 0.5 bpp	.920	.958	**.963**
Goldhill, 0.25 bpp	.842	.870	**.871**
Goldhill, 0.125 bpp	.747	**.783**	.781
Barbara, 0.5 bpp	.932	.955	**.959**
Barbara, 0.25 bpp	.861	.894	**.902**
Barbara, 0.125 bpp	.739	.767	**.785**
Airfield, 0.5 bpp	.922	**.939**	.937
Airfield, 0.25 bpp	.857	.871	**.878**
Airfield, 0.125 bpp	.766	.790	**.803**

Although WDR is simple, competitive with SPIHT in PSNR values, and often provides better perceptual results, there is still room for improvement. We now turn to a recent enhancement of the WDR algorithm.

6.3.5 ASWDR Algorithm

One of the most recent image compression algorithms is the adaptively scanned wavelet difference reduction (ASWDR) algorithm of Walker [66]. The adjective adaptively scanned refers to the fact that this algorithm modifies the scanning order used by WDR in order to achieve better performance.

ASWDR adapts the scanning order so as to predict locations of new significant values. If a prediction is correct, then the output specifying that location will just be the sign of the new significant value — the reduced binary expansion of the number of steps will be empty. Therefore a good prediction scheme will significantly reduce the coding output of WDR.

The prediction method used by ASWDR is the following: if $w(m)$ is significant for threshold T, then the values of the children of m are predicted to be significant for half-threshold $T/2$. For many natural images, this prediction method is a reasonably good one. As an example, Fig. 6.17 shows two vertical subbands for a Daub 9/7 wavelet transform of the "Lena" image. The image in Fig. 6.17(a) is of those significant values in the second level vertical subband for a threshold of 16 (significant values shown in white). In Fig. 6.17(b), we show the *new* significant values in the first vertical subband for the half-threshold of 8. Notice that there is a great deal of similarity in the two images. Since the image in Fig. 6.17(a) is magnified by two in each dimension, its white pixels actually represent the predictions for the locations of new significant values in the first vertical subband. Although these predictions are not

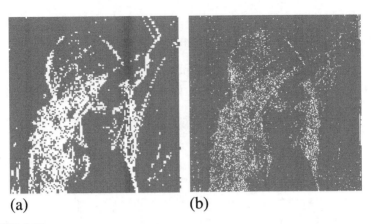

(a)   (b)

FIGURE 6.17
(a) Significant values, second vertical subband, threshold 16. **(b)** *New* **significant values, first vertical subband, threshold** 8. **Reproduced by Special Permission of** *Playboy* **magazine. Copyright ©1972, 2000 by Playboy.**

perfectly accurate, there is a great deal of overlap between the two images. Notice also how the locations of significant values are highly correlated with the location of edges in the "Lena" image. The scanning order of ASWDR dynamically adapts to the locations of edge details in an image, and this enhances the resolution of these edges in ASWDR compressed images.

Table 6.4   Number of Significant Values Encoded, *No* Arithmetic Coding

Image\Method	WDR	ASWDR	% increase
Lena, 0.125 bpp	5,241	5,458	4.1%
Lena, 0.25 bpp	10,450	11,105	6.3%
Lena, 0.5 bpp	20,809	22,370	7.5%
Goldhill, 0.125 bpp	5,744	5,634	−1.9%
Goldhill, 0.25 bpp	10,410	10,210	−1.9%
Goldhill, 0.5 bpp	22,905	23,394	2.1%
Barbara, 0.125 bpp	5,348	5,571	4.2%
Barbara, 0.25 bpp	11,681	12,174	4.2%
Barbara, 0.5 bpp	23,697	24,915	5.1%
Airfield, 0.125 bpp	5,388	5,736	6.5%
Airfield, 0.25 bpp	10,519	11,228	6.7%
Airfield, 0.5 bpp	19,950	21,814	9.3%

A complete validation of the prediction method just described would require assembling statistics for a large number of different subbands, thresholds, and images.

Rather than attempting such an a priori argument (see [6, 66]), we instead argue from an a posteriori standpoint. We present statistics that show that the prediction scheme employed by ASWDR does, in fact, encode more significant values than are encoded by WDR for a number of different images. As the pseudocode presented below shows, the only difference between ASWDR and WDR is in the predictive scheme employed by ASWDR to create new scanning orders. Consequently, if ASWDR typically encodes more values than WDR does, this must be due to the success of the predictive scheme.

Table 6.4 shows the numbers of significant values encoded by WDR and ASWDR for four different images. In almost every case, ASWDR was able to encode more values than WDR. This gives an a posteriori validation of the predictive scheme employed by ASWDR.

We now present the pseudocode description of ASWDR encoding. Notice that the significance pass portion of this procedure is the same as the WDR significance pass described above, and that the refinement pass is the same as for bit-plane encoding (hence the same as for WDR). The one new feature is the insertion of a step for creating a new scanning order.

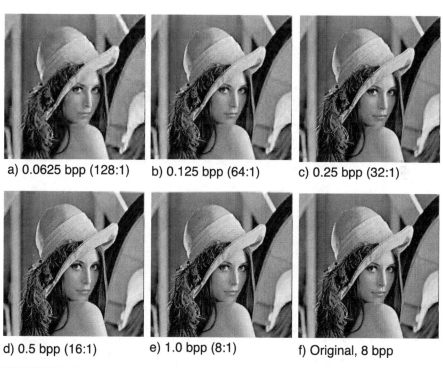

a) 0.0625 bpp (128:1)   b) 0.125 bpp (64:1)   c) 0.25 bpp (32:1)

d) 0.5 bpp (16:1)   e) 1.0 bpp (8:1)   f) Original, 8 bpp

FIGURE 6.18
ASWDR compressions of "Lena image." PSNR values: (a) 27.73 dB. (b) 30.61 dB. (c) 33.64 dB. (d) 36.67 dB. (e) 39.90 dB. Reproduced by Special Permission of *Playboy* magazine. Copyright ©1972, 2000 by Playboy.

ASWDR encoding —

Step 1: *Initialize.* Choose initial threshold, $T = T_0$, such that *all* transform values satisfy $|w(m)| < T_0$ and at least one transform value satisfies $|w(m)| \geq T_0/2$. Set the initial scan order to be the baseline scan order.

Step 2: *Update threshold.* Let $T_k = T_{k-1}/2$.

Step 3: *Significance pass.* Perform the following procedure on the insignificant indices in the scan order:

```
Initialize step-counter C = 0
Let  C_old = 0
Do
        Get next insignificant index m
        Increment step-counter C by 1
        If |w(m)| ≥ T_k then
                Output sign w(m) and set w_Q(m) = T_k
                Move m to end of sequence of significant indices
                Let  n = C - C_old
                Set  C_old = C
                If n > 1 then
                        Output reduced binary expansion of n
        Else if |w(m)| < T_k then
                Let  w_Q(m) retain its initial value of 0.
Loop until end of insignificant indices
Output end-marker as per WDR Step 3
```

Step 4: *Refinement pass.* Scan through significant values found with higher threshold values T_j, for $j < k$ (if $k = 1$ skip this step). For each significant value $w(m)$, do the following:

```
If |w(m)| ∈ [w_Q(m), w_Q(m) + T_k),  then
        Output bit 0
Else if |w(m)| ∈ [w_Q(m) + T_k, w_Q(m) + 2T_k),  then
        Output bit 1
        Replace value of  w_Q(m) by w_Q(m) + T_k .
```

Step 5: *Create new scan order.* For the highest-scale level (the one containing the all-lowpass subband), use the indices of the remaining insignificant values as the scan order at that level. Use the scan order at level j to create the new scan order at level $j - 1$ as follows. The first part of the new scan order at level $j - 1$ consists of the insignificant children of the significant values at level j. The second part of the new scan order at level $j - 1$ consists of the insignificant children of the insignificant values at level j. Use this new scan order for level $j - 1$ to create the new scan order at level $j - 2$, until all levels are exhausted.

Step 6: *Loop.* Repeat steps 2 through 5.

The creation of the new scanning order only adds a small degree of complexity to the original WDR algorithm. Moreover, ASWDR retains all of the attractive features of WDR: simplicity, progressive transmission capability, and ROI capability.

a) SPIHT   b) WDR   c) ASWDR

d) SPIHT   e) WDR   f) ASWDR

FIGURE 6.19
SPIHT, WDR, and ASWDR compressions of "Lena" at low bpp. (a)–(c) 0.0625 bpp, 128:1. (d)–(f) 0.125 bpp, 64:1. Reproduced by Special Permission of *Playboy* **magazine. Copyright ©1972, 2000 by Playboy.**

Fig. 6.18 shows how ASWDR performs on the Lena image. The PSNR values for these images are slightly better than those for WDR, and almost as good as those for SPIHT. More importantly, the perceptual quality of ASWDR compressions are better than SPIHT compressions and slightly better than WDR compressions. This is especially true at high compression ratios. Fig. 6.19 shows magnifications of 128:1 and 64:1 compressions of the "Lena" image. The ASWDR compressions better preserve the shape of Lena's nose and details of her hat, and show less distortion along the side of her left cheek (especially for the 0.125 bpp case). These subjective observations are borne out by the edge correlations in Table 6.3. In almost every case, the ASWDR compressions produce slightly higher edge correlation values.

As a further example of the superior performance of ASWDR at high compression ratios, in Fig. 6.20 we show compressions of the "airfield" image at 128:1. The WDR and ASWDR algorithms preserve more of the fine details in the image. Look especially along the top of the images: SPIHT erases many fine details such as

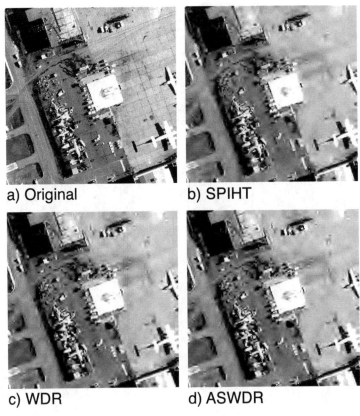

a) Original   b) SPIHT

c) WDR   d) ASWDR

FIGURE 6.20
Comparisons of 128:1 compressions of "airfield" image. (From Walker, James S., A lossy image codec based on adaptively scanned wavelet difference reduction, in *Optical Engineering*, July 2000. With permission.)

the telephone pole and two small square structures to the right of the thin black rectangle. These details are preserved, at least partially, by both WDR and ASWDR. The ASWDR image does the best job in retaining some structure in the telephone pole. ASWDR is also superior in preserving the structure of the swept-back winged aircraft, especially its thin nose, located to the lower left of center. These are only a few of the many details in the airplane image which are better preserved by ASWDR.

As quantitative support for the superiority of ASWDR in preserving edge details, we show in Table 6.5 the values for three different edge correlations γ_k, $k = 3$, 4, and 5. Here k denotes how many levels in the Daub 9/7 wavelet transform were used. A higher value of k means that edge detail at lower resolutions was considered in computing the edge correlation. These edge correlations show that ASWDR is superior over several resolution levels in preserving edges in the "airfield" image at the low bit rate of 0.0625 bpp.

Table 6.5 Edge Correlations for 128:1
Compressions of "Airfield" Image

Corr./Method	SPIHT	WDR	ASWDR
γ_3	.665	.692	**.711**
γ_4	.780	.817	**.827**
γ_5	.845	.879	**.885**

High compression ratio images like these are used in reconnaissance and in medical applications, where fast transmission and ROI (region selection) are employed, as well as multiresolution detection. The WDR and ASWDR algorithms do allow for ROI while SPIHT does not. Furthermore, their superior performance in displaying edge details at low bit rates facilitates multiresolution detection.

Further research is being done on improving the ASWDR algorithm. One important enhancement will be the incorporation of an improved predictive scheme, based on weighted values of neighboring transform magnitudes as described in Buccigrossi and Simoncelli [6].

6.3.6 Lossless Compression

A novel aspect of the compression/decompression methods diagrammed in Figs. 6.3 and 6.4 is that *integer-to-integer* wavelet transforms can be used in place of the ordinary wavelet transforms (such as Daub 9/7) described so far. An integer-to-integer wavelet transform produces an integer-valued transform from the gray-scale, integer-valued image [7]. Since n loops in bit-plane encoding reduces the quantization error to less than $T_0/2^n$, it follows that *once 2^n is greater than T_0, there will be zero error.* In other words, the bit-plane encoded transform will be exactly the same as the original wavelet transform; hence lossless encoding is achieved (with progressive transmission as well). Of course, for many indices, the zero error will occur sooner than with the maximum number of loops n. Consequently, some care is needed in order to efficiently encode the minimum number of bits in each binary expansion. A discussion of how SPIHT is adapted to achieve lossless encoding can be found in Said and Pearlman [47]. The algorithms WDR and ASWDR can also be adapted in order to achieve lossless encoding (public versions of these adaptations are available [16].)

6.3.7 Color Images

Following the standard practice in image compression research, we have concentrated here on methods of compressing gray-scale images. For color images, this corresponds to compressing the *intensity* portion of the image. That is, if the color image is a typical RGB image, with 8 bits for red, 8 bits for green, and 8 bits for blue, then the intensity I is defined by $I = (R + B + G)/3$, which rounds to an 8-bit gray-scale image. The human eye is most sensitive to variations in intensity, so the most difficult part of compressing a color image lies in the compressing of the intensity. Usually, the two color channels are denoted Y and C and are derived from the R, G, and B values [43]. Much greater compression can be done on the Y and C versions of the image since the human visual system is much less sensitive to variations in these two variables. Each of the algorithms described above can be modified so as to compress color images. For example, the public domain SPIHT coder [46] does provide programs for compressing color images. For reasons of space, we cannot describe compression of color images in any more detail.

6.3.8   Other Compression Algorithms

There is a wide variety of wavelet-based image compression algorithms besides the ones that we focused on here. Some of the most promising are algorithms that minimize the amount of memory which the encoder and/or decoder must use [20, 29]. A new algorithm which is embedded and which minimizes PSNR is described by Li and Lei [24]. Many other algorithms are cited in the review article by Davis and Nosratinia [14]. In evaluating the performance of any new image compression algorithm, one must take into account not only PSNR values but also the following factors: (1) perceptual quality of the images (edge correlation values can be helpful here); (2) whether the algorithm allows for progressive transmission; (3) the complexity of the algorithm (including memory usage); and (4) whether the algorithm has ROI capability.

6.3.9   Ringing Artifacts and Postprocessing Algorithms

As observed from the simulation for low bit rate (high compression ratio) compression, the decompressed image has ringing artifacts at the strong edges in the image. This is caused by the quantization process and by the overlapping nature of the wavelet transform. Edges have significant coefficients in all detailed subbands (horizontal, vertical, and diagonal subbands) and at low bit rate compression, these subband coefficients are quantized heavily. The ringing artifact at an edge is the linear combination of the (overlapping) wavelets and the quantization errors. Several postprocessing algorithms are proposed to reduce the ringing artifacts [22, 36, 72] and improve the perceptual quality of the decompressed image. Simulation results, software, and further details can be found at http://mmsplab.ece.wisc.edu/post/index.html.

References

[1]   Antonini, M., Barlaud, M., Mathieu, P., and Daubechies, I., Image coding using the wavelet transform, *IEEETrans. on Image Processing*, 1, 205–220, 1992.

[2] Bamberger, R.H., Eddins, S.L., and Nuri, V., Generalized symmetric extension for size-limited multirate filter banks, *IEEETrans. on Image Processing*, 3, 82–86, 1994.

[3] Brislawn, C., Classification of symmetric wavelet transforms, *LosAlamos Tech. Report*, 1993.

[4] Brislawn, C., A simple lattice architecture for even-order linear-phase perfect reconstruction filter banks, *Proc. IEEE-SP Intl. Symp. Time-Frequency and Time-Scale Analysis*, Philadelphia, PA, 124–127, 1994.

[5] Brower, B.V., Low-bit-rate image compression evaluations, *Proc. SPIE*, Orlando, FL, April 4–9, 1994.

[6] Buccigrossi, R.W. and Simoncelli, E.P., Image compression via joint statistical characterization in the wavelet domain, *IEEE Trans. on Image Processing*, 8(12), 1999.

[7] Calderbank, A.R., Daubechies, I., Sweldens, W., and Yeo, B.-L., Wavelet transforms that map integers to integers, *Applied and Computational Harmonic Analysis*, 5(3), 332–369, 1998.

[8] Cohen,A., *Ondelettes, analyses multirésolutions et traitement numérique du signal*, Ph.D. thesis, Universite Paris IX, Dauphine, 1990.

[9] Cohen, A., Daubechies, I., and Feauveau, J.-C., Biorthogonal bases of compactly supported wavelets, *Comm. Pure Appl. Math.*, 45, 1992.

[10] Compression with Reversible Embedded Wavelets, RICOH Company Ltd. submission to ISO/IEC JTC1/SC29/WG1 for the JTC1.29.12 work item, 1995. Can be obtained on the World Wide Web, address: http://www.crc.ricoh.com/CREW.

[11] Daubechies,I., *Ten Lectures on Wavelets*, CBMS Conference Series, SIAM, Philadelphia, 1992.

[12] Daubechies, I., Orthonormal bases of compactly supported wavelets, *Comm. Pure Appl. Math.*, 41, 909–996, 1988.

[13] Daubechies, I. and Lagarias, J., Two-scale difference equations I. Existence and global regularity of solutions, *SIAMJ. Math. Anal.*, 22, 1388–1410, 1991.

[14] Davis, G.M. and Nosratinia, A., Wavelet-based image coding: an overview, *Applied and Computational Control, Signals and Circuits*, 1(1), 1998.

[15] Eirola,T., Sobolev characterization of solutions of dilation equations, *SIAM J. Math. Anal.*, 23, 1015–1030, 1992.

[16] WDR and ASWDR compressors are part of the FAWAV software package at http://www.crcpress.com/edp/download/fawav/fawav.htm/.

[17] Gopinath, R.A., Odegard, J.E., and Burrus, C.S., Optimal wavelet representation of signals and the wavelet sampling theorem, *IEEE Transaction on Circuits & Systems II*, 41, 262–277, 1994.

[18] Heller, P.N. and Wells, Jr., R.O., Spectral theory of multiresolution operators and applications, in *Wavelets: Theory, Algorithms, and Applications*, Chui, C.K., Ed., AcademicPress, San Diego, CA, 13–31, 1994.

[19] Go to `ftp://ipl.rpi.edu/pub/image/still/usc/gray/` for "Lena," "Goldhill," and "Barbara." Go to `http://www.image.cityu.edu.hk/imagedb/` for "airfield."

[20] Islam, A. and Pearlman, W.A., An embedded and efficient low-complexity hierarchical image coder, *Proc. SPIE 3653, Visual Communications and Image Processing '99*, San Jose, CA, Jan. 1999.

[21] Kiya, H., Nishikawa, K., and Iwahashi, M., A development of symmetric extension method for subband image coding, *IEEE Trans. on Image Processing*, 3, 78–81, 1994.

[22] Shen, M. and Jay Kuo, C.C., Artifact removal in low bit rate wavelet coding with robust nonlinear filtering, *MMSP98*, 480–485, 1998.

[23] Lawton,W., Necessary and sufficient conditions for construction orthonormal wavelet bases, *J. Math. Phys.*, 32, 57–61, 1991.

[24] Li, J. and Lei, S., An embedded still image coder with rate-distortion optimization, *IEEE Trans. on Image Processing*, 8(7), 913–924, 1999.

[25] Majani, E. and Lightstone, M., Biorthogonal wavelets for image compression, *Proc. 1994 Data Compression Conference*, Snowbird, Utah, 462, 1994.

[26] Mallat, S., *A Wavelet Tour of Signal Processing*, Academic Press, New York, 1998.

[27] Mallat, S., A theory for multiresolution signal decomposition: the wavelet representation, *IEEE Trans. PAMI*, 11, 674–693, 1989.

[28] Mallat, S., Multifrequency channel decomposition of images and wavelet models, *IEEE Trans. on Acoust. Speech and Signal Processing*, 37(12), 2091–2110, 1989.

[29] Malvar, H., Progressive wavelet coding of images, *Proc. of IEEE Data Compression Conference*, Salt Lake City, UT, 336–343, March 1999.

[30] Marr, D., *Vision*, W.H. Freeman, San Francisco, CA, 1982.

[31] Recommendation H.262, ISO/IEC 13818. Generic coding of moving picture and associates audio, Draft International Standard of MPEG-2.

[32] Mintzer, F., Filters for distortion-free two-band multirate filter banks, *IEEE Trans. on ASSP*, 626–630, 1985.

[33] Nayebi, K., Barnwell, III, T.P., and Smith, M.J.T., Time-domain filter bank analysis: a new design theory, *IEEE Trans. on Signal Processing,* 40, 1992.

[34] Nguyen, T.Q. and Vaidyanathan, P.P., Two-channel perfect-reconstruction FIR QMF structures which yield linear-phase analysis and synthesis filters, *IEEE Trans. on ASSP,* 37, 676–690, 1989.

[35] Nguyen,T.Q., A quadratic constrained least-squares approach to the design of digital filter banks, *Proc. IEEE ISCAS,* San Diego, 1344–1347, May 1992.

[36] Oguz, S.H., Hu, Y.H., and Nguyen, T.Q., Morphological post-filtering of ringing and lost data concealment in generalized lapped orthogonal transform based image and video coding, Ph.D. thesis, University of Wisconsin, 1999. Additional informations can be found at http://mmsplab.ece.wisc.edu/post/index.html.

[37] Orchard, M. and Ramchandran, K., An investigation of wavelet-based image coding using an entropy-constrained quantization framework, *Proc. Data Compression Conf.,* Snowbird, Utah, 341–350, 1994.

[38] Pennebaker, W.B. and Mitchell, J.L., *JPEG: Still Image Compression Standard,* Van Nostrand Reinhold, NewYork, 1993.

[39] Ramchandran, K. and Vetterli, M., Best wavelet packet bases in a rate-distortion sense, *IEEE Trans. on Image Processing,* 2, 160–175, 1993.

[40] Ramos, M.G. and Hemami, S.S., Activity selective SPIHT coding, *Proc. SPIE 3653, Visual Communications and Image Processing '99,* San Jose, CA, Jan. 1999. See also errata for this paper at http://foulard.ee.cornell.edu/marcia/asspiht2.html.

[41] Rioul, O., A discrete-time multiresolution theory, *IEEE Trans. on Signal Processing,* 41, 2591–2606, 1993.

[42] Rioul, O. and Vetterli, M., Wavelets and signal processing, *IEEE Signal Processing Magazine,* 8(3), 14–38, 1991.

[43] Russ,J.C., *The Image Processing Handbook,* CRC Press, Boca Raton, FL, 1995.

[44] Said, A. and Pearlman, W.A., Image compression using the spatial-orientation tree, *IEEE Int. Symp. on Circuits and Systems,* Chicago, IL, 279–282, 1993.

[45] Said, A. and Pearlman, W.A., A new, fast, and efficient image codec based on set partitioning in hierarchical trees, *IEEE Trans. on Circuits and Systems for Video Technology,* 6(3), 243–250, 1996.

[46] SPIHT programs can be downloaded from ftp://ipl.rpi.edu/pub/.

[47] Said, A. and Pearlman, W.A., An image multi-resolution representation for lossless and lossy image compression, *IEEE Trans. Image Processing,* 5(9), 1303–1310, 1996.

[48] Shapiro, J.M., Embedded image coding using zerotrees of wavelet coefficients, *IEEE Trans. on Signal Processing,* 41, 3445–3462, 1993.

[49] Shoham, Y. and Gersho, A., Efficient bit allocation for an arbitrary set of quantizers, *IEEE Trans. on ASSP,* 36, 1445–1453, 1988.

[50] Smith, M.J.T. and Barnwell, III, T.P., Exact reconstruction techniques for tree-structured suband coders, *IEEE Trans. ASSP,* 434–441, 1986.

[51] Smith, M.J.T. and Eddins, S., Analysis-synthesis techniques for subband image coding, *IEEE Trans. ASSP,* 38, 1446–1456, 1990.

[52] Strang, G., Wavelets and dilation equations, *SIAM Review,* 31, 614–627, 1989.

[53] Strang, G. and Nguyen, T., *Wavelets and Filter Banks,* Wellesley-Cambridge Press, Wellesley, MA, 1997.

[54] Vaidyanathan, P.P., *Multirate Systems and Filter Banks,* Prentice-Hall, Englewood Cliffs, NJ, 1993.

[55] Vaidyanathan, P.P. and Hoang, P.Q., Lattice structures for optimal design and robust implementation of two-channel perfect-reconstruction QMF banks, *IEEE Trans. on ASSP,* 36, 81–94, 1988.

[56] Tian, J. and Wells, Jr., R.O., A lossy image codec based on index coding, *IEEE Data Compression Conference, DCC '96,* 456, 1996.

[57] Tian, J. and Wells, Jr., R.O., Embedded image coding using wavelet-difference-reduction, in *Wavelet Image and Video Compression,* Topiwala, P., Ed., 289–301, Kluwer Academic, Norwell, MA, 1998.

[58] Tian, J. and Wells, Jr., R.O., Image data processing in the compressed wavelet domain, *3rd International Conference on Signal Processing Proc.,* Yuan, B. and Tang, X., Eds., 978–981, Beijing, China, 1996.

[59] Vetterli, M., A theory of multirate filter banks, *IEEE Trans. on ASSP,* 35, 356-372, 1987.

[60] Vetterli, M., Multidimensional subband coding: some theory and algorithms, *Signal Processing,* 6, 97–112, 1984.

[61] Vetterli, M. and Herley, C., Wavelets and filter banks, *IEEE Trans. on Signal Processing,* 40, 2207-2233, 1992.

[62] Vetterli, M. and LeGall, D., Perfect reconstruction FIR filter banks: some properties and factorization, *IEEE Trans. on ASSP,* 37, 1057–1071, 1989.

[63] Villasenor, J.D., Belzer, B., and Liao, J., Wavelet filter evaluation for image compression, *IEEE Trans. on Image Processing,* 4, 1053–1060, 1995.

[64] Villemoes,L., Energy moments in time and frequency for two-scale difference equation solutions and wavelets, *SIAM J. Math. Anal.,* 23, 1519–1543, 1992.

[65] Volkmer, H., On the regularity of wavelets, *IEEE Trans. on Information Theory,* 38, 872–876, 1992.

[66] Walker, J.S., A lossy image codec based on adaptively scanned wavelet difference reduction, *Optical Engineering,* in press.

[67] Wallace, G.K., The JPEG still picture compression standard, *Comm. of the ACM,* 34(4), 30–44, 1991.

[68] Witten, I., Neal, R., and Cleary, J., Arithmetic coding for compression, *Comm. of the ACM,* 30(6), 1278–1288, 1986.

[69] Woods, J. and O'Neil, S.D., Subband coding of images, *IEEE Trans. on ASSP,* 34, 1278–1288, 1986.

[70] Wavelet scalar quantization gray scale fingerprint image compression specification, Criminal Justice Information Services, FBI, Washington, DC, 1993.

[71] Xiong, Z., Ramchandran, K., and Orchard, M., Joint optimization of scalar and tree-structured quantization of wavelet image decomposition, *27th Asilomar Conf.,* Pacific Grove, CA, November 1993.

[72] Yang, S., Tull, D., Hu, Y.H., and Nguyen, T., Maximum a posteriori parameter estimation for image ringing artifact removal, submitted to the *IEEE Transaction on Image Processing,* 1999. Additional information can be found at http://mmsplab.ece.wisc.edu/post/index.html.

[73] Zettler, W.R., Huffman, J., and Linden, D., The application of compactly supported wavelets to image compression, *Proc. SPIE,* 1244, 150–160, 1990.

[74] Zhu, B., Tewfik, A.H., Colestock, M.A., Gerek, O.N., and Cetin, A.E., Image coding with wavelet representations, edge information and visual masking, *Proc. IEEE ICIP,* Washington, DC, 1995.

Chapter 7

Fractal-Based Image and Video Compression

Guojun Lu
Monash University

7.1 Introduction

This chapter describes image and video compression techniques based on affine transforms or iterated function systems (IFS) [2, 4]. These techniques are fundamentally different from techniques based on other transforms, such as discrete cosine transform (DCT). In DCT-based techniques, image data are transformed from the spatial domain to the frequency domain where quantization and entropy coding are carried out to achieve data compression. In IFS-based techniques, we exploit the fact that part of an image is similar to another part of the image after certain affine transforms called IFS. Data compression is achieved by determining these transforms and storing parameters representing them.

The chapter is organized as follows. Section 7.2, describes the basic properties of fractals and the basic principle of fractal-based image compression. Section 7.3 describes concepts of contractive affine transforms, iterated function systems, and the fractal image generation process. Section 7.4 discusses how to find IFS directly from images. Section 7.5 describes how to compress images based on a library of known IFS.

Techniques introduced in Sections 7.4 and 7.5 can achieve very high compression ratios. However, it is difficult to find IFS in a natural image. To solve this problem, image coding methods using partitioned IFS (PIFS) have been developed. The difference between IFS and PIFS is that IFS consists of affine transforms that map an entire image to parts of the image while PIFS consists of transforms that map parts of an image to other parts of the image. In these methods, an image to be coded is divided into nonoverlapping blocks. For each block, a transformation is found that converts part of the image into a block similar to this block. The combination of all transforms found for each block is called PIFS. PIFS corresponds to a unique image. The compression performance depends on the contents of the image and how the image is partitioned. There are many types of partitions: fixed size partitioning, quadtree partitioning, horizontal-vertical (HV) partitioning, and triangular partitioning. We discuss PIFS-based coding using fixed size partition and quadtree partition in Sections 7.6 and 7.7, respectively.

One property of fractals is scalability: fractals have fine detail in any scale. Section 7.8 explores how we can use this property to reduce the required compression time and improve performance.

It is well known that there is much redundancy among neighboring frames of a video sequence. It is very likely that part of a current frame is the transformation of a part in the previous frame. Based on this observation, we can apply quadtree partition techniques to video sequence coding, which is discussed in Section 7.9.

There are other image compression techniques based on various properties of fractals. Section 7.10 briefly describes two techniques based on fractal dimension and fractal approximation, respectively. Section 7.11 concludes the chapter.

7.2   Basic Properties of Fractals and Image Compression

The word fractal was coined by Mandelbrot from the Latin word *fractus,* meaning broken, to describe objects that were too irregular to fit into traditional geometrical settings [1]. Several definitions have been proposed. Mandelbrot defined a fractal to be a set with Hausdorff dimension strictly greater than its Euclidean dimension, i.e., a set for which the only consistent description of its metric properties requires a "dimension" value larger than our standard, intuitive definition of the set's "dimension." A fractal has a fractional dimension; thus some people say we get the word *fractal* from *fractional dimension.* According to Barnsley [2], a fractal is a geometric form whose irregular details recur at different scales and angles which can be described by affine or fractal transforms. Various other definitions have been proposed, but they are not complete in that they exclude a number of sets that clearly ought to be regarded as fractals.

Falconer [3] proposed that it is best to regard a fractal as a set that has properties such as those listed below, rather than to look for a precise definition which will almost certainly exclude some interesting cases. Typical properties of a fractal are

- It has a fine structure, i.e., details on arbitrarily small scales.

- It is too irregular to be described in traditional geometrical language, both locally and globally.

- It usually has some form of self-similarity, perhaps approximate or statistical.

- Its fractal dimension (Hausdorff dimension) is usually higher than its Euclidean dimension.

- In most cases of interest, a fractal is defined in a very simple way, perhaps recursively.

Fig. 7.1 shows the construction of one of the common fractals called the Sierpinski gasket or triangle [3]. It is constructed by repeatedly replacing an equilateral triangle with three equilateral triangles of half the height. This construction process can be represented by a set of fractal transforms (see Section 7.4). Fractal transforms have been used to generate complicated fractal images. Fractal image compression is the inverse of fractal image generation; instead of generating an image from a given formula, fractal image compression searches for sets of fractals in a digitized image which describe and represent the entire image. Once the appropriate sets of fractals are determined, they are represented by very compact fractal transform formulae. These formulae are the rules for reproducing the various sets of fractals which, in turn, regenerate the entire image. Because fractal transform formulae require a very small amount of data to be represented and stored, fractal compression can result in very high compression ratios.

FIGURE 7.1
Construction of the Sierpinski triangle or gasket.

7.3 Contractive Affine Transforms, Iterated Function Systems, and Image Generation

Since fractal image compression based on iterated function systems is the inverse of image generation, in this section, we see how fractals are generated from an iterated function system. We introduce contractive affine transforms and iterated function systems.

Contractive Affine Transforms

A two-dimensional affine transform W maps points in the Euclidean plane into new points in the Euclidean plane, according to the formula

$$W \begin{bmatrix} x \\ y \end{bmatrix} = \begin{bmatrix} a & b \\ c & d \end{bmatrix} \begin{bmatrix} x \\ y \end{bmatrix} + \begin{bmatrix} e \\ f \end{bmatrix}. \tag{7.1}$$

It consists of a linear transform, represented by the 2×2 matrix with entries a, b, c, and d, followed by a shift or translation, represented by the vector with entries e and f. An example for an affine transform is

$$W \begin{bmatrix} x \\ y \end{bmatrix} = \begin{bmatrix} 0.5 & 0 \\ 0 & 0.5 \end{bmatrix} \begin{bmatrix} x \\ y \end{bmatrix} + \begin{bmatrix} 0 \\ 0 \end{bmatrix}.$$

Applying the above transform to all points in the triangle F of Fig. 7.2(a) results in a smaller triangle in Fig. 7.2(b). Notice that the cross and the star in the transformed triangle $W(F)$ are closer than in F. We say that the transform is contractive if it always moves pairs of points closer together. Formally, a transform W is said to be contractive if for any two points P_1 and P_2, the distance

$$d\left(W\left(P_1\right), W\left(P_2\right)\right) < sd\left(P_1, P_2\right) \tag{7.2}$$

where $s \in (0, 1)$, and is called the contractive factor.

Contractive affine transforms have the property that when they are repeatedly applied, they converge to a point which remains fixed upon further iterations. For example, applying the transform

$$W \begin{bmatrix} x \\ y \end{bmatrix} = \begin{bmatrix} 0.5 & 0 \\ 0 & 0.5 \end{bmatrix} \begin{bmatrix} x \\ y \end{bmatrix}$$

repetitively to any initial point (x_0, y_0) will yield the sequence of points $(x_0/2, y_0/2)$, $(x_0/4 \cdot y_0/4), \ldots$, which can be seen to converge to the point $(0, 0)$, in the limit.

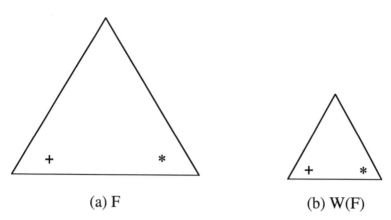

(a) F (b) W(F)

FIGURE 7.2
Effect of applying a contractive affine transform to a shape.

Iterated Function Systems

An iterated function system (IFS) is a collection of contractive affine transforms. The following is an IFS consisting of four transforms.

$$W_1 \begin{bmatrix} x \\ y \end{bmatrix} = \begin{bmatrix} 0.5 & 0 \\ 0 & 0.16 \end{bmatrix} \begin{bmatrix} x \\ y \end{bmatrix} + \begin{bmatrix} 0 \\ 0 \end{bmatrix} ; \qquad (7.3)$$

$$W_2 \begin{bmatrix} x \\ y \end{bmatrix} = \begin{bmatrix} 0.2 & 0.26 \\ 0.23 & 0.22 \end{bmatrix} \begin{bmatrix} x \\ y \end{bmatrix} + \begin{bmatrix} 0 \\ 0.2 \end{bmatrix} ; \qquad (7.4)$$

$$W_3 \begin{bmatrix} x \\ y \end{bmatrix} = \begin{bmatrix} -0.15 & 0.28 \\ 0.26 & 0.24 \end{bmatrix} \begin{bmatrix} x \\ y \end{bmatrix} + \begin{bmatrix} 0 \\ 0.2 \end{bmatrix} ; \qquad (7.5)$$

$$W_4 \begin{bmatrix} x \\ y \end{bmatrix} = \begin{bmatrix} 0.85 & 0.04 \\ 0.04 & 0.85 \end{bmatrix} \begin{bmatrix} x \\ y \end{bmatrix} + \begin{bmatrix} 0 \\ 0.2 \end{bmatrix} . \qquad (7.6)$$

Actually, this is the IFS of a fern leaf. Later we see how this IFS generates a fern image.

A fundamental theorem of fractal geometry is that each IFS, that is, each set of contractive transforms, defines a unique fractal image. It is called the attractor of the IFS. The attractor of an IFS is unique; for each IFS there is only one attractor. This is the contractive mapping fixed-point theorem. The attractor of an IFS has the following property: if the IFS is made up of N affine transforms which are denoted by $W_1, W_2, \ldots, W_N$, then the corresponding attractor A obeys

$$A = W_1(A) \cup W_2(A) \cup \cdots \cup W_N(A) .$$

This says that the attractor of the IFS is the same as the union of the transforms of the attractor. Now let us see how to generate the attractor of an IFS using the chaos game algorithm [2].

The Chaos Game Algorithm

Suppose an IFS contains N affine transforms $W_1, W_2, \ldots, W_N$. Let these transforms have associated probabilities $p_1, p_2, \ldots, p_N$, respectively. They obey

$$p_1 + p_2 + \cdots + p_N = 1 \quad \text{and} \quad p_i > 0 \quad \text{for} \quad i = 1, 2, \ldots, N \, .$$

They are the probabilities with which each transform is to be selected and applied in the chaos game algorithm.

Here is how the chaos game is played: choose any point (x_0, y_0) in a Euclidean plane. Select one transform in the IFS according to its probability and apply it to point (x_0, y_0) to get a new point (x_1, y_1). Select another transform according to its probability and apply it to point (x_1, y_1) to get a new point (x_2, y_2). Repeat this process to obtain a long sequence of points:

$$(x_0, y_0), (x_1, y_1), (x_2, y_2), (x_3, y_3), \ldots$$

A basic result of the IFS theory is that this sequence of points will converge, with 100% probability, to the attractor of the IFS.

The following is pseudocode showing how the chaos game algorithm is applied in general:

(i) Let $x = 0; y = 0$.

(ii) Choose k to be one of the numbers $1, 2, \ldots, N$, with probability p_k.

(iii) Apply transform W_k to point (x, y) to obtain a new point $(x_{\text{new}}, y_{\text{new}})$.

(iv) Let $x = x_{\text{new}}; y = y_{\text{new}}$.

(v) Plot (x, y).

(vi) Return to step (ii) and repeat until a preset number of iterations is reached.

Using the above algorithm and the four transforms in Eqs. (7.3)–(7.6), we can generate a fern leaf as shown in Fig. 7.3.

Transforms (7.3)–(7.6) can be represented compactly by just 24 parameters, although the fern generated from them is quite complicated and requires large amounts of data if stored in a bit-mapped format. So, if we can do the reverse of image generation and find the IFS of a given image, we can achieve a very high compression ratio. The next section describes how to find the IFS for a given image.

7.4   Image Compression Directly Based on the IFS Theory

The direct method of image compression using IFS is based on the collage theorem [2, 6]. Loosely speaking, it states that if we can find IFS W for an image B so that

FIGURE 7.3
The fern image generated from the transforms (7.3)–(7.6).

B and $W(B)$ are very similar, then the attractor of W will also be very similar to image B. Thus, we can store W instead of image B to achieve a very high compression ratio. Hence, to compress an image is to find the IFS of that image.

To find an IFS for an image, based on the collage theorem and the property of IFS attractors, we split the whole image into nonoverlapping segments whose union covers the entire image. If each segment is a transformed copy of the entire image or is very close to it, the combination of these transforms is the IFS of the original image. In other words, to encode an image into IFS is to find a set of contractive affine transforms, $W_1, W_2, \ldots, W_N$, so that the original image B is the union of the N subimages:

$$B = W_1(B) \cup W_2(B) \cup \cdots \cup W_N(B) .$$

We use an example to show how to find an IFS for an image. As shown in Fig. 7.4, the Sierpinski triangle is the union of three small triangles: the top, bottom left, and bottom right triangles. Each small triangle is a copy of the transformed original Sierpinski triangle. If we can find these transforms, the IFS of the Sierpinski triangle is their combination.

Let us find the transform for the top triangle first. We know that the general format of the affine transform is Eq. (7.1). To determine the transform, we have to find six variables: a, b, c, d, e, and f. The first thing to do is to find corresponding points in the original Sierpinski triangle and the top triangle. Since it is clear, from Fig. 7.4(a), that point (x_1, y_1) is transformed to (x'_1, y'_1), (x_2, y_2) to (x'_2, y'_2) and

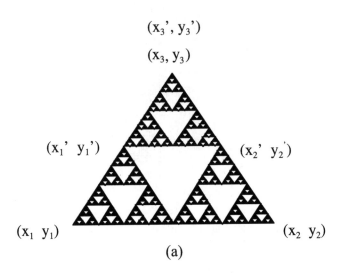

(x_3', y_3')

(x_3, y_3)

$(x_1'\ y_1')$ $(x_2'\ y_2')$

$(x_1\ y_1)$ $(x_2\ y_2)$

(a)

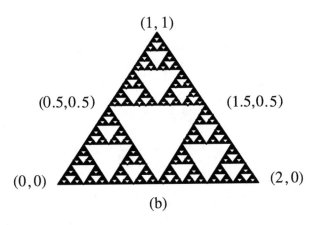

$(1, 1)$

$(0.5, 0.5)$ $(1.5, 0.5)$

$(0, 0)$ $(2, 0)$

(b)

FIGURE 7.4
An example to find IFS for a given image.

(x_3, y_3) to (x_3', y_3'), we have the following six equations:

$$
\begin{aligned}
ax_1 + by_1 + e &= x_1' \\
ax_2 + by_2 + e &= x_2' \\
ax_3 + by_3 + e &= x_3' \\
cx_1 + dy_1 + f &= y_1' \\
cx_2 + dy_2 + f &= y_2' \\
cx_3 + dy_3 + f &= y_3'
\end{aligned}
$$

Using the coordinates in Fig. 7.4(b), we can solve the above equations to get $a = 0.5$, $b = 0$, $c = 0$, $d = 0.5$, $e = 0.5$, and $f = 0.5$.

In a similar way, we can find six variables for each of the transforms for the bottom left and bottom right triangles, respectively. Combining these transforms we have the IFS of the Sierpinski triangle as follows:

$$W_1 \begin{bmatrix} x \\ y \end{bmatrix} = \begin{bmatrix} 0.5 & 0 \\ 0 & 0.5 \end{bmatrix} \begin{bmatrix} x \\ y \end{bmatrix} + \begin{bmatrix} 0.5 \\ 0.5 \end{bmatrix}$$

$$W_2 \begin{bmatrix} x \\ y \end{bmatrix} = \begin{bmatrix} 0.5 & 0 \\ 0 & 0.5 \end{bmatrix} \begin{bmatrix} x \\ y \end{bmatrix} + \begin{bmatrix} 0 \\ 0 \end{bmatrix}$$

$$W_3 \begin{bmatrix} x \\ y \end{bmatrix} = \begin{bmatrix} 0.5 & 0 \\ 0 & 0.5 \end{bmatrix} \begin{bmatrix} x \\ y \end{bmatrix} + \begin{bmatrix} 1.0 \\ 0 \end{bmatrix} .$$

Readers can verify the obtained IFS by applying the chaos game algorithm (with $p_i = 1/3$) using this IFS to see whether the Sierpinski triangle is obtained. After obtaining the IFS of the Sierpinski triangle, we can store the 18 parameters of three transforms instead of the bit-mapped data to achieve significant compression.

It should be noted that although an IFS has a unique attractor, many IFS can be found for a given image. For example, the Sierpinski triangle can be thought of as the union of nine small triangles, and then the determined IFS will consist of nine transforms.

In the above example, the attractor of the IFS perfectly covers the original image, which can be reconstructed from the IFS exactly. For an arbitrary image, it may be impossible or difficult to find an IFS whose attractor perfectly covers the original image. We should then find the "collage" as close to the original image as possible. According to the collage theorem [2, 6], the attractor of the IFS determined by the collage will be close to the original image.

For images with gray-scale or color, the same principle can be used to find the IFS, but obviously it would be much more difficult and time consuming.

7.5 Image Compression Based on IFS Library

For natural images, it is difficult to find the IFS directly as discussed in the previous section. But it is possible that images consist of a number of small objects whose IFS are known. For this kind of image we can proceed as follows. Using an image segmentation technique, we segment the image into small objects. An object can be a fern, leaf, cloud, fence post, or more complex collection of pixels.

We then look up these objects in a library of fractals. The library does not contain literal fractals; that would require large amounts of storage. Instead, the library contains compact sets of IFS codes that will reproduce the corresponding fractals. The library is searched to find fractals that approximate each segmented object. The corresponding IFS codes of these fractals are stored instead of the original image to achieve high compression. This image compression approach using the IFS library is very similar to vector quantization in which each image block is represented with the index of the codeword that is most similar to a particular image block [15].

IFS codes in the library are obtained using the direct approach as discussed in the previous section. Since the same object can be contracted, rotated, and translated, it is not practical for the library to contain the same object in many different scales and different angles. Instead, fractals in the library are transformed to match the objects in the image. This matching process is similar to the matching process used in PIFS coding described in the next section.

The main problem with the library searching method is how to automatically and accurately segment an image into meaningful objects. There is no single effective method to do this.

7.6 Image Compression Based on Partitioned IFS

The direct and library-based approaches can compress images with a very high compression ratio. However, it is very difficult to find IFS automatically in natural images. To solve this problem, an alternative method has been developed [5, 6]. A natural image, such as a face, does not contain the type of self-similarity that can be found in the fractals. The image does not appear to contain affine transformations themselves. However images do, in fact, contain a different sort of similarity. Part of the image is similar to another part of the image. The distinction from fractal self-similarity is that rather than forming the image from copies of its whole self (under appropriate affine transformation), here the image is formed from copies of properly transformed parts of itself. Experimental results suggest that most images that one would expect to see can be compressed by taking advantage of this type of self-similarity [6].

Based on the above observation, fractal-based block coding, or PIFS-based coding, was developed [5]–[8]. The PIFS-based approach is as follows. To encode an image f, we divide it into range blocks $R_1, R_2, \ldots, R_i \ldots R_N$, such that

$$f = R_1 \cup R_2 \cup \cdots \cup R_N$$

and

$$R_i \cap R_j = 0 \quad \text{when} \quad i \neq j \ .$$

That is, the range blocks cover the whole image and do not overlap.

The image is also divided into overlapping domain blocks $D_1, D_2, \ldots, D_j, \ldots, D_M$. For each range block R_i, we find a contractive transform W_i and a domain block D_j in the image, so that

$$R_i \approx W_i \left(D_j \right) \ .$$

The combination of $W_1, W_2, \ldots, W_i, \ldots, W_N$ is called PIFS W. If W is simpler than the original image, we can encode f into W and achieve certain compression. When decoding, according to the contractive mapping fixed point theorem, so long as W is contractive, application of W to an arbitrary image repeatedly will result in a fixed image. When $W(f)$ is close to f, the fixed image will be close to the original image f.

The three main issues involved in the design and implementation of a fractal block-coding system based on the above idea are (i) how the image is partitioned, (ii) the choice of a distortion measure between two images, and (iii) types of contractive affine transformations to be used. We now discuss these issues.

7.6.1 Image Partitions

The simplest partition of an image is fixed size partitioning; an image is divided into nonoverlapping square range blocks of fixed size ($B \times B$ pixels). For each range block, the entire image is searched for a square domain block which when suitably transformed is similar to the range block. The domain block is larger than the range block. Typical size is $2B \times 2B$ pixels, and they overlap every B pixels in both x-direction and y-direction.

The selection of sizes for range blocks is a compromise between compression ratio and reconstructed image quality. It is easy to find good matching domain blocks for small range blocks (4×4 and below), leading to high decoded image quality. But the achievable compression ratios will not be very high. On the other hand, it is generally harder to find good matching domain blocks for large range blocks (8×8 and above). But they allow a good exploitation of the redundancy in smooth image areas, leading to high compression ratios.

Fixed-size partitioning is the simplest. It ignores the contents of the image. To take advantage of image contents, other partitioning techniques, such as quadtree partitioning, horizontal-vertical (H-V) partitioning, have been proposed [6, 9]. In Section 7.7, we discuss the quadtree partitioning technique.

7.6.2 Distortion Measure

We use a distortion measure to determine the closeness of two image blocks: the smaller the distortion measure the more alike the two image blocks. There are many kinds of distortion measures. The common one used is the root-mean-square (RMS) distortion. For two square image blocks u and v of size $B \times B$ pixels, it is defined as

$$d(u, v) = \sqrt{\sum_{i,j}(u(i, j) - v(i, j))^2}$$

where summation is for $i = 0$ to $B - 1$ and $j = 0$ to $B - 1$.

In the search and mapping process, the RMS distortion is used to determine the closeness between a range block and a transformed domain block. The domain block causing the least distortion after certain transformation is deemed as a matching domain block to the range block.

We use the peak signal-to-noise ratio (PSNR) to measure the decoded image quality relative to the original image. Note that PSNR is not an exact measure of picture quality. It is used here for comparison purposes only.

7.6.3   A Class of Discrete Image Transformations

For each R_i, we must find D_j and W_i. Since we want to compress images with gray levels or color, we have to extend the basic form of affine transform to include pixel depth. We define the pixel depth at position (x, y) as $z = f(x, y)$. The extended affine transform becomes

$$W_i \begin{bmatrix} x \\ y \\ z \end{bmatrix} = \begin{bmatrix} a_i & b_i & 0 \\ c_i & d_i & 0 \\ 0 & 0 & s_i \end{bmatrix} \begin{bmatrix} x \\ y \\ z \end{bmatrix} + \begin{bmatrix} e_i \\ f_i \\ o_i \end{bmatrix}.$$

For ease of reference, we simplify this as

$$V_i \begin{bmatrix} x \\ y \end{bmatrix} = \begin{bmatrix} a_i & b_i \\ c_i & d_i \end{bmatrix} \begin{bmatrix} x \\ y \end{bmatrix} + \begin{bmatrix} e_i \\ f_i \end{bmatrix}.$$

V_i determines how the partitioned domain blocks of an original image are mapped to range blocks, while s_i and o_i determine the contrast scaling and brightness shift of the transform, respectively. The operation of W_i to transform D_j to R_i can be decomposed into the following stages: geometric contraction of D_j, contrast scaling, brightness shift, and rotation and flip operations [5, 7]. If all pixels in R_i have the same (or similar) pixel values, these operations are not required. We just need to store a single pixel value for this range block. This operation is called absorption.

Geometric Contraction

The domain blocks must be spatially contracted to the size of the range block. In the simple case where the domain block is twice the size of the range block, the pixel values of the contracted domain block are the average values of four neighbouring pixels in the domain block.

Contrast Scaling

A contrast scaling factor s_i must be found to make the contrast among pixels in the domain block match the contrast among pixels in the range block. s_i is defined as the greatest brightness difference among pixels in the range block divided by the greatest brightness difference among pixels in the domain block.

Brightness Shift

The brightness shift o_i must be found to make the brightness of the domain block and the range block the same. o_i is defined as the difference between average pixel value in the range block and average pixel value in the domain block.

Rotation and Flip Operations

The rotation and flip operations do not modify pixel values; they simply shuffle pixels within a block, in a deterministic way — we call them isometries. There are many isometries. The following eight are commonly used [5, 7]:

- identity (no rotation or flip operation),

- orthogonal reflection about mid-vertical axis of block,

- orthogonal reflection about mid-horizontal axis of block,

- orthogonal reflection about first diagonal of block,

- orthogonal reflection about second diagonal of block,

- rotation around center of block, through $+90°$,

- rotation around center of block, through $+180°$,

- rotation around center of block, through $-90°$.

In effect, these operations are able to generate, from a single block, a whole family of geometrically related transformed blocks, which provides a pool in which matching blocks will be sought during the encoding. More complex transformations can be used. But more bits will be required to identify each transformation.

7.6.4 Encoding and Decoding Procedures

An image is divided into nonoverlapping range blocks. For each range block R_i, we first determine whether it is an absorption block. If it is not, a domain block D_j is sought which matches the range block best after certain transformation. The search starts at the domain block closest to the range block and extends in a spiral fashion until a satisfactory match is found.

For each nonabsorption range block, we must store information on the position of the corresponding domain block, the contrast scaling factor, the brightness shift, and the rotation and flip operations. Table 7.1 shows the number of bits required to encode a range block, assuming that the image size is $I \times I$ pixels and the range block size is $R \times R$ pixels.

During decoding, starting from an arbitrary image of the same size as the original image, each range block is computed from the corresponding domain block using the encoding information. Computing all the range blocks once is one iteration. After several iterations, the reconstructed image will be very close to the original image. The PSNR is used to determine whether further iterations can improve the quality of the reconstructed image. If there is no improvement in PSNR, the iterative process ends. According to the collage theorem, how close the reconstructed image can be to the original image is determined by the accuracy of the mapping from domain blocks to range blocks at the encoding stage.

Table 7.1 Number of Bits Required for Coding a Fixed-Size Range Block

Type of Coding	Parameters	Number of bits
Absorption	Identifier	4
	Absorption factor	8
Isometric	Identifier	4
	Scaling factor	3
	Shifting factor	8
	Domain block coordinates	$2\log_2(I/R)$

7.6.5 Experimental Results

Table 7.2 shows experimental results obtained on the image "Lena" (512×512 pixels) and image "Lena" (256×256 pixels) with 8 bits per pixel (bpp), using different range block sizes [9]. It can be seen that decoded image quality is determined by the range block size used. The smaller the range block size, the easier it is to find a closer matching domain block, thus the higher the decoded image quality. Fig. 7.5 shows the original and compressed images "Lena" of 512×512 pixels with a compression ratio of 19 to 1.

Table 7.2 Experimental Results on Image "Lena" Using Fixed Size Range Blocks

Image size	Range block size	Compression ratio	PSNR (dB)
512×512	32×32	361.4	22.41
512×512	16×16	84.3	25.73
512×512	8×8	19.0	29.65
512×512	4×4	4.4	34.98
256×256	16×16	88.6	23.08
256×256	8×8	20.5	27.24
256×256	4×4	4.8	32.99

7.7 Image Coding Using Quadtree Partitioned IFS (QPIFS)

The weakness of the fixed-size partition is that the image is partitioned without considering the image contents. There are regions of the image that are difficult to cover well using fixed size range blocks. Similarly, there are regions that could be covered well with larger size range blocks, thus reducing the total number of maps needed (and increasing the compression ratio). This observation leads to the use of the quadtree partitioning technique [6, 9]. In a quadtree partition, range blocks as large as possible are used to code the image. When no good matching domain

FIGURE 7.5
(a) The original image "Lena" of 512×512 pixels, 8 bpp. (b) Compressed image using fixed size range block of 8×8 pixels, at 0.42 bpp with PSNR of 29.65 dB. Reproduced by Special Permission of *Playboy* magazine. Copyright ©1972, 2000 by Playboy.

block (of one size bigger than the range block) can be found for a range block in the image, it is divided into four equally sized child range blocks as shown in Fig. 7.6. These four children are named according to their direction in the parent range block – Northwest (NW), Northeast (NE), Southwest (SW), and Southeast (SE). This process repeats, starting from the whole image and continuing until the range blocks are small enough to be matched within some specified RMS tolerance. Small range blocks can be matched better than large ones because contiguous pixels in an image tend to be highly correlated. Therefore, one important issue in quadtree partitioning is to select a suitable RMS tolerance used in the matching process for range blocks of different sizes.

In fixed-size partitioning, the range block size is stored only once for the entire image, and the range block positions are implied if coded information of each range

Parent range block

Four equally sized
child range blocks

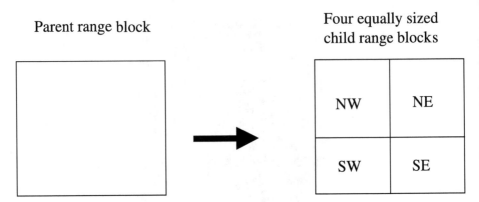

FIGURE 7.6
Partitioning of a parent range block.

block is stored sequentially. In quadtree partitioning, range block size varies. Information regarding range block size and position must be stored somehow. If this
information is stored straightforwardly, we need 3 bits to store range block size and
18 bits for position for each block, assuming image size is 512×512 pixels. This total
21-bit overhead would likely offset the possible compression improvement gained by
taking advantage of the image contents. So it is important to design a scheme to store
this information compactly. In the next two subsections, we discuss RMS tolerance
threshold selection and a compact storage scheme.

7.7.1   RMS Tolerance Selection

One of the main criteria for partitioning a parent range block into four equally
sized child range blocks lies in the RMS tolerance value. Although a large tolerance
value will lead to a high compression ratio, the quality of the decoded image will
be low. Conversely, a small tolerance value will ensure that the decoded image is of
high quality but the compression ratio will be compromised. Thus the selection of
a suitable tolerance threshold plays a major role in the achievable compression ratio
and decoded image quality.

PSNR is used to determine the decoded image quality. Consider a range block of
size $R \times R$ pixels and pixel values range form 0 to 255,

$$PSNR = 10 \log_{10}[(255 \times 255 \times R \times R)/(d(f, g) \times d(f, g))]$$

where $d(f, g)$ is the RMS of the pixel difference between the range block and the
transformation of the corresponding matching domain block.

Reorganizing the above equation, we have

$$d(f, g) = \frac{255R}{10^{PSNR/20}}.$$

This function determines the maximum allowable RMS difference used in the matching process given the size of the range block and required PSNR. This method has the advantage that the user can control the quality of the decoded image by choosing the required PSNR. The encoder, in turn, will calculate the maximum allowable distortion for the range blocks of different sizes.

7.7.2 A Compact Storage Scheme

Our aim is to use as few bits as possible to store the information about the range block sizes and positions, so that high compression ratios can be achieved.

Fig. 7.7 shows a simplified quadtree. The root node corresponds to the entire image of size $2^n \times 2^n$ pixels. We call this layer level n. The root has four children, which are on level $n - 1$. The next level is called level $n - 2$, and so on.

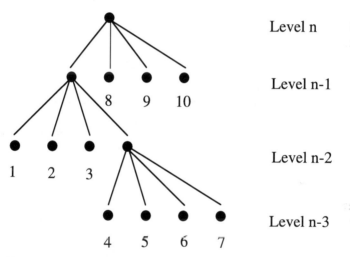

FIGURE 7.7
A simple quadtree.

The quadtree shows that the size and coordinates of four children can be calculated given the size and coordinates of the parent range block and the direction of children in the parent block. In other words, the range block size and coordinates are stored implicitly in the quadtree. The following recursive functions can be used for the decoder to calculate the sizes and coordinates of range blocks in a $2^n \times 2^n$ image given the level number and direction in the partition.

$$R(L) = 2^n \qquad\qquad \text{if } L = n$$
$$R(L) = R(L + 1)/2 \qquad \text{otherwise}$$

where $R(L)$ is the size of a range block at level L of the quadtree,

$$x(L, D) = 0, y(L, D) = 0$$
$$\text{if} \quad L = n$$
$$x(L, D) = x(L + 1, PD), y(L, D) = y(L + 1, PD) + R(L)$$
$$\text{if} \quad L \neq n \quad \text{and} \quad D = NW$$
$$x(L, D) = x(L + 1, PD) + R(L), y(L, D) = y(L + 1, PD) + R(L)$$
$$\text{if} \quad L \neq n \quad \text{and} \quad D = NE$$
$$x(L, D) = x(L + 1, PD), y(L, D) = y(L + 1, PD)$$
$$\text{if} \quad L \neq n \quad \text{and} \quad D = SW$$
$$x(L, D) = x(L + 1, PD) + R(L), y(L, D) = y(L + 1, PD)$$
$$\text{if} \quad L \neq n \quad \text{and} \quad D = SE$$

where $x(L, D)$ and $y(L, D)$ are the x and y coordinates of a range block at level L, D is the direction of the range block with respect to its parent range block, and PD is the direction of the parent block with respect to the grandparent range block.

From the above discussion, it is clear that if we know the quadtree level number of each block and store the compressed information of blocks in a certain order, the decoder will be able to work out the block sizes and positions. Block information is stored in depth-first order. Take quadtree in Fig. 7.7 as an example; we store range block (node) information in the order as indicated by the number under each node. For each range block, we can simply store the level number followed by the information presented in Table 7.1.

A close look reveals that it may be redundant to store the level number for each block because there is a good chance that several consecutive blocks share the same level number. To eliminate this redundancy, we reserve a bit (called a *run bit*) for each block to indicate whether this block is at the same level as the previous block. If the block is at the same level as the previous block, this bit is set to 1 and no level number is stored. Otherwise, this bit is set to 0 and the level number is stored following it.

Experimental results show that for an image of $2^n \times 2^n$ pixels, the useful maximum range block size is $2^{n-2} \times 2^{n-2}$ pixels; the useful maximum range block size is one sixteenth of the original image size [9]. The minimum block size is 4×4 pixels. So the total number of quadtree levels is $n - 3$. The number of bits required to store this level information is $\log_2(n - 3)$ rounded to the next integer.

To summarize, the compressed file contains a list of compressed block information stored in depth-first order. For each block, we store a run bit, followed by transformation information (if run bit is set) or by level information and transformation information (if the run bit is off). The format of transformation information is the same as that used for the fixed-size partitioning method discussed in the previous section (Table 7.1).

7.7.3   Experimental Results

Tables 7.3 and 7.4 show the results of experiments carried out on the image "Lena" of 512×512 and image "Lena" of 256×256 pixels, respectively [9]. The maximum range block size is set to 128×128 pixels while the minimum is set to 4×4 pixels. It can be seen that more large range blocks are used when the required decoded image quality is low, and few large blocks are used when the required decoded image quality is high. This is because when the required decoded image quality increases, the distortion thresholds used for various range blocks decrease. As a result, large range blocks are not able to encode the region to within the acceptable threshold, and they are broken down further by quadtree partition. Fig. 7.8 shows the compressed image of 0.44 bpp using the quadtree partitioning method with PSNR of 30.30 dB.

FIGURE 7.8

Compressed image of 512×512 pixels using QPIFS at 0.44 bpp with PSNR of 30.3 dB. Reproduced by Special Permission of *Playboy* magazine. Copyright ©1972, 2000 by Playboy.

Table 7.3   Test Results on "Lena" 512×512 Using QPIFS

Number of range Blocks						Compression	Decoded image quality
128×128	64×64	32×32	16×16	8×8	4×4	Ratio	(PSNR in dB)
0	23	90	196	345	220	84.8	23.56
0	11	107	223	558	920	39.5	25.90
0	6	89	295	676	1728	25.4	28.27
0	4	63	325	866	2664	18.0	30.30
0	2	49	287	1025	4044	13.0	32.17
0	0	42	215	1113	5804	9.8	33.37
0	0	13	188	1111	8100	7.4	34.27

Table 7.4   Test Results on "Lena" 256 × 256 Using QPIFS

Number of range Blocks					Compression	Decoded image quality
64 × 64	32 × 32	16 × 16	8 × 8	4 × 4	Ratio	(PSNR in dB)
0	25	90	196	272	33.0	23.82
0	11	109	255	628	18.9	26.50
0	7	88	320	960	13.7	28.76
0	4	72	353	1276	11.0	30.39
0	3	52	349	1676	9.0	31.81
0	1	48	283	2132	7.6	32.42
0	0	37	237	2556	6.6	32.73

By comparing the results obtained using fixed-size partitioning (Table 7.2) and those of quadtree partitioning (Tables 7.3 and 7.4), we can make the following observations:

1. When the required decoded image quality is low (below 25 dB), the fixed size partitioning provides a better compression ratio. For example, compressing image "Lena" of 256 × 256 pixels using range block size 16 × 16 pixels, a compression ratio of 88.6 is achieved, with a corresponding decoded image quality of 23.08 dB. In comparison, the same image compressed using quadtree partitioning gives a compression ratio of 33.0, although the quality is slightly higher at 23.82 dB. This is because the overhead information of range block size and position (on average about 2 bits for each block) is required for quadtree partitioning. Also, at low required decoded image quality, the image can be encoded using larger range blocks. In other words, the use of range blocks of 16 × 16 pixels in fixed-size partitioning is sufficient to generate a decoded image quality of 23 dB. This fact is evidenced in the quadtree partitioning method, in which approximately 75% of the image is encoded by range blocks of size 16 × 16 pixels or larger. Although only 25% of the image is encoded by smaller range blocks, the number of smaller ranger blocks used (total of 468, the sum of range blocks of 8 × 8 and 4 × 4) is much higher than the number of blocks of 16 × 16 or larger (total of 115, the sum of range blocks of 32 × 32 and 16 × 16). This is why the compression ratio decreases so much, although the decoded quality is increased by 0.8 dB due to usage of these smaller range blocks.

2. When the required decoded image quality is high, 30 dB and above, the quadtree partitioning method gives a better compression ratio and quality compared to the fixed-size partitioning. For example, we achieve a compression ratio of 4.8 with decoded image quality of 32.99 dB when image "Lena" of 256 × 256 pixels is compressed with fixed-size range blocks of 4 × 4 pixels. In comparison, the quadtree method is able to achieve a compression ratio of 6.6 with similar decoded image quality of 32.73 dB. This demonstrates the strength of the

quadtree partitioning method as it is able to encode certain regions with large range blocks, thus reducing the total number of range blocks to 2830. In fixed size partitioning, a total of 4096 range blocks of 4 × 4 pixels are used.

3. The results also show the flexibility of the quadtree partitioning method. The user is able to select the required decoded image quality. The encoder is then able to calculate the respective distortion tolerances for range blocks of different sizes. As a result, the method is able to use a combination of range blocks of different sizes to reproduce the decoded image with required quality. Fixed size partitioning does not have this flexibility. The achievable decoded image quality is determined by the range block size used in the encoding process. For example, in Table 7.2, when 4 × 4 range blocks are used, the resulting decoded image quality is 34.98 dB. When 8 × 8 range blocks are used, the resulting decoded image quality is 29.65 dB. As it is difficult to use range block sizes between 4 × 4 and 8 × 8, the fixed-size partitioning will not be able to produce a decoded image quality between 34.98 dB and 29.65 dB with the highest possible compression ratio.

4. The larger the image size, the higher the achievable compression ratio because larger images normally contain more spatial redundancy.

7.8 Image Coding by Exploiting Scalability of Fractals

Fractals are scalable in the sense that they have fine details at any scale. Based on this property, many researchers [4, 6, 7] have indicated that a fractal-encoded image can be decoded to any size. But this is true only to a certain extent with natural image compression. The original image is not a fractal, but the reconstructed image from PIFS is a fractal. Although we can display fractals at any size without losing fine details, it will not be very close to the original image if we enlarge it too many times.

Image coding using PIFS is time consuming. The complexity of computation is $O(n^4)$, assuming the image size is $n \times n$ and fixed-size partition is used. Thus it takes much longer to compress a larger image. This problem leads to the development of a scheme to reduce required compression time by using the scalability of fractals [9].

The basic idea is as follows. Given an image to compress, we reduce the original image size. Then we find the PIFS for this reduced image. During decoding, we use the scalability property to display the decoded image in its original size. As long as the decoded image has acceptable quality compared to the original one, we have achieved image compression using much less time and possibly with a higher compression ratio because a smaller image is coded. The important issues are how to reduce the original image size and how to decode the PIFS found on the reduced image to the original size.

7.8.1   Image Spatial Sub-Sampling

To reduce the original image, we use basic spatial subsampling techniques. Given an original image of $2^n \times 2^n$. If we want to reduce it to $2^{n-m} \times 2^{n-m}$, we represent each $2^m \times 2^m$ square block using one pixel. The value of the pixel is the average of these $2^m \times 2^m$ pixel values. For example, if we want to reduce the image size to one quarter of the original, we use one pixel to represent each square block of four pixels.

7.8.2   Decoding to a Larger Image

Decoding a compressed image of size $I \times I$ pixels represented by PIFS to $2^m I \times 2^m I$ pixels, where m is a nonnegative integer, involves decoding all the range blocks to size $2^m R \times 2^m R$ pixels instead of $R \times R$. The same decoding algorithm described in Section 7.6 is used, but special consideration must be taken to change block sizes and coordinates as shown below.

For a range block of size $R \times R$ with lower left corner at (x, y) that is encoded using absorption, to decode it to size $2^m R \times 2^m R$, the decoder will simply replace the block of $2^m R \times 2^m R$ pixels defined by lower left corner at $(2^m x, 2^m y)$ with the absorption factor.

To decode a range block with lower left corner at (x, y) that has been encoded by a contractive transformation, the decoder must be able to extract the matching domain block from the enlarged image. Thus the coordinates of the domain block (x_d, y_d) indicated in the compressed file must be changed to $(2^m x_d, 2^m y_d)$. A block of $2^m 2R \times 2^m 2R$ pixels with lower left corner at $(2^m x_d, 2^m y_d)$ is obtained from enlarged image. The appropriate contractive affine transformation is then applied to the block, to obtain the $2^m R \times 2^m R$ range block.

7.8.3   Experimental Results

Given the image "Lena" of 512×512 pixels, we first reduce it to 256×256 pixels. We find QPIFS for the reduced images. During decoding, we decode the QPIFS to image size of 512×512 pixels using the scalability property. Table 7.5 shows the experimental results obtained when images are compressed at 256×256 and then decoded to 512×512 using fractal scalability. Note that, in the table, both compression ratio and decoded image quality are calculated relative to the original image of 512×512 pixels. Table 7.6 shows results obtained when images are compressed at 512×512 pixels and decoded at the same size. Note that the results shown were obtained on a slow processor (SGI Indigo with a 50 MHz IP20 processor). With a faster processor, the time shown should be much shorter. Fig. 7.9 shows a reconstructed image using fractal scalability.

From the table we can make the following observations:

1. At similar decoded image qualities, coding time required to compress images of 256×256 is much less compared with compressing 512×512 directly without reducing its size first. Although not shown in the table, the decompression times required for decoding are similar for all cases.

FIGURE 7.9
Image "Lena" of 512×512 **pixels reconstructed from QPIFS found in images of** 256×256 **pixels using fractal scaling, at 0.22 bpp with PSNR of 28.77 dB. Reproduced by Special Permission of** *Playboy* **magazine. Copyright ©1972, 2000 by Playboy.**

Table 7.5 Test Results Using Combination of Subsampling and Scaling Technique

QPIFS coded at	Compressing time (s)	Decoded to	Compression ratio	Decoded image quality (PSNR in dB)
256×256	777	512×512	131.5	23.10
256×256	1392	512×512	75.4	25.26
256×256	1957	512×512	54.7	26.91
256×256	2444	512×512	44.0	27.96
256×256	3006	512×512	35.9	28.77
256×256	9538	512×512	30.0	29.08
256×256	11008	512×512	26.4	29.24

2. At similar decoded image qualities, the compression ratio achieved using scaling is much higher than that using direct encoding and decoding at the same size. For example, we achieved a compression ratio of 35.9 with decoded image quality of 28.77 dB using the scaling property, compared with a compression ratio of 25.4 with image quality of 28.27 dB achieved with direct encoding and decoding.

3. Since our purpose is to obtain a decoded image at the original image size, in the case of 512×512 pixels, the decoded image quality is calculated relative to the original image of 512×512 pixels. There is a quality degradation when decoding QPIFS found in the image 256×256 pixels to an image of 512×512 pixels. For example, in the last row of Table 7.5, the QPIFS found

Table 7.6   Test Results Using QPIFS Directly on a 512×512 "Lena" Image

Compressed at	Compressing time (s)	Compression ratio	Decoded image quality (PSNR in dB)
512×512	4645	84.8	23.56
512×512	10235	39.5	25.90
512×512	16075	25.4	28.27
512×512	22789	18.0	30.30
512×512	31797	13.0	32.17
512×512	42639	9.8	33.37
512×512	56415	7.4	234.27

for the image of 256×256 pixels should be able to generate an image of 256×256 pixels at 32.73 dB (relative to the uncompressed image of 256×256 pixels). However, when it is decoded to 512×512 pixels, the PSNR drops to 29.24 (relative to the original image of 512×512 pixels). This proves that scalability of fractals can be used only to a certain extent in natural image compression. Enlarging too often will result in unacceptable decoded image quality.

7.9   Video Sequence Compression using Quadtree PIFS

There are high temporal correlations among images in a video sequence. To achieve high video compression performance, these temporal redundancies must be removed or reduced. In this section, we describe a technique to remove temporal redundancy using quadtree partitioning [9].

The basic idea of still image compression using PIFS is to find similarities among parts of an image. We can borrow this idea to compress video sequences. However, instead of finding similarities among parts of an image, we find similarities among neighboring images in a video sequence. In this case, quadtree partitioning will be very efficient because there are large areas that are unchanged between consecutive frames; thus many large range blocks can be used. Based on whether or not there is a change in a block of pixels relative to the corresponding block in the previous image and the type of changes (translation or rotation, etc.), we identify different types of blocks in the image and code them differently to achieve better compression performance.

7.9.1   Definitions of Types of Range Blocks

If a block in the current frame is identical or very similar to the corresponding block in the previous frame, we call this block a *type one block*. If a block in the current frame is a translation of a block in the previous frame, we call it a *type two block*.

If a block in the current frame is an affine transformation of a block in the previous frame, we call it a *type three block*.

Type One Range Blocks

In a typical video sequence, a large portion of the current frame is similar to the previous frame except for areas where moving objects are located. We can use type one blocks to code this unchanged portion. Since the only information to be stored is the position of this block, a high compression ratio can be achieved. To identify a type one range block, a block in the current frame is compared with the corresponding range block in the previous frame. If the error between the two range blocks is below a preset threshold value, the block is deemed as a type one range block.

Type Two Range Blocks

For the area containing moving objects, if the movement of objects can be traced from one frame to the next, a high compression ratio can be achieved. This observation leads to the implementation of type two range blocks. Each type two range block in the current frame is produced by a translation of a same size block in the previous frame. A range block is identified as a type two block if the distortion between itself and a block obtained from the search region defined in the previous frame is smaller than a preset threshold.

The search region is defined based on the observation that objects generally do not move too far from one frame to the next. Thus the search region is a small area in the previous frame instead of the whole image, saving bits required to represent the x and y offsets between the range block and the matching block in the previous frame. A search range of M is used. The search region is defined as in Fig. 7.10. The size of the search region is $(2M + R) \times (2M + R)$, assuming the range block size is $R \times R$. When the range block is near the boundaries of the image, the search region is constrained by the boundaries. During the search process, the matching block will be searched exhaustively in the previous frame by changing the block position one pixel at a time within the search region. After the exhaustive search process, the block giving the minimum error between itself and the range block to be coded is deemed as a type two matching block if the distortion is below the preset distortion threshold.

Type Three Range Blocks

For the area that cannot be encoded using type one or type two blocks, type three blocks are used. Type three range blocks are encoded in a similar way as the range block in the still image as discussed in Sections 7.6 and 7.7. The only difference is that the matching domain blocks are searched in the previous frame. This is based on the observation that parts of the current frame may possibly be parts in the previous frame after rotation, intensity change, etc. Thus, it would be easier to find matching blocks in the previous frame than in the current frame.

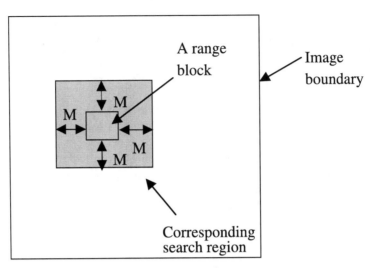

FIGURE 7.10
Definition of the search region.

Distortion Tolerance

The distortion tolerance for all three types of blocks is determined by the same function as described in Section 7.7.

7.9.2   Encoding and Decoding Processes

We treat the first image in a video sequence as a still image (intraframe) and code it using quadtree PIFS as described in Section 7.7. There are two ways to code the subsequent frames. One way is to code them relative to the previous original images. The advantage of this method is that the similarity between the current frame and the previous original image will be high, resulting in the use of more type one and type two blocks, thus achieving higher compression ratios. The disadvantage is that the decoder does not have the original images. It has to decode images based on the previous decoded image. Since we are using a lossy coding method, the decoded image is different from the original image. If we decode images based on previous decoded image, the differences will accumulate and eventually the decoded image quality will no longer be acceptable unless we take remedial measures. The other way to code the images is to code them relative to the previous decoded image. The advantage and disadvantage are the opposite of the first method. In the following, we describe the implementation of the first method. Measures are taken to prevent error accumulation to an unacceptable level.

To solve the error accumulation problem, we divide a video sequence into layers (Fig. 7.11). The highest layer is the video sequence itself. It is divided into a number of fixed length subsequences. The first image in each subsequence is intraframe coded. That is, it is treated as a still image and coded using quadtree PIFS. Each subsequence is further divided into a number of fixed-length groups of images or frames. The first image in each group of images is coded relative to the first image in the previous group of images. By doing so, the rate of quality degradation is reduced. The encoding process for all images in a group of images except the first is the same; they are coded relative to the immediately previous images. Therefore, except for the first image in a subsequence, all images are interframe coded.

In the following discussion, we use the general term *reference image* to describe the image based on which the current image is encoded. The reference image is the previous image if the current image is not the first in a group of images. Otherwise, the reference is the first image in the previous group of images.

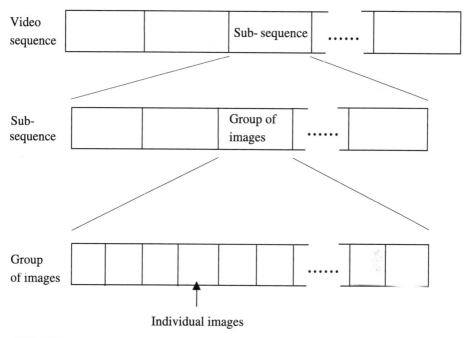

FIGURE 7.11
The hierarchy of a video sequence.

The reference image is used to search for type one and type two range blocks, and for matching domain blocks to encode type three range blocks. The quadtree partitioning method with extension to take care of the use of type one and type two range blocks is utilized to encode each image except the first image in a subsequence.

An image to be interframe coded is divided into 16 equal square range blocks. Each range block to be encoded is first tested to determine whether or not it can be coded using type one block relative to the reference image. It is encoded using type one block if the distortion is below the preset threshold. Otherwise, it is tested to determine if it can be encoded using type two block. If the current range block cannot be encoded using type one or type two blocks, a matching domain block is searched from the reference image. If the difference between the range block and a domain block, after it has undergone an appropriate transformation, is smaller than a preset threshold, the range block is encoded using type three block. Otherwise, the range block must be partitioned into four smaller range blocks using quadtree partitioning. For each smaller range block, the above encoding process is repeated until all range blocks are encoded.

During decoding, the quadtree used in the encoding process is rebuilt as described in Section 7.7. The range blocks encoded using type one or type two blocks are reconstructed from the reference image. The areas encoded using type three blocks are initialized with the contents at the corresponding location in the reference image and are reconstructed by applying stored PIFS. By initializing these areas with the contents of the reference image, very few iterations are required to obtain the converging image. Experiments show that in most cases only one iteration of applying the PIFS is required to achieve convergence to the final image.

7.9.3   Storage Requirements

Table 7.7 shows the storage requirements for different types of blocks. In an actual implementation, additional bits are required to store range block sizes and coordinates as described in Section 7.7. By using quadtree partitioning, we expect that unchanged areas will be coded using larger range blocks, leading to higher compression ratios.

Table 7.7   Storage Requirements for Different Types of Range Blocks

Type of encoding	Parameters	Number of bits
Type one	Identifier	4
Type two	Identifier	4
	Coordinates	$2\log_2(2M)$
Type three with absorption	Identifier	4
	Absorption	8
Type three with isometric	Identifier	4
	Scaling factor	3
	Shifting factor	8
	Domain coordinates	$2\log_2(I/R)$

7.9.4   Experimental Results

In the reported experiments, the size of a subsequence used is 100 images and the size of a group of images is 10. These numbers are chosen to achieve optimal compromise between compression ratios and effects of accumulated error.

The "salesman" video sequence of 90 frames is compressed using the algorithm described above. The compression ratio achieved is 57.9 with an average decoded image quality of 29.17 dB. Using software-only decoding, a decoding rate of 3 frames per second with image size of 256 × 256 pixels was achieved on an SGI Indigo workstation with 50 MHz IP20 processor. With optimization of the code and faster processors, real-time video decoding is possible.

7.9.5 Discussion

The interframe coding method discussed in this section is similar, to a certain extent, to the motion estimation and compensation techniques used in the MPEG standard [14]. The QPIFS-based method has two advantages. First, it uses quadtree partitioning instead of fixed-block size used in MPEG. This leads to better compression performance because nonmoving areas can be coded using larger blocks. Second, the QPIFS-based method not only estimates and compensates for translation but also considers object rotation, brightness changes, etc. by using affine transformations.

7.10 Other Fractal-Based Image Compression Techniques

The techniques discussed so far are all based on iterated function systems. In this section, we briefly describe two techniques that are not based on IFS but make use of other properties of fractals.

7.10.1 Segmentation-Based Coding Using Fractal Dimension

In most cases, images are meant to be viewed by the human eye. The human visual system (HVS) is not perfect, and it is less sensitive to certain frequencies than to others. The less sensitive components can be coded coarsely without much perceived quality loss. Therefore, if a coding system can take advantage of the properties of HVS, a high compression ratio can be achieved.

One such technique is segmentation-based image coding [10]. Images are segmented into homogeneous regions with similar features, and each region is coded using different techniques based on their visual importance. However, there are limitations in the traditional segmentation-based coding. The main limitation is due to the fact that the image is normally segmented into regions of constant intensity. In complicated texture areas, a trade-off must be made. Good representation of texture requires many small segments. In order to get low bit rates or high compression ratios, however, the number of segments should be small. This problem can be solved by segmenting images into texturally homogeneous regions with respect to the degree of roughness perceived by HVS.

A characteristic of a fractal is the fractal dimension that provides a good measure of perceptual roughness of texture, with increasing values in fractal dimension representing perceptually rougher texture. Techniques to calculate the fractal dimension can be found in Falconer [3] and Peleg, Naor, Hartley, and Avnir [11]. After the fractal dimension is calculated, an image is segmented into several texture classes according to the fractal dimension. After image segmentation, an efficient image coding technique is chosen to encode each texture class according to visual importance. It was reported that a compression ratio of 40 was achieved for gray scale images with good quality reconstructed images [10].

7.10.2   Yardstick Coding

This method is based on fractal geometry to measure the length of a curve using a yardstick of fixed-length (see Fig. 7.12). We want to measure the curve drawn in thin dotted lines using a yardstick. The thick lines are covered by the yardstick travelling along the curve. It is obvious that the shorter the yardstick, the closer the measurement result will be to the true length of the curve, and the better the curve will be covered by the yardstick travelling along the curve.

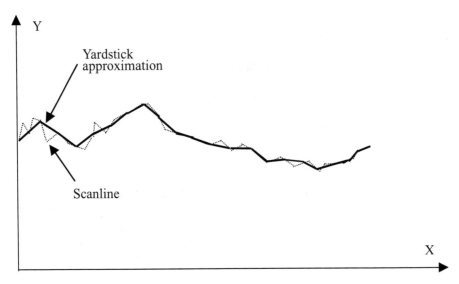

FIGURE 7.12
Coding (approximation) of a scan line.

Walach and Karnin [12] proposed the so called "yardstick travelling" mechanism for coding each line of pixels in an image. A scan line of pixels can be thought of as a curve: the x-coordinate is the pixel number along the line, and the y-coordinate is the intensity of pixels. The curve can be approximated with a set of straight lines covered by the running yardstick. These straight lines can be represented by points coinciding with the ends of the yardstick when it is running along the curve. Since the number of end points will be smaller than the pixel numbers when the length of the yardstick is appropriately chosen (normally between 8 and 24 pixels), compression can be achieved by storing these points instead of pixels [12, 13].

The yardstick method is similar to the traditional subsampling and interpolation method, both exploiting the correlation among neighbouring pixels.

7.11   Conclusions

This chapter has described a number of image and video compression techniques based on fractal properties. Techniques based on IFS and PIFS are most promising. The compression performance of PIFS-based techniques is similar to that of DCT-based techniques.

The PIFS-based compression technique has an advantage in that the compression achievable at a given signal to noise ratio scales with image size: higher compression ratios can be achieved with larger images. This shows the potential of this technique in applications involving large images. IFS-based fractal image coding is highly asymmetric in that significantly more processing is required for encoding than for decoding. It is highly suitable for information dissemination applications where images are encoded once and decoded many times. At the decoding site, no sophisticated hardware is needed to achieve high decoding speed.

IFS-based image/video compression techniques are potentially suitable for interactive multimedia applications where indexing, retrieving, and browsing of images are required. When an image is encoded into an IFS, one image or object is just one IFS. It is easy to index and search based on IFS.

Pointers to Further Reading and Available Software

This chapter has introduced basic concepts and techniques of fractal-based image and video compression. Many papers and much software are available online. Fisher maintains a Web site (`http://inls.ucsd.edu/y/Fractals/`) where one can find papers, books, software, and other resources. The University of Waterloo has an active research group working on fractal compression. Its home page (`http://links.uwaterloo.ca/`) has links to many papers and software. Institut fuer Informatik of Universitaet Freiburg, Germany, has an FTP site (`ftp://ftp.informatik.uni-freiburg.de/documents/papers/fractal/`) that contains many papers. One can find the latest products and developments in fractal compression and applications from the home page of Iterated Incorporated (`http://www.iterated.com/`).

References

[1] Mandelbrot, B.B., *The Fractal Geometry of Nature,* Freeman, San Francisco, 1982.

[2] Barnsley, M.F., *Fractals Everywhere,* Academic Press, Boston, 1988.

[3] Falconer, K., *Fractal Geometry — Mathematical Foundations and Applications,* John Wiley & Sons, New York, 1990.

[4] Barnsley, M.F. and Sloan, A.D., A better way to compress images, *Byte,* 215–223, 1988.

[5] Jacquin, A.E., *A Fractal Theory of Iterated Markov Operators with Applications to Digital Image Coding,* Ph.D. thesis, Georgia Institute of Technology, August 1989.

[6] Fisher, Y., Fractal image compression, SIGGRAPH'92, course notes.

[7] Jacquin, A.E., A novel fractal block-coding technique for digital images, ICASSP'90, 2225–2228, Albuquerque, NM, 1990.

[8] Jacquin, A.E., Fractal image coding based on a theory of iterated contractive image transformations, *SPIE,* vol. 1360, *Visual Communications and Image Processing,* 227–239, 1990.

[9] Lu, G. and Yew, T.L., Applications of partitioned iterated function systems in image and video compression, *J. Visual Communication and Image Representation,* 7(2), 144–154, 1996.

[10] Jang, J. and Rajala, S.A., Segmentation-based image coding using fractals and the human visual system, ICASSP'90, 1957–1960, Albuquerque, NM, 1990.

[11] Peleg, S., Naor, J., Hartley, R., and Avnir, D., Multiple resolution texture analysis and classification, *IEEE Transactions on Pattern Analysis and Machine Intelligence,* 6(4), 518–523, 1984.

[12] Walach, E. and Karnin, E., A fractal based approach to image compression, ICASSP'86, 529–532.

[13] Zhang, N. and Yan, H., Hybrid image compression method based on fractal geometry, *Electronics Letters,* 27(5), 406–408, 1991.

[14] MPEG home page, `http://drogo.cselt.stet.it/mpeg/`.

[15] Gersho, A. and Gray, R.M., *Vector Quantization and Signal Compression,* Kluwer Academic Publishers, Boston, 1992.

[16] Forte, B. and Vrscay, E.R., Theory of generalized fractal transforms, in *Fractal Image Encoding and Analysis,* Fisher, Y., Ed., Springer Verlag, Heidelberg, 1998.

[17] Mendivil, F. and Vrscay, E.R., Correspondence between fractal-wavelet transforms and iterated function systems with gray-level maps, in *Fractals in Engineering: From Theory to Industrial Applications,* Levy Vehel, J., Lutton, E., and Tricot, C., Eds., Springer Verlag, London, 1997.

[18] Vrscay, E.R., A generalized class of fractal-wavelet transforms for image representation and compression, *Canadian J. of Electrical and Computer Engineering,* 23(1–2), 69–83, 1998. (Special issue on Visual Computing and Communications.)

[19] Saupe, D. and Vrscay, E.R., Can one break the "collage barrier" in fractal image coding?, *Fractals in Engineering Conference,* June 14–15, 1999, Delft University, The Netherlands.

[20] Zhao, Y. and Yuan, B., A new affine transformation: its theory and application to image coding, *IEEE Transactions on Circuits and Systems for Video Technology,* 8(3), 1998.

Chapter 8

Compression of Wavelet Transform Coefficients

Xiaolin Wu

University of Western Ontario

8.1 Introduction

Mathematical transforms are widely used in signal compression, particularly in compression of sensory data such as audio, image, and video. Although sensory signals are typically sampled and presented to users in the spatial/time domain, a direct signal representation in the spatial/time domain creates a huge volume of data with excessive redundancy. Clearly, signals in original sample form are unsuitable for transmission or storage. Transform coding is a proven paradigm for signal compression. In transform coding, signal samples are mapped from spatial/time domain into another space, typically a frequency or joint time-frequency domain in which statistical and subjective redundancies in the samples can be better understood, exploited, and removed. Transformed samples are thus more amenable to compression. This paradigm of transform-based signal compression is exemplified by the current and commercially successful industrial standards for image compression (JPEG standard [19]) and video compression (MPEG standards [4, 5]) [21]. A schematic description of a typical transform coding system is given in Fig. 8.1. The compression (encoding) process is completed in three major steps: transform of signal samples, quantization of transform coefficients, and entropy coding of quantized coefficients. The decompression (decoding) process is a reverse of the compression process.

The JPEG and MPEG standards use discrete cosine transform (DCT) in the transform step of the compression system. The acronyms JPEG and MPEG stand for the Joint Photographic Experts Group and the Moving Picture Experts Group. The two groups consist of members from both the International Standards Organization (ISO) and the International Telecommunications Union (ITU). They are charged respectively with the missions of developing international standards for the coded representation of compressed still images, and of compressed moving pictures and associated

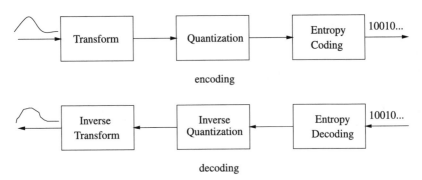

FIGURE 8.1
Schematic description of a typical transform coding system.

audio. Their efforts are instrumental for the prevalence of digital visual communications in multimedia and Internet applications. Due to the popularity of JPEG and MPEG compression standards and products, the DCT-based coding system is now considered a matured and effective technology for image and video compression. In 1988 when the JPEG members evaluated various image compression schemes and decided on the JPEG standard, the DCT-based image codecs offered the best compromise between compression performance, computational complexity (hardware complexity in particular), and coder flexibility, among other competing image compression technologies at that time, specifically vector quantization (VQ) [14] and DPCM (differential pulse coding modulation) coding.

Since the standardization of DCT-based compression technology, the past few years have seen rapidly increasing sophistication and maturity of wavelet-based image compression methods. Wavelet-based image codecs have so far delivered the best lossy compression performance in both peak signal to noise ratio (PSNR) and visual quality, over bit rates from 0.05 bits/pixel (summary quality for browsing) to 2.00 bits/pixel (visually indistinguishable from the original). During the same period, research on VQ compression and fractal compression has also advanced. But in image compression, neither VQ nor fractal compression methodology has matched the rate-distortion performance of wavelet-based image codecs at the time of this writing. Indeed, the recent success of wavelet transform in image compression has reinforced the dominance of widely practiced transform coding paradigm for signal compression. Only in the realm of lossless image compression, adaptive predictive coding has slightly (about 3%) higher compression ratio than lossless transform coding such as reversible integer wavelet codecs [34]. But this small advantage of predictive lossless coding becomes even more marginal in the presence of other unique features of wavelet lossless codecs, on which we elaborate later.

Within the transform coding family, discrete wavelet transform is threatening to unseat DCT as the transform of choice, at least for image compression applications. The current state-of-the-art wavelet image codecs significantly outperform the existing DCT-based JPEG standard in PSNR measure and subjective image quality, particularly for low bit rates at which the block effects of DCT are noticeable [25, 28, 30, 29, 36, 37]. Being encouraged by the improvements brought on by wavelet-based image compression techniques over DCT, and prompted by increasing acceptance of wavelet compression technology by industry, the JPEG committee has developed a new wavelet-based still image compression standard called JPEG 2000 [2]. Also, in 1993 the FBI chose a wavelet-based image codec to be the standard for fingerprint image compression [1].

The superior compression performance of wavelet-based image coding systems over their DCT-based counterpart might suggest that the improved performance was primarily made by replacing DCT with wavelet transform, and hence the choice of transform would matter the most to coding efficiency. However, in strict technical terms, all existing transforms used in signal compression by themselves do not lead to any data reduction. Both DCT and dyadic wavelet transforms, the two most widely used types of transforms in image and video compression, generate as many coefficients as the number of samples. Furthermore, while the original sample values of digital signals are integers, the transform coefficients are nonintegers. Therefore, without efficient coding of transform coefficients, a transform not only cannot compress but can even expand the data. The main benefit of transform to data compression is from its property of energy packing. A suitable transform can transfer the majority of signal energy into a few transform coefficients, resulting in a large number of zero and near-zero coefficients. In other words, the probability distribution of transform coefficients is much more biased than that of original samples. The more biased the distribution, the easier it is to compress signals by entropy coding. Despite the well-accepted folklore that lossy signal compression is better done via transform coding, it is the process of entropy coding that actually achieves data reduction. Informally, entropy coding refers to a family of coding techniques that uses shorter codewords for more probable symbols (smaller transform coefficients), and longer codewords for less probable symbols (larger transform coefficients). An optimal variable length code can achieve an average code length that approaches the information theoretic lower bound called entropy, hence the term entropy coding. Entropy coding is also referred to as noiseless or lossless coding since the coding process is perfectly reversible. It is a key machinery of information theory, a field fathered by Shannon [27] half a century ago and that has guided data compression engineering ever since. For background and rigorous treatment of entropy coding, we refer readers to textbooks such as Cover and Thomas [10].

In terms of energy packing capability, the principal component transform (also known as Karhunen-Loève transform [15]) is optimal in the sense that it distributes the largest amount of signal energy into the direction of the eigenvector of the largest eigenvalue (the direction of largest sample variance), and the second largest amount of signal energy into the second largest eigenvector direction, and so on. Therefore, if one is to choose only k coefficients to best approximate the original signal in L_2 metric, then the optimal choice will be the k coefficients corresponding to the eigenvectors of the k largest eigenvalues. DCT has been shown to be very close to the principal component transform when applied to the first order stationary Markov process [22]. This justifies the wide use of DCT in data compression. The energy packing capability of wavelet transform was studied by DeVore, Jawerth, and Lucier [11] who showed that wavelet bases are optimal among all possible basis functions in minmax nonlinear approximation obtained by retaining the k largest coefficients and discarding the remaining. Both DCT and wavelet transforms possess some good properties in terms of energy packing.

Wavelet transforms have two additional advantages over DCT that are important for coefficient compression. The first is the multiresolution representation of the signal by wavelet decomposition that greatly facilitates subband coding, a notion that existed long before the popularity of wavelets [32]. Fig. 8.2 shows an image pyramid associated with wavelet decomposition. It can be seen from the figure that wavelet transform preserves to some extent spatial signal features in subbands of different scales and creates self-similarities between the subbands of the same spatial orientation. This fractal-like structure reveals sample dependencies across scales to the benefit of statistical context modeling and coding of wavelet coefficients. In fact, the well-known zerotree techniques precisely exploit the self-similarity of regions of zero and near-zero coefficients. The second advantage of wavelet transform is that it reaches a good compromise between frequency and time resolutions of the signal. From the perspective of energy packing, statistically short-term signal constructs such as image edges, or transients in signal processing terminology, have much higher energy concentration in time domain; hence they can be modeled and coded far more efficiently in the time domain than in the frequency domain. However, the exact opposite is true for long-term signal constructs such as smooth shades and regular textures in images. Wavelet transforms are superior to DCT in that their basis functions offer good frequency resolution in the lower frequency range, and at the same time they yield good time resolution at a higher frequency range (see the well-preserved edge information in the three highest subbands in Fig. 8.2).

However, neither the multiresolution property nor the frequency-time character-istics of wavelets suffices for signal compression. Whether and how much signal compression can benefit from the good properties of wavelets largely depends on statistical context modeling (implicit or explicit) and entropy coding of wavelet co-efficients. The difference in rate-distortion performance between the DCT-based JPEG codec and wavelet-based image codecs is mostly caused by the differences in entropy coding of transform coefficients between the two methods. Indeed, be-fore Shapiro's zerotree technique (EZW) in 1993 [28], a landmark work on wavelet coefficient coding, wavelet transforms had not won over DCT in rate-distortion per-formance. More recently, particularly during the ongoing JPEG 2000 standardization process, further advances have been made in statistical context modeling and adaptive entropy coding of wavelet coefficients. The modern wavelet coefficient coding tech-niques [25, 36, 29, 41, 37] significantly outperform the pioneer EZW coder for any given wavelet transform. The new techniques have better rate-distortion performance over EZW because they overcome a weakness of zerotree. That is, while being an

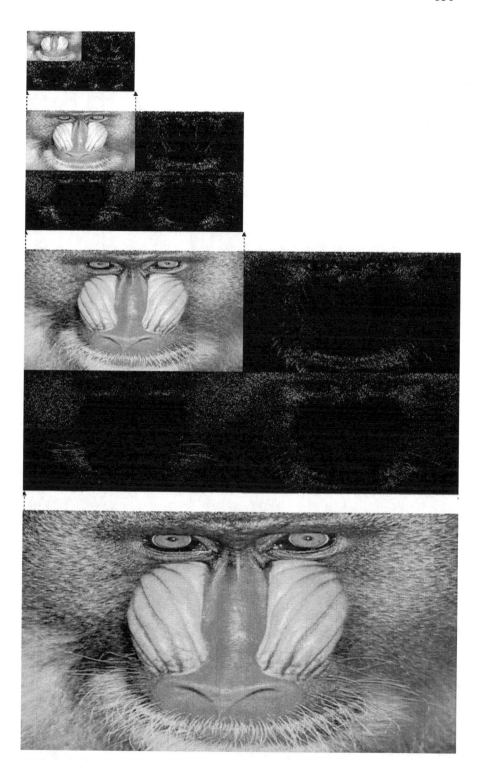

FIGURE 8.2
Dyadic wavelet decomposition of a test image.

effective technique to remove data redundancy in the form of a long-term trend, ze-rotree is less efficient to describe short-term signal constructs than the more advanced statistical modeling techniques discussed later in this chapter.

In summary, it is the increasing sophistication of coefficient coding, not the transforms alone, that contributes the most to the success enjoyed by wavelet image compression technology. Compression of coefficients is perhaps the most critical issue for any transform-based signal compression system. This chapter is dedicated to the problem of compression of transform coefficients. In order to make our discussions concrete and lucid, we focus on compression of wavelet coefficients in the setting of image coding. The general principles and techniques of this chapter, however, are applicable to compression of other transform coefficients and also effective with other types of signals, such as video and audio.

The structure of this chapter is as follows. Section 8.2 discusses the problem of compressing transform coefficients in wavelet-based image compression systems. Specifically, we introduce the popular approach of embedded bit-plane coding of quantized wavelet coefficients. Section 8.3 formulates the problem of statistical context modeling of wavelet coefficient and explains why this is the single most important issue that determines the compression performance. Since wavelet transforms cannot achieve total decorrelation between the signal samples, particularly when sample correlation is nonlinear, high-order statistical dependencies exist between wavelet coefficients. Therefore, optimum compression performance can be made possible only by high-order statistical context modeling of wavelet coefficients. However, if not treated with care, the number of Markov conditioning states can grow exponentially in the order of the model. This leads to a so-called problem of context dilution, addressed in Section 8.4. The challenge is how to maintain a modest number of conditioning states while still making high-order statistics available to aid entropy coding. Section 8.5 discusses how to discriminatingly choose modeling contexts in wavelet subbands as a means to control model cost. Section 8.6 introduces the process of context quantization to reduce drastically the number of conditioning states. The essence of context quantization is to merge different conditioning states that have similar symbol probability distributions. The subject is further pursued in Section 8.7, which investigates how to optimize context quantization for minimum code length. We borrow a common strategy of nonparametric multivariate statistical analysis to overcome high model cost: data projection in the direction of statistical dominance. Specifically, Fisher's linear discriminant [12] is used to guide context quantization. Section 8.8 presents a context quantizer design algorithm via dynamic programming that can minimize conditional entropy for a given number of conditioning states. Section 8.9 turns to the computational aspect of context modeling. Efficient algorithm techniques are developed to compute modeling contexts. We demonstrate that the time complexity of forming a modeling context is $O(1)$, independent of the order of the model, and thus high-order statistical context modeling is made computationally feasible. The chapter concludes with experimental results that provide convincing empirical evidence for the importance and effectiveness of context modeling and conditional entropy coding of wavelet coefficients in practical compression systems.

MSB   1   0   0   0   1
        0   1   0   1   0
        0   1   0   0   1   ──────▶ 1 0 0 0 1, 0 1 0 1 0, 0 1 0 0 1, 1 0 0
LSB   1   0   0   1   1

FIGURE 8.3
Embedded bit stream of coefficients.

8.2   Embedded Coefficient Coding

Like most transform coding systems a typical wavelet-based signal compression system consists of three cascaded modules, as depicted by Fig. 8.1, first wavelet transform, followed by quantization of wavelet coefficients, and finally entropy coding of quantized coefficients. To improve coding efficiency, one can perform adaptive wavelet transforms for better energy packing or optimal quantization for rate-distortion considerations. But most of the coding gains are usually made by conditional entropy coding of wavelet coefficients coupled with universal statistical context modeling. This is because transforms can remove only linear correlations between samples, whereas universal statistical context modeling can discover and remove more complex types of sample dependencies. In the ongoing JPEG 2000 standardization process, entropy coding of wavelet coefficients is by far the hottest subject being studied and debated by participating parties. It has been established empirically that the best rate-distortion results can be obtained by adaptive entropy coding of coefficients even with a fixed wavelet transform and uniform scalar coefficient quantization.

In order to focus this chapter on the last system component of coefficient entropy coding, in the following discussions we assume that the standard dyadic wavelet transform due to Mallat [17] is used in the transform module, and that simple uniform scalar quantization of wavelet coefficients is used in the quantization module. The input of entropy coder is the quantization indices of the coefficient magnitudes plus the signs of the coefficients. The quantized coefficients of dyadic transform are thus signed integers arranged in a two-dimensional layout of subbands as in Fig. 8.2. Image compression is finally achieved by lossless entropy coding of quantized coefficients.

A breakthrough of wavelet-based image compression technology is a coding scheme called embedded bit plane coding that was pioneered by Shapiro [28] in 1993 and then improved very rapidly by many other authors [25, 30, 43, 36, 37]. The idea is simple. Instead of coding all coefficients in one pass, and coding each coefficient once, we scan the coefficients in multiple passes, one bit plane per pass, from the most to the least significant bit, as illustrated in Fig. 8.3. Within a bit plane, the order of traversing wavelet coefficients in a two-dimensional subband layout can be arbitrary. A common traversal order is from the lowest frequency or the most coarse subband to the highest frequency or the most detailed subband, as depicted by Fig. 8.4. The binary sequence generated by such a traversal is called the embedded bit stream. An important property of the embedded bit stream is its scalability in

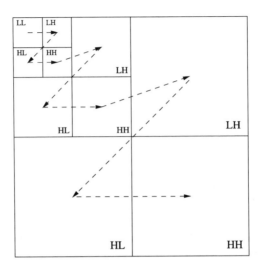

FIGURE 8.4
A traversal of subbands within a bit plane.

both spatial resolution and sample fidelity. Truncating an embedded bit stream at any point means approximating all wavelet coefficients at a certain precision; hence, the truncated bit stream can reconstruct the image at a corresponding fidelity — the longer the bit stream being used in the reconstruction, the higher the fidelity. The effect of successive refinement of a coded image via progressive transmission of an embedded bit stream is illustrated in Fig. 8.5. Scalable image and video compression allows transmission and distribution of the same source material at different quality levels to meet different bandwidth and storage capacity requirements, and the ability to do so with a single unified code stream. This feature is highly desirable in many applications, such as Internet, multimedia, medical imaging, prepress imaging, and image databases. With scalable coding one needs only to archive one master copy of the material in the database to support applications at different quality levels and under different constraints — from fast browsing to professional high quality reproduction — instead of maintaining multiple copies of the same materials at different bit rates for different bandwidths and quality trade-offs.

Scalable wavelet coding can also unify lossy and lossless compressions. If reversible integer wavelet transforms [8] are used, all coefficients are integers in the first place. No coefficient quantization is necessary; hence, there will be no quantization errors. In this case an embedded bit stream can eventually achieve perfect lossless decompression if every bit is received, while any truncation of the bit stream corresponds to a lossy decompression. The use of reversible integer wavelets for lossy to lossless scalable image compression was proposed by Zandi et al. [43] and Said and Pearlman [26]. This approach has very recently been extended to lossy and lossless compression of image sequences such as three-dimensional medical data, multi/hyperspectral remote sensing data, and video [18, 42].

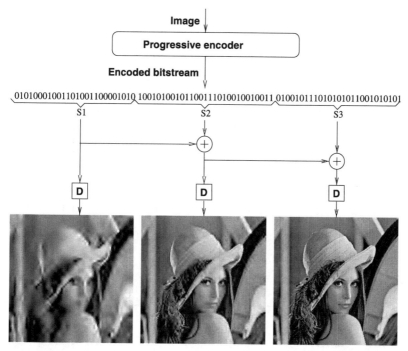

FIGURE 8.5

Progressive image reconstruction via scalable embedded code stream. Reproduced by Special Permission of *Playboy* magazine. Copyright ©1972, 2000 by Playboy.

The first published work on embedded bit plane coding of wavelet coefficients was Shapiro's zerotree algorithm [28]. Shapiro developed his embedded zerotree of wavelets (EZW) by observing that large blocks of zero coefficients exist in high frequency subbands and at bit planes of high significance. Furthermore, a block of zero coefficients statistically tends to reside in the same spatial location across different scales. If we consider a coefficient at a coarser subband as parent, and the four coefficients corresponding to the spatial location of the parent coefficient at the next finer subband as children, then coefficients of a dyadic wavelet transform can be naturally organized into quadtree data structures, as shown in Fig. 8.6. The conditional probability for all four children to be 0 given that the parent is 0 is much higher than given that the parent is 1. This statistical inheritance of 0 across different scales tends to form quadtrees of all 0 nodes, with their roots at upper levels of the multiresolution hierarchy and their leaves at the bottom level. Therefore, one can code a large number of 0 coefficients very compactly with a special code symbol for such zerotrees. This technique is very much like the zigzag technique of the existing DCT-based JPEG standard for coding long runs of 0 coefficients. In essence, the EZW technique compresses wavelet coefficients using a prior statistical model, i.e., assuming that

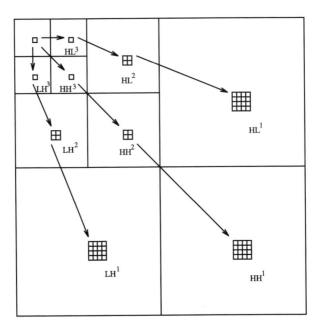

FIGURE 8.6
Coefficient quadtrees across different scales. Zerotrees are those quadtrees whose nodes are all 0.

zero and near-zero coefficients are clustered in both spatial and frequency domains, and that the regions of low sample energy are self-similar across different scales. The rate-distortion performance of the EZW technique was improved by a variant of the zerotree coder called SPIHT, proposed by Said and Pearlman [25]. Unlike the EZW algorithm that forms and codes zerotrees in a fixed spatial scanning order, SPIHT codes the zerotrees in an order that is beneficial to rate-distortion performance; those trees that are likely to generate higher reduction in distortion are coded first. The better performance of SPIHT over EZW is also due to the use of a finer tree-based classification of source symbols and the use of joint entropy (specifically, coding four binary symbols in a block).

But the best image compression results reported so far in the literature were not generated by zerotree-based methods, but rather by a sample-by-sample bit plane coding technique called ECECOW (embedded conditional entropy coding of wavelet) coefficients [36, 41]. A drawback of the zerotree, or similar quadtree type of data structures used by EZW [28] and SPIHT [25] algorithms, is that the tree imposes an artificial structure on the wavelet coefficients. Only contexts of square shape in the spatial domain can be used, whereas statistically related wavelet coefficients may form regions of arbitrary shapes. Moreover, like any run-length type codes, quadtree code cannot efficiently describe statistically short-term signal constructs, such as edges, because the implicit statistical model used by zerotree breaks down on transient sample behavior. Relative to sample-by-sample coding, zerotree can be considered as a block-based entropy code. It largely ignores the sample dependency between neighboring quadtree nodes. This limitation is particularly regrettable considering that wavelet transform represents a fundamental departure from block-based DCT. The first technique of embedded bit plane coding of wavelet coefficients without any tree constraints seems to be Taubman and Zakhor's layered zero coding (LZC) algorithm [30]. Another early wavelet image coder, called CREW (Compression via Reversible Embedded Wavelets) [43], also did not confine the formation of modeling contexts to quadtree nodes. Compared with its predecessors, the main strength of the ECECOW algorithm is its using higher-order context modeling of embedded wavelet coefficient symbol streams.

8.3 Statistical Context Modeling of Embedded Bit Stream

This section formulates the problem of entropy coding of embedded wavelet bit streams, namely, coding uniformly quantized wavelet coefficients bit plane by bit plane, scanning from the most to the least significant bits. Within each bit plane there are many possibilities of traversing different subbands, and different ways of traversing a subband other than raster scan. Flexible bit traversal can support many desirable functionalities such as region of interests, error resilience, and rate-distortion optimization [29]. The context modeling and entropy coding techniques developed in this chapter all support any traversals within a bit plane. For simplicity, we assume a raster scan in the following descriptions.

The bit plane coding deals with only two source symbols: 0 or 1. However, accompanying the most significant bit of a coefficient, its sign should also be coded. Since the sign is a binary event, we again have only two possible source symbols in this situation. Therefore, in bit plane coding, all wavelet coefficients of an image can be conveniently converted into a sequence of binary symbols: $x_1, x_2, \ldots, x_n$, $x_i \in \{0, 1\}$. The minimum code length of the binary sequence in bits is given by

$$- \log_2 \prod_{i=1}^{n} P\left(x_i | x^{i-1}\right), \tag{8.1}$$

where x^{i-1} denotes the sequence $x_{i-1}, x_{i-2}, \ldots, x_1$. If the conditional probability $P(x_i | x^{i-1})$ is known, then arithmetic coding can approach this minimum rate. Arithmetic coding is a powerful entropy coding technique with an arbitrarily high coding efficiency (limited only by the precision of arithmetic operations). It was pioneered by Rissanen and Langdon [23] and popularized by Witten, Neal, and Cleary [31]. Since embedded wavelet symbol sequence is binary, it can be compressed by adaptive binary arithmetic coding, the simplest and fastest version of adaptive arithmetic coding. Efficient, good approximation algorithms, such as QM coder [20], for adaptive binary arithmetic coding have been well studied and can be easily implemented by both software and hardware. Indeed, QM coder and other variants of binary arithmetic coding are used in many image compression standards, such as the new lossless JPEG standard JPEG-LS (JPEG-LS high-performance extension, LS mean lossless) [3], the JBIG (Joint Binary Image Group) lossless binary image compression standard [7], the JPEG 2000 standard [2], and others [21]. In addition to facilitating binary arithmetic coding, the binarization of the wavelet coefficients also offers great operational advantages for high-order context modeling, as appreciated in subsequent sections.

With arithmetic coding, we can separate the entropy coding completely from statistical context modeling, i.e., the problem of estimating $P(x_i|x^{i-1})$. Given a probability estimate $\hat{P}(x_i|x^{i-1})$, arithmetic coding can achieve the code length $-\log_2 \prod_{i=1}^{n} \hat{P}(x_i|x^{i-1})$. The remaining problem, also a far more difficult one, is how to reach a good estimate $\hat{P}(x_i|\underline{x}^{i-1})$ of $P(x_i|x^{i-1})$, where $\underline{x}^{i-1}$ denotes a subsequence of x^{i-1} that consists of past samples of statistical significance to x_i. Note that the most relevant past subsequence $\underline{x}^{i-1}$ is not necessarily a prefix of x^{i-1}. In image coding, $\underline{x}^{i-1}$ or a causal template for x_i consists of adjacent symbols in both time and frequency. The estimated conditional probability mass function $\hat{P}(x_i|\underline{x}^{i-1})$ serves as a statistical model of the source. The modeling context is the set of past observations $\underline{x}^{i-1}$ on which the probability of the current symbol is conditioned.

In fact, statistical context modeling in the form of probability estimation lies at the heart of any compression system. Ultimately it is the model quality, or the precision of probability estimate, that determines the rate-distortion performance. The true magic of wavelet transforms to compression is in their support of context modeling of sample dependencies via the localization of signal energy in both frequency and time/spatial domains. Specifically, wavelet coefficients of similar magnitudes statistically cluster in frequency subbands and in time/spatial locations. Large wavelet coefficients in different frequency subbands tend to register at the same spatial locations. This localization property makes statistical context modeling of the image signals much easier in wavelet domain than in time/spatial domain or other transform spaces. Specifically, the choice of relevant modeling context $\underline{x}^{i-1}$ becomes easier in the wavelet domain, as explained below.

We take a universal source coding approach to compression of the binary sequence x^n, assuming no prior knowledge about $P(x_i|x^{i-1})$. The central task is to estimate the conditional probability $P(x_i|\underline{x}^{i-1})$ "on the fly" based on the past coded bits and to use the estimate $\hat{P}(x_i|\underline{x}^{i-1})$ to drive an adaptive binary arithmetic coder. For easy reference to individual samples x_i in the binary sequence x^n, we denote the b-th bit of a coefficient c by c_b, the i-th through j-th bits of c, $j > i$, by $c_{j..i}$, and all the bits of c that are above the b-th bit by $c_{..b+1}$. In the sequel, the notation $c_{j..i}$ always refers to the bits in the binary encoding of coefficient magnitude $|c|$. The sign of c is denoted by $\tilde{c}$. Note that the bits of $c_{j..i}$ are not consecutive in an embedded wavelet bit stream but are scattered around. If the most significant bit of c is lower than b, then $c_{..b}$ is considered to be 0. We use directional notations N, W, S, E, NW, NE, NN, WW, and so on, to denote the coefficients to the north, west, south, east, northwest, northeast, northnorth, and westwest of the current coefficient c. Similarly, we denote the parent coefficient by P, and those coefficients in the parent subband to the north, west, south, and east of P by PN, PW, PS, and PE.

In coding of the b-th bit plane, we may condition c_b on

$$\text{C}_{..b+1}, \text{N}_{..b}, \text{W}_{..b}, \text{S}_{..b+1}, \text{E}_{..b+1}, \text{NW}_{..b}, \text{NE}_{..b},$$

$$\text{NN}_{..b}, \text{WW}_{..b}, \text{P}_{..b}, \text{PN}_{..b}, \text{PS}_{..b}, \text{PW}_{..b}, \text{PE}_{..b}, \ldots \qquad (8.2)$$

We treat all the known bits, up to the moment of coding c_b, of the neighboring coefficients in current and parent subbands as potential feature events in modeling context $\underline{x}^{i-1}$ of $x_i = c_b$. Unlike in the EZW and SPIHT algorithms, our modeling context of c_b contains some future information if one considers that the octave-raster scanning of coefficients produces a time series. Specifically, this refers to the use of $\text{S}_{..b+1}$, $\text{E}_{..b+1}$, $\text{PS}_{..b}$, $\text{PE}_{..b}$, and the like in context modeling of c_b. The ability of looking into the future in a time series significantly reduces the uncertainty of c_b.

8.4 Context Dilution Problem

The modeling context of Eq. (8.2) leads to a statistical model

$$P\left(c_b | \text{N}_{..b}, \text{W}_{..b}, \text{S}_{..b+1}, \text{E}_{..b+1}, \text{NW}_{..b}, \text{NE}_{..b}, \text{P}_{..b} \cdots \right) \qquad (8.3)$$

of very high order or long memory. High-order context modeling is necessary for optimal compression performance because image features such as edges can involve pixels that are spatially far apart. Given a modeling context $(\text{N}_{..b}, \text{W}_{..b}, \text{S}_{..b+1}, \text{E}_{..b+1}, \text{NW}_{..b}, \text{NE}_{..b}, \text{P}_{..b} \cdots)$, the average code length of c_b is bounded from below by the conditional entropy

$$H\left(c_b | \text{N}_{..b}, \text{W}_{..b}, \text{S}_{..b+1}, \text{E}_{..b+1}, \ldots \right)$$

$$= -E\left\{ \log P\left(c_b | \text{N}_{..b}, \text{W}_{..b}, \text{S}_{..b+1}, \text{E}_{..b+1}, \ldots \right)\right\} . \qquad (8.4)$$

The fact that conditional entropy is monotonically nonincreasing [10] seems to suggest that the higher the order of the context model, the shorter the code length. But this is not necessarily true.

In universal source coding we do not have prior knowledge of the source. The model itself must be either explicitly sent to the decoder or learned on the fly from the samples. In the former case, we need to add side information to the total description length of the source. In the latter case, the learning requires a large number of samples to fit a statistical model to the source. The number of possible conditioning states grows exponentially with the order of the context, an image of finite resolution may not provide sufficient samples to reach a robust estimate of the underlying conditional probability

$$\hat{P}\left(c_b | \text{N}_{..b}, \text{W}_{..b}, \text{S}_{..b+1}, \text{E}_{..b+1}, \text{NW}_{..b}, \text{NE}_{..b}, \text{P}_{..b} \cdots \right) . \qquad (8.5)$$

In order words, too high an order of modeling context spreads sample statistics too thin among all possible modeling states to yield statistical significance. The code length will actually increase when the order of modeling context gets too high. Thus, from an implementation point of view, high order of context modeling is more than a problem of high time and space complexity. It can reduce coding efficiency as well. This problem is commonly known as *context dilution* and formulated by Rissanen analytically as *model cost* [24]. Intuitively, the higher the model complexity (i.e., the more model parameters), the longer the time the model takes to *learn* from the samples to set the parameters right. Before the model converges to the underlying statistics via online learning, entropy coding cannot achieve the minimum code length of Eq. (8.1). Therefore, the context model has an inherent cost to the total description length, either in the form of side information to describe the model, as in two-pass coding, or in the form of extra code length due to model mismatch in the beginning of the learning process, as in one-pass coding.

By now, one may appreciate an advantage of turning the wavelet coefficients into a binary sequence. Since a conditional binary probability has only two parameters, we do not need nearly as many samples to obtain a good probability estimate as for a large symbol alphabet. But even with c_b being binary, we still have to reduce Eq. (8.2) to a modest number of conditioning states; otherwise the benefits of context modeling will be negated by high model cost. Indeed, in his original EZW paper [28], Shapiro remarked that Markov conditioning did not offer significant coding gains over "single histogram strategy" (entropy coding based on symbol probability without context modeling). In their original SPIHT paper [25], Said and Pearlman also implied that in their experiments high-order context modeling made only marginal coding gains over the simple Huffman coding. But their observations did not mean the lack of high-order statistical dependencies between samples in the wavelet domain. Their experimental results with context modeling were somewhat disappointing only because the problem of context dilution was not considered. The challenge is to reduce the model cost and still capture statistically significant structures of high orders between the samples.

8.5   Context Formation

One way to reduce the number of model parameters, and thus to reduce the model cost, is to include into the modeling context only those past samples that are statistically related to the current sample being coded. For one-dimensional sources, such as text, speech, and audio, the modeling context selection criterion can be some prefix of the current sample because the amount of sample dependency is proportional to the distance between the samples. Similarly, for image and video sources the general practice is to choose a spatial and temporal neighborhood to form the modeling context. However, as we pointed out in the previous section, the resulting context can be of a very high order. A more selective rule than k nearest neighbors, where k is the size of context template, should be used if we have any prior knowledge about sample structures.

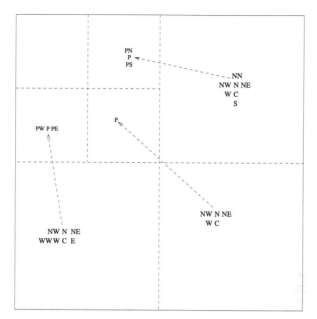

FIGURE 8.7
Modeling contexts in different subbands.

The feature orientations of different wavelet subbands are the kind of prior knowledge that is useful for reducing the model cost. For instance, the LH subband exhibits predominantly vertical sample structures, while the HL subband exhibits predominantly horizontal sample structures. Therefore, we choose a subset of Eq. (8.2):

$$S_{LH} = \{N_{..b}, W_{..b}, NW_{..b}, NE_{..b}, NN_{..b}, S_{..b+1}, P_{..b}, PN_{..b}, PS_{..b}\} \qquad (8.6)$$

to be the modeling context of c_b in LH subbands. This choice of conditioning events forms a vertically prolonged modeling context. Similarly, we use a horizontally prolonged modeling context

$$S_{HL} = \{N_{..b}, W_{..b}, NW_{..b}, NE_{..b}, WW_{..b}, E_{..b+1}, P_{..b}, PW_{..b}, PE_{..b}\} \qquad (8.7)$$

in modeling of c_b in HL subbands.

Note that we include corner samples $NW_{..b}$ and $NE_{..b}$ in the northwest and northeast directions, but not $SW_{..b+1}$ and $SE_{..b+1}$ in the southwest and southeast directions. The reason is that the former two samples have one bit more precision and are therefore statistically more significant than the latter two samples in the raster scan of bit planes. In our experiments, including two more samples at the southwest and southeast corner did not bring any compression gains, and in some cases it could even increase the code length due to the effect of context dilution. In practice, when choosing a modeling context one can monitor the resulting code length as the order of modeling context increases. This will empirically detect the point where the increased model cost just starts to have negative impact on compression. Thus one can choose an appropriate order of the model by not adding to the modeling context samples of less statistical significance to c.

In HH subbands, sample structures tend to be much weaker than in LH and HL subbands. A smaller modeling context can be used without reducing compression performance. In our experiments, we found that maximum coding gains can be made by conditioning a c_b in an HH subband on the following set of samples:

$$S_{HH} = \{N_{..b}, W_{..b}, NW_{..b}, NE_{..b}, S_{..b+1}, E_{..b+1}, P_{..b}, C_{HL}, C_{LH}\} \qquad (8.8)$$

where C_{HL} and C_{LH} are two sister coefficients of c that are at the same spatial location in the HL and LH subbands of the same scale. The different shapes and orientations of modeling contexts used in different subbands are illustrated in Fig. 8.7.

Due to the use of ubiquitous L_2 metric in wavelet approximation, samples in all subbands except the one in the lowest frequency are drawn from zero mean processes. The coefficient sign $\tilde{c}$ has equal probability to be positive and negative. Consequently, we have $H(\tilde{c}) = 1$; i.e., the self entropy of coefficient sign is at the maximum. But this does not necessarily mean that the signs are uncompressible. In fact, the conditional entropy of the signs can be significantly lower than 1. The waveform structures of the input image are often exhibited by sign patterns of wavelet coefficients. In Fig. 8.8 we plot the spatial distributions of signs for parts of two popular test images, "Barb" and "Lena" that have high textures. The clearly visible structures of signs suggest that the sign bits of wavelet coefficients can be modeled as a Markov process and compressed by conditional entropy coding.

During embedded bit plane coding, the sign $\tilde{c}_{..b}$ of a wavelet coefficient c has three states: +, −, and 0. At the b-th bit plane, $\tilde{c}_{..b}$ is still unknown to the decoder if $c_{..b} = 0$; i.e., the most significant bit of c is below b. In this case the coder assigns state 0 to $\tilde{c}_{..b}$; otherwise it assigns + or − to $\tilde{c}_{..b}$ by the conventional meanings of sign. Here the state 0 is a dynamic concept; it may change to + or − as the coding process advances to deeper bit planes. We distinguish 0 from + and − because such a distinction yields a more revealing modeling context for the signs. The use of three states of signs in context modeling exploits the correlation between the signs and the magnitudes of neighboring wavelet coefficients because $\tilde{c}_{..b} = 0$ indicates a relatively small $|c|$. The modeling context for $\tilde{c}$ commonly consists of sign status of c's four immediate neighboring samples, namely it is the set

$$\tilde{S} = \{\tilde{N}_{..b}, \tilde{W}_{..b}, \tilde{S}_{..b}, \tilde{E}_{..b}\} \ . \qquad (8.9)$$

8.6 Context Quantization

Careful selection of past samples to be used in modeling context based on subband orientations is only a screening process. The number of possible conditioning states of

Sign map of barb

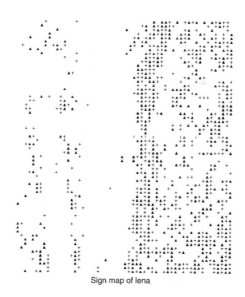

Sign map of lena

FIGURE 8.8

Sign patterns in parts of "Barb" (top) and "Lena" (bottom). The triangles are for negative signs, + for positive signs, and spaces for insignificant coefficients up to the current bit plane.

the chosen context is still far too large. Context dilution remains a serious problem. A rule of thumb for the right number of conditioning states in embedded wavelet image coding is about 64. The use of more than 100 conditioning states hardly makes any compression gain, and in many cases it can even increase the bit rate. A common technique of reducing the number of conditioning states for entropy coding is context quantization. The idea is to merge conditioning states in which the sample probability distributions are close in terms of Kullback-Leibler distance or relative entropy [10].

A simple scheme is scalar quantization of samples in the modeling context. In a modeling context that consists of eight or so samples, such as those in Eqs. (8.6) and (8.7), scalar quantization has to be very coarse in order to bring the number of conditioning states under 100. Indeed, many of the wavelet image coders reported in the literature use binary quantization of feature samples [43, 29]. In other words, feature samples $N_{..b}$, $W_{..b}$, $S_{..b}$, $E_{..b}$, etc. are entered into the context as either 1 (already significant at the current bit plane) or 0 (not yet significant at the current bit plane). Such a coarse quantization can obscure some subtleties in correlations between c and the energy level of the neighboring coefficients.

In order to capture the correlation between c and its neighbors in the wavelet domain, we use a linear estimator Δ of the magnitude of c, one for each of three orientations (LH, HL, and HH) of subbands:

$$\Delta_\theta = \sum_{z_i \in S_\theta} \alpha_{\theta,i} z_i, \quad \theta \in \{LH, HL, HH\}, \tag{8.10}$$

where the terms z_i are conditioning events in the context chosen for the given subband of c as described above. The parameters $\alpha_{\theta,i}$ are determined by linear regression so that Δ_θ is the least-squares estimate of c in the given subband orientation. The linear regression can be done offline for a general set of training images, a given class of images, and even for a given image. Of course in the last case, the optimized parameters have to be sent as side information.

For each of Δ_θ we can design an optimal quantizer Q_θ to minimize the conditional entropy

$$H\left(c|Q_\theta\left(\Delta_\theta\right)\right) = E\left\{\log P\left(c|Q_\theta\left(\Delta_\theta\right)\right)\right\}. \tag{8.11}$$

Since Δ_θ is a scalar random variable, the optimal quantizer Q_θ to achieve minimum conditional entropy can be computed via a standard dynamic programming process [33]. The optimization is carried out offline using a training set, and the quantizer parameters are stored and available at both the encoder and decoder. (In order not to interrupt the flow of our presentation we defer the details of dynamic programming process for designing minimum conditional entropy quantizers to Section 8.8.)

Besides the correlation between c and the local energy level Δ, the wavelet coefficient c also has dependence on spatial patterns of its neighboring coefficients, particularly at locations of strong edges or high textures. This dependence is due to the fact that a wavelet transform offers certain time resolution of the signal at the expense of frequency resolution. Therefore it is necessary to model the spatial sample patterns in the wavelet domain to maximize coding gains. Again the required statistical modeling has to be done without drastically increasing the number of conditioning states. To achieve this, we quantize the sample spatial pattern around c into a binary vector (bit pattern) $T_b = t_4 t_3 t_2 t_1 t_0$ by

$$
\begin{aligned}
t_0 &= \text{N}_{..b} > \text{C}_{..b+1}?0:1 ; \\
t_1 &= \text{W}_{..b} > \text{C}_{..b+1}?0:1 ; \\
t_2 &= \text{S}_{..b+1} > \text{C}_{..b+1}?0:1 ; \\
t_3 &= \text{E}_{..b+1} > \text{C}_{..b+1}?0:1 ; \\
t_4 &= \begin{cases} \text{P}_{..b} + \text{PN}_{..b} + \text{PS}_{..b} > 6\text{C}_{..b+1}?0:1 & \text{in LH subbands} ; \\ \text{P}_{..b} + \text{PW}_{..b} + \text{PE}_{..b} > 6\text{C}_{..b+1}?0:1 & \text{in HL subbands} . \end{cases}
\end{aligned}
\tag{8.12}
$$

The type of binary context quantization as in Eq. (8.12), as we mentioned earlier, is directly used to form conditioning states in many embedded wavelet image/video coders [29, 30, 43]. But significantly higher compression gains can be made by combining quantized energy level $Q_\theta(\Delta_\theta)$ and the spatial pattern T_b of c's neighboring coefficients to form conditioning states in entropy coding of c. Specifically, c is coded by an adaptive binary arithmetic coder driven by probability estimate

$$
\hat{P}\left(c_b | Q_\theta\left(\Delta_\theta\right), T_b\right) .
\tag{8.13}
$$

8.7 Optimization of Context Quantization

The previous section introduced context quantization as a necessary component for statistical modeling and entropy coding of wavelet coefficients and presented some context quantization techniques. However, these techniques are largely based on heuristics, albeit being proven to be useful in practice. This section investigates the problem of context quantization in a multivariate analysis approach of statistics and develops algorithms for designing optimum context quantizer for minimum conditional entropy.

Context quantization is a special form of vector quantization whose criterion should ideally be minimum conditional entropy. It is well-known that optimal vector quantization is NP-complete — a problem is said to be NP-complete if its exact solution requires an amount of computations that increases exponentially in the input size [13]. In other words, for a large training set which is required if the derived VQ solution is to have any statistical significance, designing the globally optimal vector is computationally intractable. Thus, we necessarily resort to alternative techniques that are computationally feasible. Since high dimensionality is the main cause for the complexity of the problem, we would naturally like to reduce the dimensionality of the problem. A classical approach in multivariate analysis is to project sample vectors of high dimensions onto a lower dimensional space that contains most of the statistical variations.

A high-order modeling context such as the one in Eq. (8.2) can be viewed as a modeling event vector $\mathbf{v} = (v_1, v_2, \ldots, v_d)$, where v_i is a modeling event. Let $V = \{\mathbf{v}_1, \mathbf{v}_2, \ldots, \mathbf{v}_k\}$ be a training set of event vectors. V can be the set of all event vectors observed so far in an online learning process, or an offline training set. The former is necessary if the context quantizer is designed on the fly in one-pass coding, whereas the latter is for offline context quantizer design. We partition V into V_0 and V_1, where subset V_0 (V_1) contains all the modeling event vectors associated with $c_b = 0$ ($c_b = 1$). If there exists a hyperplane or some other surface in the d-dimensional event space that can completely separate V_0 and V_1, then the binary symbol c_b to be coded is uniquely determined by its modeling context. In this ideal case the conditional entropy of c_b is 0. In reality, however, the two point subsets V_0 and V_1 are mingled in the event space in a complicated way. To simplify the problem, we can project all training event vectors onto a line and hope that V_0 and V_1 form distinct clusters along the line. This approach is due to Fisher [12]. Let the projection be

$$u_i = \mathbf{a}^T \mathbf{v}_i, \quad i = 1, 2, \ldots, |V| .$$
(8.14)

Given a training set V, we want to determine the projection direction $\mathbf{a}$ such that

$$G(\mathbf{a}) = \frac{(\mu_0 - \mu_1)^2}{\sigma_0^2 + \sigma_1^2}$$
(8.15)

is maximized, where

$$\mu_j = E\left\{u_i | \mathbf{v}_i \in V_j\right\} = E\left\{\mathbf{a}^T \mathbf{v}_i | \mathbf{v}_i \in V_j\right\}, \quad j = 0, 1$$
(8.16)

and

$$\sigma_j^2 = E\left\{\left(\mathbf{a}^T \mathbf{v}_i - \mu_j\right)^2 | \mathbf{v}_i \in V_j\right\}, \quad j = 0, 1 .$$
(8.17)

The criterion of maximum $G(\mathbf{a})$ can be intuitively understood as maximum separation of V_0 and V_1. The numerator demands maximum distance between the projected means of V_0 and V_1 in direction $\mathbf{a}$, whereas the denominator requires minimum overlap of V_0 and V_1 along the projection line.

We use a well-known procedure in multivariate analysis literature for maximizing $G(\mathbf{a})$ based on sample event vectors. We rewrite Eq. (8.15), by scaling, in terms of sample scatter matrices S_0 and S_1 for V_0 and V_1, respectively,

$$G(\mathbf{a}) = \frac{(\mu_0 - \mu_1)^2}{\mathbf{a}^T (S_0 + S_1) \mathbf{a}} .$$
(8.18)

The scatter matrix is defined by

$$S_i = \sum_{\mathbf{v} \in V_i} (\mathbf{v} - \mathbf{m}_i)(\mathbf{v} - \mathbf{m}_i)^T, \quad i = 0, 1$$
(8.19)

where

$$\mathbf{m}_i = \frac{1}{|V_i|} \sum_{\mathbf{v} \in V_i} \mathbf{v}, \quad i = 0, 1 . \tag{8.20}$$

We also express the numerator of Eq. (8.15) in terms of sample means:

$$(\mu_0 - \mu_1)^2 = \mathbf{a}^T (\mathbf{m}_0 - \mathbf{m}_1)(\mathbf{m}_0 - \mathbf{m}_1)^T \mathbf{a} . \tag{8.21}$$

Letting $M = (\mathbf{m}_0 - \mathbf{m}_1)(\mathbf{m}_0 - \mathbf{m}_1)^T$ and $S = S_0 + S_1$, we have

$$G(\mathbf{a}) = \frac{\mathbf{a}^T M \mathbf{a}}{\mathbf{a}^T S \mathbf{a}} . \tag{8.22}$$

Differentiating $G(\mathbf{a})$ and setting $\partial G / \partial \mathbf{a} = 0$ to determine the direction $\hat{\mathbf{a}}$ that maximizes $G(\mathbf{a})$, we arrive at

$$\frac{\hat{\mathbf{a}}^T M \hat{\mathbf{a}}}{\hat{\mathbf{a}}^T S \hat{\mathbf{a}}} S \hat{\mathbf{a}} = M \hat{\mathbf{a}} . \tag{8.23}$$

Now the underlying optimization problem reduces to one of an eigenvalue with the scalar term $\lambda = (\hat{\mathbf{a}}^T M \hat{\mathbf{a}})/(\hat{\mathbf{a}}^T S \hat{\mathbf{a}})$. If S^{-1} exists, the direction of $\hat{\mathbf{a}}$ is given by

$$\hat{\mathbf{a}} = S^{-1} M \hat{\mathbf{a}} . \tag{8.24}$$

Since $M \hat{\mathbf{a}}$ has the direction of $\mathbf{m}_0 - \mathbf{m}_1$, it follows that

$$\hat{\mathbf{a}} = S^{-1} (\mathbf{m}_0 - \mathbf{m}_1) . \tag{8.25}$$

The simple solution above is made possible by the binarization of source symbols via embedded bit plane coding. The binarization conveniently lends Fisher's linear classifier with two classes to our context quantization problem. In Fisher's original work, the objective is to find a linear discriminant to classify between V_0 and V_1 for minimum classification error. But in reality, the projected samples of V_0 and V_1 in the direction of $\hat{\mathbf{a}}$ can be intermingled in such complicated ways that Fisher's discriminant leaves a significant degree of uncertainty. Much finer context quantization is required to further resolve the uncertainty and to approach rate-distortion optimality.

8.8 Dynamic Programming for Minimum Conditional Entropy

Once the direction of maximum separation $\hat{\mathbf{a}}$ is determined, we project all training event vectors onto a line in this direction. On this line the projection establishes an order of training event vectors by their projection values $u_i = \hat{\mathbf{a}}^T \mathbf{v}_i, i = 1, 2, \ldots, |V|$, namely $\mathbf{v}_i \le \mathbf{v}_j$ if $u_i \le u_j$. This linear ordering enables a constrained optimization approach of dynamic programming to design a K-level context quantizer. The constraint is that all quantizer cells are perpendicular to direction $\hat{\mathbf{a}}$. Under the constraint, the K-level context quantizer can be globally optimized for minimum conditional entropy, which is better than a gradient descent method that may be trapped in a local minimum. It is easy to see that in Section 8.6, the least-square estimator Δ of Eq. (8.10) also corresponds to a projection in high-dimensional feature space and establishes an order of training event vectors via the projection. Therefore, the same dynamic programming process to be developed in this section can be used to solve the optimization problem posted around Eq. (8.11) as well.

Let $\underline{u} = \min_i u_i$, $\bar{u} = \max_i u_i$, and denote by $Q(\tau, k)$ the set of all possible k-dimensional vectors $\mathbf{q} = (q_1, q_2, \ldots, q_k)$ such that

$$\underline{u} \equiv q_0 < q_1 < q_2 < \cdots < q_{k-1} < q_k = \tau < \bar{u} . \tag{8.26}$$

In designing the context quantizer, we associate each modeling event vector $\mathbf{v} \in V$ with the random variable $c_b \in \{0, 1\}$ being modeled. Then the optimal context quantizer that minimizes conditional entropy is given by

$$\hat{\mathbf{q}} = \arg \min_{\mathbf{q} \in Q(\bar{u}, K)} \sum_{k=1}^{K} P\left(u_i \in (q_{k-1}, q_k]\right) H\left(c_b | u_i \in (q_{k-1}, q_k]\right) \tag{8.27}$$

where

$$H\left(c_b | u_i \in (q_{k-1}, q_k]\right) = -E\left\{\log P\left(c_b | u_i \in (q_{k-1}, q_k]\right)\right\} . \tag{8.28}$$

In the formulation, the k-th quantizer cell corresponds to a subset $\Psi_k = \{\mathbf{v}_i | q_{k-1} < u_i \leq q_k\}$ of training event vectors. Denote by $n_0(q_{k-1}, q_k]$ the number of modeling event vectors in Ψ_k that are associated with $c_b = 0$, and by $n_1(q_{k-1}, q_k] = |\Psi_k| - n_0(q_{k-1}, q_k]$ the number associated with $c_b = 1$. Also let

$$
\begin{aligned}
L_0\left(q_{k-1}, q_k\right] &= n_0\left(q_{k-1}, q_k\right] \log n_0\left(q_{k-1}, q_k\right] \\
L_1\left(q_{k-1}, q_k\right] &= n_1\left(q_{k-1}, q_k\right] \log n_1\left(q_{k-1}, q_k\right] \\
L\left(q_{k-1}, q_k\right] &= |\Psi_k| \log |\Psi_k| .
\end{aligned}
\tag{8.29}
$$

When working with the training set V and using the notations above, the minimization problem of Eq. (8.27) becomes

$$\hat{\mathbf{q}} = \arg \min_{\mathbf{q} \in Q(\bar{u}, K)} \sum_{k=1}^{K} \left(L\left(q_{k-1}, q_k\right] - L_0\left(q_{k-1}, q_k\right] - L_1\left(q_{k-1}, q_k\right]\right) . \tag{8.30}$$

The optimal K-level context quantizer $\hat{\mathbf{q}}$ as given by Eq. (8.30) can be efficiently computed by observing the following recursion:

$$
\min_{\mathbf{q} \in Q(r,j)} \sum_{k=1}^{j} \left(L(q_{k-1}, q_k] - L_0(q_{k-1}, q_k] - L_1(q_{k-1}, q_k] \right)
$$

$$
= \min_{\tau < r} \left\{ \min_{\mathbf{q} \in Q(\tau,j-1)} \sum_{k=1}^{j-1} \left(L(q_{k-1}, q_k] - L_0(q_{k-1}, q_k] - L_1(q_{k-1}, q_k] \right) \right.
$$

$$
\left. + L(\tau, r] - L_0(\tau, r] - L_1(\tau, r] \right\}. \tag{8.31}
$$

The recursion means that the solution for the problem of size j can be constructed on the solutions of subproblems of size $j - 1$. Because of this property (called the principle of optimality, in optimization literature), we can use a straightforward dynamic programming algorithm to solve Eq. (8.30). The primitive operations in the dynamic programming process are those in Eq. (8.29). We can precompute and store L_0, L_1, and L for all possible subsets in $O(|V|^2)$ time. The expensive logarithmic computations can be done via table lookup. After the preprocessing, the dynamic programming algorithm takes $O(K|V|^2)$ time.

8.9 Fast Algorithms for High-Order Context Modeling

High-order context modeling is indispensable for good rate-distortion performance of wavelet image coders. But if care is not taken in algorithm design and implementation, the formation of high-order modeling contexts can be both CPU and memory intensive, creating a computation bottleneck for wavelet coding systems. Indeed, our earlier research prototype of ECECOW, a high-order embedded conditional entropy coder of wavelet coefficients, spent 70% of its execution time on context modeling. It is unacceptable for most applications that a module of a wavelet image codec is six times more expensive than the wavelet transform itself. In this section, we focus on the operational aspects of high-order statistical context modeling and introduce some fast algorithmic techniques that can drastically reduce both time and space complexities of high-order context modeling in the wavelet domain.

Two computationally intensive parts in context formation are the linear combination Eq. (8.10) of neighboring samples and the texture pattern extraction Eq. (8.12) from neighboring samples. Once Δ is computed, its quantization is very fast via table lookup. Although Eq. (8.10) and Eq. (8.12) involve only basic arithmetic and logic operations — namely additions, comparisons, and bit manipulations — straightforward computations of Eq. (8.10) and Eq. (8.12) require a large number of operations per binary symbol. Furthermore, forming a high-order context that spans over several scan lines needs to access data (modeling events) stored in distant memory locations. This activity can cause excessive cache misses on modern hardware architecture, slowing down the computation. The high computational complexity is seemingly inherent in high-order context modeling. In order to speed up context formation we have to question if the computational complexity of statistical context modeling is necessarily proportional to the order of the model. The answer is pleasantly, if somewhat surprisingly "no," as we will see shortly.

8.9.1   Context Formation via Convolution

By tracing the major causes of high computational complexity, we come to the following key observation. High-order modeling contexts for neighboring samples have large overlaps in the wavelet domain. This means that samples are accessed and operated on repetitively. We can improve computational efficiency by eliminating repetitive arithmetic, logic, and memory operations in spatially overlapped modeling contexts. This idea leads to an incremental algorithm to compute Δ in $O(1)$ time independent of the order of modeling context. Given that the wavelet coefficients are coded in raster scan order at a given bit plane and in a given subband, we denote the coefficient vector in the current row by $x_0[t]$, where $t = 0, 1, \ldots$ represents spatial locations. The coefficient vector in the next row is denoted by $x_1[t]$, and likewise in the previous two rows by $x_{-1}[t]$ and $x_{-2}[t]$, respectively. The x values are up to the current decoded precision in bit plane coding, i.e., in the notations of previous sections, $x_{-2}[\cdot]$, $x_{-1}[\cdot]$, and $x_0[\tau]$, $\tau < t$, are $c_{..b}$, while $x_0[\tau]$, $\tau \geq t$, and $x_1[\cdot]$ are $c_{..b+1}$. We drop the subscripts for bit ranges $..b$, $..b + 1$ because they are clearly implied in spatial locations of the wavelet coefficients x.

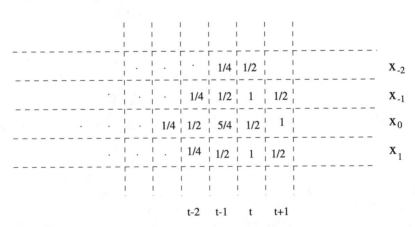

FIGURE 8.9
Convolution kernel effected by the incremental Δ computation of Eq. (8.32).

In sequential coding of $x_0[t]$ for increasing t, we compute incrementally

$$
\begin{aligned}
\alpha_t &= x_{-1}[t+1] + x_1[t+1] \\
\beta_t &= \frac{\beta_{t-1}}{2} + \alpha_{t-1} + x_0[t-1] + x_0[t+1] + \frac{x_{-2}[t]}{2} \\
\Delta_t &= \frac{\alpha_t}{2} + \beta_t .
\end{aligned}
\tag{8.32}
$$

Expanding the recursion above reveals

$$
\begin{aligned}
\Delta_t &= x_{-1}[t] + x_1[t] + x_0[t-1] + x_0[t+1] \\
&+ \frac{x_0[t]}{2} \frac{x_{-2}[t] + x_{-1}[t-1] + x_{-1}[t+1] + x_1[t-1] + x_1[t+1] + x_0[t-2]}{2} \\
&+ \frac{\beta_{t-2} + x_{-2}[t-1]}{4} .
\end{aligned}
\tag{8.33}
$$

This corresponds to a high-order linear filter whose kernel is graphically depicted in Fig. 8.9. Note that Fig. 8.9 illustrates only the part of the filter kernel with coefficients larger than $1/4$ — the 14 most important modeling events with respect to $x_0[t]$. We can see that Δ_t is a weighted sum of N, W, S, E, NW, NE, SW, SE, NN, WW, and many other past observations with the weights proportional to their distances to $x_0[t]$. Therefore, Δ_t offers a modeling context of $x_0[t]$ of order higher than 14. But Δ_t can be computed by the incremental algorithm of Eq. (8.32) in only six additions, five memory accesses, and two bit shifts — less than half the number of operations required by a direct implementation of Eq. (8.10). In fact, the computational complexity of the proposed incremental algorithm is independent of the order of modeling contexts. Indeed, we can rewrite the second line of Eq. (8.32) as

$$
\beta_t = \lambda \beta_{t-1} + \cdots
\tag{8.34}
$$

where λ is a forgetting factor. Increasing λ gives higher weights to the past observations and hence increases the order of modeling context. Therefore, we derived an $O(1)$ time algorithm for computing Δ_t that can increase the order of context modeling for a fixed number of operations. The optimal value of λ is determined by the length of memory in the source. In practice, we empirically found that $\lambda = 1/2$ gave very close to optimal compression results on natural images while avoiding divisions. The incremental computations of Eq. (8.32) have a simple convolution structure, and hence particularly suitable for hardware implementation.

8.9.2 Shared Modeling Context for Signs and Textures

Next we consider efficient computations of Eq. (8.12) and Eq. (8.9) and introduce algorithmic techniques to greatly reduce the amount of computation and memory accesses to set up spatial texture patterns T_b and sign contexts. As we did for Δ, we dropped the references to parent subband in T_b, i.e., not using t_4 in Eq. (8.12). Then there are four status bits $t_3 t_2 t_1 t_0$ to be set depending on the outcomes of four comparisons between $C_{..b+1}$ and $N_{..b}$, $W_{..b}$, $S_{..b+1}$, and $E_{..b+1}$. By a careful organization of computations in Eq. (8.12) and Eq. (8.9), we can save the comparison and bit setting operations. The basic idea is to let sign modeling and texture modeling share as much context information as possible.

For each coefficient C, we introduce a *syndrome byte* $S_b = s_7 s_6 \cdots s_1 s_0$:

$$
\begin{aligned}
s_0 &= N_{..b} > 0?1:0, & s_4 &= \tilde{N} ; \\
s_1 &= W_{..b} > 0?1:0, & s_5 &= \tilde{W} ; \\
s_2 &= S_{..b+1} > 0?1:0, & s_6 &= \tilde{S} ; \\
s_3 &= E_{..b+1} > 0?1:0; & s_7 &= \tilde{E} .
\end{aligned}
\tag{8.35}
$$

where N, W, S, E are the four neighbors of C. Syndrome bytes for all coefficients are initialized to 0 and updated if necessary for decreasing bit planes. Syndrome bytes S_b support context modeling of signs by allowing three dynamic states of signs in embedded bit plane coding. Specifically, in S_b if bit $s_i = 0$, $i = 0, 1, 2, 3$, then the status bit s_{i+4} is not used or is only a "don't care" bit (although the bit is physically set to 0 at initialization). In sign modeling for $x_0[t]$, the coder needs to fetch only the syndrome byte S_b of $x_0[t]$ and then uses it as the modeling context for signs.

Note that in the embedded bit plane coding, the most significant bit and the sign of a wavelet coefficient are set at the same time. It is then immediate from Eq. (8.12) and Eq. (8.35) that

$$
T_b = t_3 t_2 t_1 t_0 = s_3 s_2 s_1 s_0 = S_b, \quad \text{if } C_{..b+1} = 0. \tag{8.36}
$$

Therefore, syndrome byte S_b can be used not only directly for modeling signs, but also for modeling textures. The entropy coder simply extracts the last four bits of S_b and sets texture pattern $T_b = s_3 s_2 s_1 s_0$, if $C_{..b+1} = 0$. All the computations of Eq. (8.12) become unnecessary and can be saved.

Each bit in syndrome byte S_b for $x_0[t]$ is set at most once. Two bits, s_i and s_{i+4}, are set when the most significant bit and the sign of one of the four neighbors N, W, S, E of $x_0[t]$ are scanned and coded. This means that the sign of a coefficient will never be accessed and examined more than once in embedded bit plane coding. Likewise, no coefficients will be accessed and tested more than once for their significance in setting T_b for decreasing b. The proposed algorithm has therefore minimized the number of arithmetic operations and memory accesses in context formation. This optimality in time complexity is achieved by eliminating repetitive computations in spatially overlapped contexts, and it is operationally realized by the use of syndrome bytes S_b. Clearly, the number of syndrome bytes is the same as the number of wavelet coefficients in the buffer to be coded. This working memory is very modest in size and well justified by the great savings in computation.

Before leaving the subject discussed above, we would like to point out that the JPEG 2000 verification model also uses a collection of status bits for each wavelet coefficient which have similar roles as the syndrome bytes [29].

The algorithmic techniques for fast context formation presented above have led to an efficient wavelet-based image coder [38] that is 20% faster than the popular SPIHT image coder (its arithmetic coded version), and at the same time it outperforms SPIHT in rate-distortion performance, as we see in the following section.

8.10 Experimental Results

In order to demonstrate the effects of different techniques for context modeling and conditional entropy coding of wavelet coefficients on coding efficiency, we present compression results of some well-known and recently published wavelet image compression algorithms and compare them with the algorithms that are described in this chapter. We evaluate both lossy and lossless compression performance of these wavelet image coders.

8.10.1 Lossy Case

For the sake of common references, we use in our evaluation two JPEG test images, "Lena" and "Barbara," that are widely used for rate-distortion comparisons in the image compression literature. Image qualities, measured by PSNR, of various wavelet image coding algorithms at different bit rates are tabulated in Tables 8.1 and 8.2. In the tables, EZW and SPIHT algorithms are well-known and were introduced earlier in this chapter. SFQ is the space-frequency quantization method by Xiong, Ramchandran, and Orchard [40], EQ is an *estimation-quantization* method by LoPresto, Ramchandran, and Orchard [16], and C/B is a context-based entropy coding method by Chrysafis and Ortega [9]. ECECOW is a coder based on techniques presented in Sections 8.5 and 8.6, and also in Wu [36]. The best results were obtained by replacing the context quantizer of ECECOW with the context quantization scheme guided by Fisher's linear discriminant and via dynamic programming [37], as described in Sections 8.7 and 8.8. This algorithm is identified as "NEW" in the tables.

Table 8.1 Rate(bpp)/PSNR(dB) Results for "Lena"

rate	EZW	SPIHT	SFQ	EQ	C/B	ECECOW	NEW
0.25	33.17	34.13	34.33	34.57	34.31	34.81	34.89
0.50	36.28	37.24	37.36	37.68	37.52	37.92	38.02
1.00	39.55	40.45	40.52	40.88	40.80	40.85	41.01

The NEW method outperforms all others in terms of rate-distortion performance, although by smaller margins against ECECOW. We need to stress that the good performances of the NEW and ECECOW methods are solely due to high-order adaptive context modeling. In our experiments, both coders used dyadic wavelet transform of the popular bi-orthogonal 9/7 filter [17]. Neither filter kernel nor coefficient quantizer was optimized for specific images. On the other hand, some of the other methods in

Table 8.2   Rate(bpp)/PSNR(dB) Results for "Barbara"

rate	EZW	SPIHT	SFQ	C/B	ECECOW	NEW
0.25	26.77	27.57	27.20	28.48	28.85	29.21
0.50	30.53	31.39	31.33	32.63	32.69	33.06
1.00	35.14	36.41	36.96	37.61	37.65	38.05

our comparison group used much longer filter kernels. Furthermore, ECECOW is an embedded coding scheme, like EZW and SPIHT, while SFQ, EQ, and C/B are not. Our experimental results clearly demonstrate the importance of high-order context modeling in compressing wavelet coefficients.

8.10.2   Lossless Case

Since entropy coding of coefficients is independent of wavelet transforms and coefficient quantization, the NEW method can be readily applied to invertible wavelet transforms [8] for lossless image compression. Invertible wavelet transforms map integer pixel values to integer wavelet coefficients. Thus no coefficient quantization is required prior to coefficient coding. Because of the absence of quantization distortion, entropy decoding of wavelet coefficients followed by inverse transform leads to lossless reconstruction. Table 8.3 compares the lossless compression performance of the NEW method with other state-of-art lossless image coders on an ISO set of test images. In the comparison group JPEG-LS is the new lossless JPEG standard [6]. CALIC is a well-known lossless image coder that seems to have the best compression performance among practical lossless image coders [34, 35]. But one needs to keep in mind that both JPEG-LS and CALIC are predictive coding schemes without progressive transmission capability. The S+P algorithm, by Said and Pearlman [26], is a pioneer work in wavelet-based, embedded lossless image compression. On average, the NEW method obtains only about 2% less lossless compression than CALIC, but it outperforms all others.

8.11   Summary

For typical transform-based signal compression systems, data reduction is mostly achieved by entropy coding of transform coefficients. If context-based adaptive arithmetic coding is used to compress the transform coefficients, then the pivotal issue that determines the compression performance is statistical context (Markov) modeling of the coefficients, or, more specifically, how to estimate the underlying conditional probability of the coefficients. In this chapter, we introduced a number of modern techniques for context modeling and adaptive entropy coding of wavelet coefficients.

Table 8.3 Lossless Rates (bits/pixel) of ECECOW
Compared with Other Lossless Image Coders on an ISO
Set of Test Images

Image	NEW	ECECOW	S+P	CALIC	JPEG-LS
Balloon	2.85	2.86	2.97	2.78	2.90
Barb 1	4.30	4.34	4.53	4.33	4.69
Zelda	3.69	3.71	3.84	3.72	3.89
Hotel	4.36	4.38	4.53	4.22	4.38
Barb 2	4.53	4.57	4.71	4.49	4.69
Board	3.61	3.62	3.82	3.50	3.68
Girl	3.79	3.81	3.96	3.71	3.93
Gold	4.41	4.42	4.56	4.38	4.48
Boats	3.85	3.86	4.03	3.77	3.93
Lena	4.07	4.09	4.16	4.04	4.24
Average	**3.96**	**3.98**	**4.12**	**3.89**	**4.08**

These techniques are used in some state-of-the-art wavelet image codecs. Although the chapter mostly relates to wavelet-based image compression for the concreteness of the discussions, the principles and techniques described here can be used in conjunction with other transforms, such as DCT, and are also readily applicable to compression of other types of signals, such as audio and video.

References

[1] Criminal Justice Information Services, *WSQ Gray-Scale Fingerprint Image Compression Specification* (ver. 2.0), Federal Bureau of Investigation, Feb. 1993.

[2] ISO/IEC WD15444-1, JPEG 2000 — Lossless and lossy compression of continuous-tone and bi-level still images, ISO, Dec. 1999.

[3] ISO/IEC JTC 1/SC 29/WG 1 WD14495, JPEG LS image coding system, July, 1996.

[4] ISO/IEC JTC1 CD 11172, Coding of moving pictures and associated audio for digital storage media up to 1.5 Mbits/s, ISO, 1992.

[5] ISO/IEC JTC1 CD 13818, Generic coding of moving pictures and associated audio, ISO, 1994.

[6] ISO/IEC JTC 1/SC 29/WG 1, JPEG LS image coding system, *ISO Working Document ISO/IEC JTC1/SC29/WG1 N399 - WD14495,* July 1996.

[7] Arps, R.B. and Truong, T., Comparison of international standards for lossless still image compression, *Proc. of the IEEE,* 82(6), 889–899, 1994.

[8] Calderbank, A.R., Daubechies, I., Sweldens, W., and Yeo, B.L., Wavelet transforms that map integers to integers, *J. Applied and Computational Harmonic Analysis,* 5(3), 332–369, 1998.

[9] Chrysafis, C. and Ortega, A., Efficient context-based entropy coding for lossy wavelet image compression, *Proc. 1997 Data Compression Conference,* 241–250, Mar. 1997.

[10] Cover, T.M. and Thomas, J.A., *Elements of Information Theory,* John Wiley & Sons, New York, 1991.

[11] DeVore, R.A., Jawerth, B.J., and Lucier, B.J., Image compression through wavelet transform coding, *IEEE Trans. Info. Theory,* 38(2), 719–746, 1992.

[12] Fisher, R.A., The use of multiple measurements in taxonomic problems, in *Contributions to Mathematical Statistics,* John Wiley & Sons, New York, 1950.

[13] Garey, M.R. and Johnson, D.S., *Computers and Intractability, A Guide to the Theory of NP-Completeness,* Freeman, New York, 1979.

[14] Gersho, A. and Gray, R.M., *Vector Quantization and Signal Compression,* Kluwer Academic Publishers, Boston, 1992.

[15] Loeve, M., *Probability Theory,* 2nd ed., Van Nostrand Reinhold, Princeton, NJ, 478, 1960.

[16] LoPresto, S.M., Ramchandran, K., and Orchard, M.T., Image coding based on mixture modeling of wavelet coefficients and a fast estimation-quantization framework, *Proc. 1997 Data Compression Conf.,* 241–250, Mar. 1997.

[17] Mallat, S., A theory for multiresolution signal decomposition: a wavelet representation, *IEEE Trans. Patt. Anal. Machine Intelligence,* 11(7), 674–693, 1989.

[18] Pearlman, W.A., Kim, B.-J., and Xiong, Z., Embedded video subband coding with 3D SPIHT, *Wavelet Image and Video Compression,* Topiwala, P., Ed., Kluwer Academic Publishers, Boston, 1998.

[19] Pennebaker, W.B. and Mitchell, J.L., *JPEG Still Image Data Compression Standard,* Van Nostrand Reinhold, New York, 1993.

[20] Pennebaker, W.B., Mitchell, J.L., Langdon, G.G., and Arps, R.B., An overview of the basic principles of the Q-coder adaptive binary arithmetic coder, *IBM J. Res. & Devel.,* 32(6), 717–726, 1988.

[21] Rao, K.R. and Hwang, J.J., *Techniques and Standards for Image, Video and Audio Coding,* Prentice Hall, Englewood Cliffs, NJ, 1996.

[22] Rao, K.R. and Yip, P., *Discrete Cosine Transform,* Academic Press, New York, 1990.

[23] Rissanen, J. and Langdon, G.G., Universal modeling and coding, *IEEE Trans. Info. Theory,* 27, 12–23, 1981.

[24] Rissanen, J., Universal coding, information, prediction, and estimation, *IEEE Trans. Info. Theory,* 30, 629–636, 1984.

[25] Said, A. and Pearlman, W.A., New, fast, and efficient image codec based on set partitioning in hierarchical trees, *IEEE Trans. Circ. & Sys. Video Tech.,* 6(3), 243–249, June 1996.

[26] Said, A. and Pearlman, W.A., An image multiresolution representation for lossless and lossy compression, *IEEE Trans. on Image Proc.,* 5(9), 1303–1310, 1996.

[27] Shannon, C.E., A mathematical theory of communication, *Bell Syst. Tech. J.,* 27, 379–423, 623–656, 1948.

[28] Shapiro, J.M., Embedded image coding using zerotrees of wavelet coefficients, *IEEE Trans. Signal Processing,* 41(12), 3445–3462, 1993.

[29] Taubman, D., EBCOT: Embedded block coding with optimized truncation, ISO/IEC JTC 1/SC 29/WG 1, No. 1020, Oct. 1998.

[30] Taubman, D. and Zakhor, A., Multirate 3-D subband coding of video, *IEEE Trans. Image Processing,* 3(5), 572–588, 1994.

[31] Witten, I.H., Neal, R.M., and Cleary, J.G., Arithmetic coding for data compression, *Communications of the ACM,* 30, 520–540, 1987.

[32] Wood, J.W. and O'Neil, S.D., Subband coding of images, *IEEE Trans. Acoustics, Speech, and Signal Processing,* 34(5), 1278–1288, 1986.

[33] Wu, X., Optimal quantization by matrix-searching, *J. Algorithms,* 12(4), 663–673, 1991.

[34] Wu, X., Lossless compression of continuous-tone images via context selection and quantization, *IEEE Trans. on Image Proc.,* 6(5), 656–664, 1996.

[35] Wu, X. and Memon, N.D., Context-based adaptive lossless image coding, *IEEE Trans. Comm.,* 45(4), 437–444, 1997.

[36] Wu, X., High-order context modeling and embedded conditional entropy coding of wavelet coefficients for image compression, *Proc. of 31st Asilomar Conf. on Signals, Systems, and Computers,* 1378–1382, 1997.

[37] Wu, X., Context quantization with fisher discriminant for adaptive embedded wavelet image coding, *Proc. of 1999 Data Compression Conference,* 102–111, Mar. 1999.

[38] Wu, X., Low complexity high-order context modeling of embedded wavelet bit streams, *Proc. of 1999 Data Compression Conference,* 112–120, Mar. 1999.

[39] Wu, X. and Xiong, Z., An empirical study of high-order context modeling and entropy coding of wavelet coefficients, ISO/IEC JTC 1/SC 29/WG 1, No. 771, Feb. 1998.

[40] Xiong, Z., Ramchandran, K., and Orchard, M.T., Space frequency quantization for wavelet image coding, *IEEE Trans. Image Processing,* 6, 677–693, 1997.

[41] Xiong, Z. and Wu, X., Wavelet image coding using trellis coded space-frequency quantization, *IEEE Signal Processing Letters,* 6(7), 158–161, 1999.

[42] Xiong, Z., Wu, X., Yun, D.Y., and Pearlman, W.A., Progressive coding of medical volumetric data using three-dimensional integer wavelet packet transform, *Proc. of 1998 IEEE Workshop on Multimedia Signal Processing,* 553–558, Dec. 1998.

[43] Zandi, A., Allen, J.D., Schwartz, E.L., and Boliek, M., CREW: compression by reversible embedded wavelets, *Proc. of Data Compression Conf.,* 212–221, IEEE Press, Piscataway, NJ, 1995.

Index